Mathematical and Statistical Methods for Actuarial Sciences and Finance

Marco Corazza · Claudio Pizzi

Editors

Mathematical and Statistical Methods for Actuarial Sciences and Finance

 Springer

Editors
Marco Corazza
Ca' Foscari University of Venice
Department of Economics
Venice, Italy

Claudio Pizzi
Ca' Foscari University of Venice
Department of Economics
Venice, Italy

ISBN 978-3-319-02498-1   ISBN 978-3-319-02499-8 (eBook)
DOI 10.1007/978-3-319-02499-8
Springer Cham Heidelberg New York Dordrecht London

Library of Congress Control Number: 2013945795

Cover-Design: Simona Colombo, Milan, Italy
Typesetting: PTP-Berlin, Protago TeX-Production GmbH, Germany
Printing and Binding: GECA Industrie Grafiche, San Giuliano Milanese (MI), Italy
Printed in Italy

Springer is part of Springer Science+Business Media (www.springer.com)

Preface

This volume is a collection of referred papers selected from the more than one hundred and twenty presented at the *International MAF Conference 2012 – Mathematical and Statistical Methods for Actuarial Sciences and Finance*.

The conference was held in Venice (Italy), from April 10 to 12, 2012, at the prestigious Cavalli Franchetti palace of the *Istituto Veneto di Scienze, Lettere ed Arti*, on the Grand Canal, very near to the Rialto bridge. It was organized by the Department of Economics of the University Ca' Foscari of Venice (Italy), with the collaboration of the Department of Economics and Statistics of the University of Salerno (Italy).

This conference was the fifth in an international biennial series, which began in 2004. It was born out of a brilliant idea by colleagues – and friends – of the Department of Economics and Statistics of the University of Salerno: the idea was that a cooperation between mathematicians and statisticians working in actuarial sciences, in insurance and in finance could improve the research on these topics.

The proof of the merits of this idea is the wide participation in all the conferences. In particular, with reference to the 2012 event, there were:

- about 180 attendants, including academics, professionals, researchers and students;
- more than 120 accepted communications, organized in 40 parallel sessions;
- attendants and authors from more than 20 countries: Australia, Austria, Belgium, Canada, Denmark, Egypt, France, Germany, Great Britain, Greece, Israel, Italy, Japan, Mexico, New Zealand, Portugal, Republic of Djibouti, Sierra Leone, Spain, Switzerland and the USA;
- 4 prestigious plenary keynote lectures delivered by:
  - Professor Giuseppe Cavaliere of the University of Bologna (Italy): "Unit roots in bounded financial time series";
  - Professor Paul Embrechts of the ETH Zurich (Switzerland): "Extreme-quantile tracking for financial time series";
  - Professor Dominique Guégan of the University Paris1, Panthéon, Sorbonne (France): "A quantitative finance and actuarial framework for risk management";
  - Professor Wolfgang Runggaldier of the University of Padua (Italy): "On stochastic filtering applications in finance";

- an instructive plenary lesson, mainly addressed to Ph.D. students and young researchers, delivered by Professor Chris Adcock of the University of Shefield (Great Britain): "Doing research and getting it published".

Generally, the papers published in this volume present theoretical and methodological contributions and their applications in real contexts.

With respect to the theoretical and methodological contributions, some of the considered areas of investigation include: actuarial models; alternative testing approaches; behavioral finance; clustering techniques; coherent and no-coherent risk measures; credit-scoring approaches; data envelopment analysis; dynamic stochastic programming; financial contagion models; financial ratios; intelligent financial trading systems; mixture normality approaches; Monte Carlo-based methodologies; multi-criteria methods; nonlinear parameter estimation techniques; nonlinear threshold models; particle swarm optimization; performance measures; portfolio optimization; pricing methods for structured and non-structured derivatives; risk management; skewed distribution analysis; solvency analysis; stochastic actuarial valuation methods; variable selection models; time series analysis tools.

As regards the applications, they are related to real problems associated, among the others, to: banks; collateralized fund obligations; credit portfolios; defined-benefit pension plans; double-indexed pension annuities; efficient-market hypothesis; exchange markets; financial time series; firms; hedge funds; non-life insurance companies; returns distributions; socially responsible mutual funds; unit-linked contracts.

Of course, success of this conference would not have been possible without the valuable help of our sponsors (in alphabetical order):

- AMASES: Associazione per la Matematica Applicata alle Scienze Economiche e Sociali;
- Centro Interdipartimentale su Cultura e Economia della Globalizzazione;
- Department of Economics of the University Ca' Foscari of Venice;
- Department of Economics and Statistics of the University of Salerno;
- DIAMAN SIM S.p.A.;
- Istituto Veneto di Scienze Lettere e Arti;
- Nethun S.p.A;
- Regione del Veneto;
- VENIS S.p.A.

Further, we would also like to express our deep gratitude to the members of the Scientific and Organizing Committees, to the Center of Quantitative Economics of the Ca' Foscari University of Venice, to the Webmaster, and to all the people whose collaboration contributed to the success of the conference *MAF 2012*.

Finally, we are pleased to inform you that the organizing machine of the next edition is already working: the conference *MAF 2014* will be held in Vietri sul Mare (Italy), on the enchanting Amalfi Coast, from April 22 to 24, 2014 (for more details visit the website http://www.maf2014.unisa.it/).

We look forward to seeing you.

Venice, August 2013   Marco Corazza
                                                                     Claudio Pizzi

Contents

Weak Form Efficiency of Selected European Stock Markets: Alternative Testing Approaches

Giuseppina Albano, Michele La Rocca and Cira Perna

Abstract Modelling and forecasting financial data is an important problem which has received a lot of attention especially for the intrinsic difficulty in practical applications. The present paper investigates the weak form efficiency of some selected European markets: AEX, CAC40, DAX, FTSE100, FTSEMIB, IBEX35. In order to keep into account nonlinear structures usually found in returns time series data, a non parametric test based on neural network models has been employed. The test procedure has been structured as a multiple testing scheme in order to avoid any data snooping problem and to keep under control the familywise error rate. For sake of comparison we also discuss the results obtained by applying some classical and well known tests based on the Random Walk Hypotheses. The data analysis results clearly show that ignoring the multiple testing structure of these latter test might lead to spurious results.

1 Introduction

The efficient market hypothesis (EMH), in its weak form, states that all available information is fully and instantaneously reflected in price, so it will be not possible for investors, using past prices, to discover undervalued stocks and develop strategies to systematically earn abnormal returns. Clearly, this is a fundamental issue in fi-

G. Albano (✉)
Department of Economics and Statistics, University of Salerno, Salerno, Italy
e-mail: pialbano@unisa.it

M. La Rocca
Department of Economics and Statistics, University of Salerno, Salerno, Italy
e-mail: larocca@unisa.it

C. Perna
Department of Economics and Statistics, University of Salerno, Salerno, Italy
e-mail: perna@unisa.it

M. Corazza, C. Pizzi (eds.), *Mathematical and Statistical Methods for Actuarial Sciences and Finance*, DOI 10.1007/978-3-319-02499-8_1, © Springer International Publishing Switzerland 2014

nance since volatility, predictability, speculation and anomalies in financial markets are also related to the efficiency and are all interdependent.

The most common used implication of EMH is the Random Walk Hypothesis (RWH), which indicates that successive price changes are random and serially independent or, in a less restrictive formulation, incorrelated. Among methodologies able to test RWH, variance-ratio tests are considered powerful. Lo e Mackinlay [13] first proposed the conventional variance-ratio test. Later, Chow and Denning [4] modified Lo-Mackinlay's test to form a simple multiple variance-ratio test and Wright [18] proposed a non parametric ranks and signs based variance-ratio tests to overcome the potential limitation of Lo-Mackinlay's conventional variance-ratio test. Recently Kim [7] has proposed an automatic variance-ratio test in which the holding period is automatically chosen by means of a procedure depending on the structure of the data.

Anyway, evidence against the RWH for stock returns in the capital markets is often shown (see, for example, [6, 12, 13] and references therein). Failure of models based on linear time series techniques to deliver superior forecasts to the simple random walk model has forced researchers to use various nonlinear techniques, such as Engle test, Tsay test, Hinich bispectrum test, Lyapunov exponent test. Also in such a literature the market efficiency confirms to be a challenging issue in finance (see, for example, [1]).

Moreover, it has been widely accepted that nonlinearity exists in the financial markets and that nonlinear models, both parametric and nonparametric, can be effectively used to uncover this pattern. Our contribution follows this research path and aims to investigate the relative merits of a neural network approach for characterizing the prices of selected European markets (AEX, CAC40, DAX, FTSE100, FTSE MIB, IBEX35). The neural model framework is used to directly testing if past lags contain useful information which can be exploited for better prediction of future values. The procedure is structured as a multiple testing scheme in order to avoid any data snooping problem and to keep under control the familywise error rate. We compare the results of our test with those obtained by using some well known variance-ratio tests.

The paper is organized as follows. In Sect. 2, we discuss the test scheme based on feedforward neural while, in Sect. 3 we briefly review the most popular variance-ratio tests. In Sect. 4, we describe the data and the data analysis results. Some remarks close the paper.

2 Neural Network Test for Market Efficiency

Let P_t, $t = 1, \ldots, n$, be the price associated to a given asset or the price index of a given market at time t. Following a standard practice, we construct returns time series ($Y_t = \nabla \log P_t$) avoiding potential problems associated with estimation of nonstationary regression functions. In the case of inefficient market (in the sense of weak efficiency) past lags contain useful information which can be exploited for better prediction of future values. To check if Y_t depends on some past lags, we can introduce the following model:

$$Y_t = g(Y_{t-1}, Y_{t-2}, \ldots, Y_{t-d}) + \varepsilon_t \tag{1}$$

where $g(\cdot)$ generally is a nonlinear function and ε_t is zero-mean error term with finite variance.

The unknown function $g(\cdot)$ can be estimated by using a feedforward neural network f (see, for example, [10, 11]) defined as

$$f(\mathbf{y}, \mathbf{w}) = w_{00} + \sum_{j=1}^{m} w_{0j} \psi(\tilde{\mathbf{y}}^T \mathbf{w}_{1j}) \tag{2}$$

where $\mathbf{w} \equiv (w_{00}, w_{01}, \ldots w_{0r}, \mathbf{w}_{11}^T, \ldots, \mathbf{w}_{1r}^T)^T$ is a $m(d+2)+1$ vector of network weights, $\mathbf{w} \in \mathbf{W}$ with $\mathbf{W}$ being a compact subset of $\mathbb{R}^{m(d+2)+1}$, and $\tilde{\mathbf{y}} \equiv (1, \mathbf{y}^T)^T$ is the input vector augmented by a bias component 1. The network (2) has d input neurons, m neurons in the hidden layer and the identity function for the output layer. The (fixed) hidden unit activation function ψ is chosen in such a way that $f(\mathbf{y}, \cdot) : \mathbf{W} \rightarrow \mathbf{R}$ is continuous for each y in the support of the explanatory variables and $f(\cdot, \mathbf{w}) : \mathbf{R}^d \rightarrow \mathbf{R}$ is measurable for each $\mathbf{w}$ in $\mathbf{W}$.

In this framework, the hypothesis that the market is efficient (that is, the given set of lags has no effect on Y) can be formulated in a multiple testing framework as

$$H_j : \theta_j = 0 \quad vs \quad H_j' : \theta_j > 0, \quad j = 1, 2, \ldots, d \tag{3}$$

where $\theta_j = \mathbb{E}\left[f_j^2(Y_{t-1}, Y_{t-2}, \ldots, Y_{t-d}, \mathbf{w}_0)\right]$ and f_j is the partial derivative of f with respect Y_{t-j}. since f is known and $\mathbf{w}_0$ can be closely approximated. Each null H_j can be tested by using the statistic $T_{n,j} = n\hat{\theta}_{n,j}$ where

$$\hat{\theta}_{n,j} = n^{-1} \sum_{t=1}^{n} f_j^2(Y_{t-1}, Y_{t-2}, \ldots, Y_{t-d}; \hat{\mathbf{w}}_n) \tag{4}$$

and the vector $\hat{\mathbf{w}}_n$ is a consistent estimator of the vector of the network weigths. Clearly, large values of the test statistics indicate evidence against the hypothesis H_j [9, 10].

In order to control the familywise error rate (FWE), we use the algorithm proposed in Romano and Wolf [15, 16], suitable for joint comparison of multiple (possibly misspecified) models. The multiple testing scheme is described in Algorithm 1 (see [11] for details).

The neural network model structure and the complexity of the test procedures distinguish the asymptotic distribution of the test statistics involved from the familiar tabulated distributions. The problem can be overcome by using resampling techniques as simulation tools to approximate the unknown sampling distributions of the statistical quantities involved in the testing procedure. Here, to obtain valid asymptotic critical values for the test, we refer to the stationary bootstrap approach proposed by Politis and Romano [14].

Algorithm 1 Multiple testing algorithm for weak form efficiency

---

1: Relabel the hypothesis from H_{r_1} to H_{r_d} in redescending order of the value of the test statistics $T_{n,j}$, that is $T_{n,r_1} \geq T_{n,r_2} \geq \ldots \geq T_{n,r_d}$.
2: Set $L = 1$ and $R_0 = 0$.
3: **for** $j = R_{L-1} + 1$ to d **do**
4:　**if** $0 \notin [T_{n,r_j} - \hat{c}_L(1 - \alpha), \infty)$ **then**
5:　　reject H_{r_j}
6:　**end if**
7: **end for**
8: **if** no (further) null hypothesis are rejected **then**
9:　Stop
10: **else**
11:　R_L = number of rejected hypothesis
12:　$L = L + 1$
13:　Go to step 3
14: **end if**

It is worthwhile to stress that more complex model structures could be easily accommodated in the described framework, by adding (and testing) other explanatory variables known to have some impact on market returns. The relevance of each new explanatory (possibly) lagged variable could be determined along the same lines as those described in the paper. In any case this analysis would be beyond the scope of this paper.

3 Variance-Ratio Tests in a Nutshell

In the following, we briefly review some classical and well known tests based on the RWH that will be used for comparison with our proposed methodology. For a more exhaustive review see, for example, [2, 18]. Here we consider the Lo-MacKinlay test and its multiple testing version proposed by Chow and Denning, non-parametric variance-ratio tests using ranks and signs and the automatic variance-ratio test proposed by Choi.

The Lo-MacKinlay test exploits the property that the variance of the increments in a random walk is linear in the sampling interval. The variance-ratio with holding period q is defined as $VR(q) = \frac{\sigma^2(q)}{\sigma^2(1)}$, where $\sigma^2(k)$ is $1/k$ times the variance of $(Y_t - Y_{t-k})$, $k = 1, q$ and $t = 1, \ldots, n$. Under the RWH, it is $VR(q) = 1$. Asymptotic standard normal test statistic for variance-ratio $VR(q)$ can be derived. Moreover, an heteroscedasticity-consistent standard normal test statistic can be derived (see [13]).

Chow and Denning extend Lo-Mackinlay's variance-ratio test and provide a simple multiple variance-ratio test in order to control the test size and reduce the Type I errors in the conventional variance-ratio test. The idea is to consider as test statistic the maximum of L test statistic of Lo-Mackinlay test for L different holding period. In this case, the test statistic follows the studentized maximum modulus distribution (SMM) with L and n (the sample size) degrees of freedom. When n is large, the SMM critical values at $L = 4$ and α equal to 5% level of significance is 2.49 (see [4]).

Wright proposes an alternative non-parametric variance-ratio tests using ranks and signs of returns and demonstrates that they may have better power properties than other variance-ratio tests [18]. The rank-based version runs as follows: Let $r(Y_t)$ be the rank of Y_t among $Y_1, Y_2, \ldots, Y_n$. Define $r_{1t} = \dfrac{\left(r(Y_t) - \frac{n+1}{2}\right)}{\sqrt{\frac{(n-1)(n+1)}{12}}}$, $\quad r_{2t} = \Phi^{-1} \dfrac{r(Y_t)}{n+1}$,

where Φ^{-1} is the inverse of the standard normal cumulative distribution function. Essentially, Wright substitutes r_{1t} and r_{2t} in place of the return $(Y_t - Y_{t-q})$ in the definition of Lo-MacKinlay's variance-ratio test statistics. The exact sampling distribution of the test statistic can be approximated by using bootstrap resampling schemes. The Sign-Based runs as follows. Let $u(Y_t) = \mathbb{I}(Y_t > 0) - 0.5$, where $\mathbb{I}$ is the indicator function. Let $s_t = 2u(Y_t)$. If the series Y_t follows a random walk, each s_t is equal to 1 with probability $\frac{1}{2}$ and is equal to -1 otherwise. As in the rank-based variance-ratio test, the test statistic is obtained from the Lo-Mackinlay statistic substiting s_t in place of the return $(Y_t - Y_{t-q})$ and the exact sampling distribution of the test statistic is approximated with a bootstrap method (see [18]).

Finally, Choi [3] proposes a fully data-dependent method of estimating the optimal choice $\hat{q}$ for q, essentially based on the wild bootstrapping. Note that the small sample properties of this automatic variance-ratio test under heteroschedasticity are unknown and have not been investigated properly.

4 Empirical Results

The time series considered in this data analysis application are daily closing values of the following stock market indices: AEX, CAC40, DAX, FTSE100 and FTSEMIB. The data set spans the period from 03/06/2002 to 01/06/2012. The market returns are depicted in Fig. 1 while some descriptive statistics are reported in Table 1. The distribution of all the series is almost symmetric and, as expected, the time series show strong kurtosis and as a consequence the Jarque Bera test clearly rejects the hypothesis of normally distributed returns.

The results of the weak form efficiency test based on feedforward neural networks which is able to take into account the multiple testing structure of the problem are reported in Tables 2 and 3. The tests are based on a nonlinear autoregressive model $Y_t = g(Y_{t-1}, \ldots, Y_{t-5}) + \varepsilon_t$ where the lag structures has been limited to a maximum of five lags, following [5] whom findings were unaffected by using different lag structures.

The function g is estimated by using neural networks with different hidden layer sizes and different values for the weight decay. The "optimal" neural network model is estimated by using nonlinear least squares and it is chosen as the one which minimize the the Schwartz Information Criterion (SIC) (see Fig. 2). The values reported in Table 2 show that the hypothesis of market efficiency cannot be rejected for all

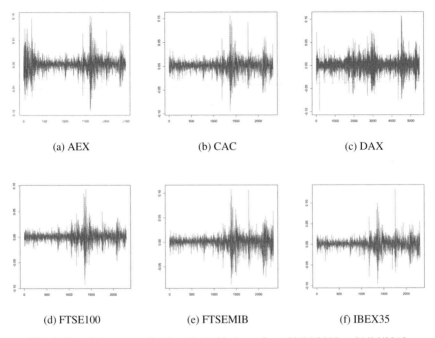

(a) AEX   (b) CAC   (c) DAX

(d) FTSE100   (e) FTSEMIB   (f) IBEX35

Fig. 1 Plot of the returns for the selected indexes from 03/06/2002 to 01/06/2012

the indexes and for all the neural models. Moreover, the results appear to be stable by changing the hidden layer sizes in a neighborhood of the optimal ones. The conclusions do not change by considering a moderate weight decay to make neural network estimates more stable (see Table 3).

For sake of comparison we also reported the results of the variance-ratio tests discussed in the previous section. In Table 4 we report the test statistics for the classical variance-ratio test, choosing for the holding period q the values $2, 4, 8, 16$, generally employed in such a literature. If we perform the variance-ratio test without taking

Table 1 Descriptive statistics for the selected daily returns from 03/06/2002 to 01/06/2012. p-values are in parenthesis

	Min.	1st Qu.	Median	Mean	3rd Qu.	Max.	Skew.	Kurt.	Jarque Bera Test
AEX	−0.096	−0.007	0.000	0.000	0.007	0.100	-0.012	5.810	3525 (< 2.2e-16)
CAC40	−0.095	−0.006	0.000	0.000	0.007	0.106	0.053	6.909	4656 (< 2.2e-16)
DAX	−0.099	−0.007	0.000	0.000	0.008	0.108	-0.096	4.593	4821 (< 2.2e-16)
FTSE100	−0.093	−0.005	0.000	0.000	0.006	0.094	-0.148	8.327	6669 (< 2.2e-16)
FTSEMIB	−0.086	−0.006	0.000	0.000	0.006	0.109	-0.082	6.492	4118 (< 2.2e-16)
IBEX35	−0.096	−0.006	0.000	0.000	0.006	0.100	0.121	7.818	5912 (< 2.2e-16)

Table 2 Neural Network tests for different hidden layer sizes. Weight decay equal to zero. Neural models which minimize the SIC are labelled with a star. The nulls are rejected, at the level 0.05, if $T_{n,r_1} > c(0.95)$

Index	m	Optim	1	2	3	4	5	T_{n,r_1}	$c(0.95)$
AEX	2		35.04	57.31	90.28	16.93	33.84	90.28	715.09
	3	*	33.05	28.37	40.20	87.84	377.22	377.22	1024.19
	4		383.71	197.16	644.86	85.86	63.62	644.86	6217.37
CAC40	1		2.52	9.35	9.32	0.32	10.52	10.52	120.26
	2	*	23.51	34.27	17.67	12.37	6.70	34.27	336.83
	3		32.22	90.98	23.27	36.98	33.80	90.98	3198.82
DAX	0	*	0.00	2.13	10.86	8.74	4.92	10.86	19.94
	1		6.00	9.34	19.57	0.01	7.23	19.57	85.24
	2		58.65	12.55	35.29	78.28	10.07	78.28	129.81
FTSE100	1		0.60	51.62	67.94	3.80	19.57	67.94	120.64
	2	*	16.13	143.45	125.87	31.18	32.91	143.45	235.38
	3		19.60	91.15	96.20	26.36	35.64	96.20	448.96
FTSEMIB	0		0.11	2.68	6.84	3.08	8.69	8.69	32.78
	1	*	1.76	40.13	29.08	2.15	19.96	40.13	110.47
	2		23.33	85.53	63.88	24.41	9.55	85.53	391.87
IBEX35	1		3.02	8.14	9.00	0.04	15.97	15.97	38.42
	2	*	15.53	49.45	36.95	12.13	16.37	49.45	870.59
	3		37.65	165.02	59.72	12.81	33.21	165.02	4133.87

Table 3 Neural network tests for different hidden layer sizes. Weight decay equal to $5e - 5$. Neural models which minimize the SIC are labelled with a star. The nulls are rejected, at the level 0.05, if $T_{n,r_1} > c(0.95)$

Index	m	Optim	1	2	3	4	5	T_{n,r_1}	$c(0.95)$
AEX	0	*	0.04	0.00	19.50	5.53	12.95	19.50	21.04
	1		0.04	0.00	19.81	5.47	12.83	19.81	21.32
	2		4.21	0.74	21.14	12.97	19.35	21.14	31.80
CAC40	0	*	5.48	6.33	11.42	3.42	8.41	11.49	24.16
	1		6.87	6.72	11.59	3.25	8.89	11.59	23.02
	2		10.77	6.46	9.97	3.31	6.11	10.77	22.07
DAX	0	*	0.00	2.13	10.86	8.74	4.92	10.86	20.26
	1		0.00	2.10	10.77	8.73	4.89	10.77	20.26
	2		7.14	2.27	15.55	11.18	4.18	15.55	26.06
FTSE100	0	*	5.72	6.81	11.06	10.68	7.82	11.06	32.44
	1		6.54	6.94	11.35	10.27	8.12	11.35	29.79
	2		6.68	6.04	11.09	11.60	4.86	11.60	29.77
FTSEMIB	0	*	0.11	2.68	6.83	3.08	8.69	8.69	33.80
	1		0.10	2.66	6.91	3.01	8.67	8.67	34.39
	2		12.78	6.64	9.02	1.77	11.24	12.78	44.99
IBEX35	0	*	0.19	6.67	5.07	0.60	6.04	6.67	22.63
	1		0.17	6.61	5.04	0.57	6.06	6.61	22.52
	2		9.02	6.25	6.23	0.47	10.43	10.43	32.20

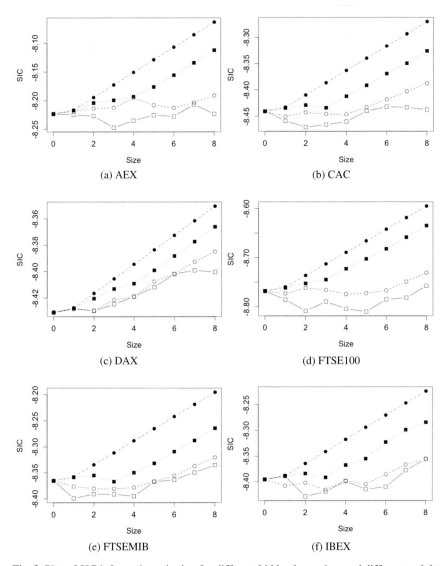

Fig. 2 Plot of SIC information criterion for different hidden layer sizes and different weight decay values ($\square = 0$, $\circ = 5E-6$, $\blacksquare = 5E-5$, $\bullet = 5E-4$)

into account the heteroscedasticity of the data and the multiple testing structure, the RWH is rejected for CAC40 and FTSE100 for all the holding periods q, and for AEX (when $q = 8$), DAX ($q = 8$) and IBEX35 ($q = 8, 16$). However, by allowing for heteroscedasticity, the variance-ratio test rejects the null hypothesis only for CAC40 (for $q = 4, 8$) and FTSE100 (for $q = 4$).

Table 4 Variance-ratio test in the presence of homoscedasticity and of heteroscedasticity of the data. Statistically significant tests at the level 5% are indicated in bold (q denotes the holding period)

Index		homoscedasticity				heteroscedasticity			
	q	2	4	8	16	2	4	8	16
AEX		−0.677	−1.519	**−2.238**	−1.311	−0.411	−0.882	−1.264	−0.736
CAC40		**−2.384**	**−3.889**	**−3.707**	**−3.266**	−1.439	**−2.279**	**−2.122**	−0.736
DAX		−0.144	−1.751	**−2.118**	−1.742	−0.099	−0.882	−1.264	−1.889
FTSE100		**−2.555**	**−4.098**	**−3.644**	**−2.995**	−1.374	**−2.140**	−1.846	−1.534
FTSEMIB		0.187	−1.371	−1.763	−1.148	0.115	−0.822	−1.023	−0.669
IBEX35		−0.410	−1.658	**−2.462**	**−2.167**	0.243	−0.984	−1.435	−1.277

Table 5 Automatic variance-ratio test. Statistically significant tests at the level 5% are indicated in bold

Index	homoschedasticity	CI (Normal)	CI (Mammen)	CI (Rademacher)
AEX	−0.325	(−2.302;2.702)	(−2.534;2.597)	(−2.273;2.558)
CAC40	**−2.402**	**(−2.232;2.736)**	**(−2.233;2.616)**	(−2.583;2.554)
DAX	−0.039	(−2.100;2.264)	(−2.214;2.203)	(−2.139;2.528)
FTSE100	**−2.778**	**(−2.642;2.958)**	(−2.932;2.898)	**(−2.799;3.204)**
FTSEMIB	0.121	(−2.383;2.220)	(−2.536;2.501)	(−2.502,2.712)
IBEX35	0.260	(-2.341;2.726)	(−2.445;2.215)	(−2.534;2.574)

The previous results appear to be sensitive to the choice of the holding period. So automatic variance-ratio tests have been performed and the results are shown in Table 5. The asymptotic normal approximation, which does not take into account the heteroscedasticity, allows to reject the RWH in the case of CAC40 and FTSE100. In order to take into account the heteroscedasticity, the wild bootstrap (both in the Mammen and in the Rademacher version) has been employed. Intervals of acceptance (CI) are identified using 1000 bootstrap replications. Also in these cases, CAC40 and FTSE100 appear to be inefficient.

By applying the non parametric-ratio test the RWH is rejected again for the indexes CAC40 and FTSE100 (see the 95% confidence intervals reported in Table 6 in the columns Rank-based test (homoschedasticity and heteroschedasticity) and identified using 1000 and 10000 replications in each case,) and possibly FTSEMIB (as reported in column Sign-based test).

Of great interest are the results of the Chow-Denning test, which takes into account the multiple testing structure of the testing scheme. By assuming homoschedasticity, again, the CAC40 and the FTSE100 appear to be inefficient. However, by allowing simultaneously heteroscedasticity and multiple testing structure (see Table 7) the efficiency hypothesis is confirmed for all the selected markets. These latter re-

Table 6 Non-parametric variance-ratio tests: ranks (homoschedasticity and heteroschedasticity) and signs (last column). Statistically significant tests at the level 5% are indicated in bold. In parentheses the confidence intervals used for the test decision

Index	q	Rank-based test		Sign-based test
		homoschedasticity	heteroschedasticity	
AEX	2	0.112 (−1.970, 1.915)	−0.333 (−1.965,1.919)	0.000 (−1.920,1.960)
	4	−0.574 (−1.984,1.937)	−1.031 (−2.001,1.930)	−0.524 (−1.935,1.999)
	8	−1.065 (−1.983,1.935)	−1.564 (−1.979,1.905)	−0.699 (−1.937,1.987)
	16	−0.495 (−1.981,1.881)	−0.854 (−1.979,1.856)	0.184 (−1.897,2.005)
CAC 40	2	−1.968 (−1.999, 1.930)	−1.908 (−2.011,1.935)	**−2.876** (−2.007,1.966)
	4	**−2.788** (−1.990,1.876)	**−3.031** (−2.027,1.921)	**−2.821** (−1.947,2.002)
	8	**−2.954** (−1.948,1.924)	**−3.159** (−1.990,1.965)	**−2.312** (−1.941,1.980)
	16	**−2.804** (−1.926,1.929)	**−3.017** (−1.949,1.991)	**−1.965** (−1.910,1.971)
DAX	2	−0.309 (−1.986, 1.974)	−0.171 (−2.014,1.954)	−0.554 (−1.961,1.961)
	4	−1.347 (−1.989,1.948)	−1.504 (−1.995,1.959)	−1.236 (−1.980,1.966)
	8	−1.847 (−1.966,1.961)	**−2.031** (−1.964,1.943)	−1.179 (−1.966,1.939)
	16	−1.714 (−1.988,1.952)	−1.813 (−1.982,1.967)	−0.529 (−1.912,1.961)
FTSE100	2	**−2.666** (−2.002, 1.963)	**−2.559** (−1.985,1.975)	**−2.377** (−2.002,1.960)
	4	**−2.863** (−1.954,1.949)	**−3.196** (−1.945,1.947)	**−2.664** (−1.950,1.995)
	8	**−2.670** (−1.995,1.939)	**−2.969** (−1.997,1.936)	−1.614 (−1.910,1.952)
	16	**−2.721** (−1.980,1.893)	**−2.833** (−1.973,1.907)	−1.247 (−1.889,2.002)
FTSEMIB	2	−0.473 (−2.021, 1.913)	0.069 (−2.047,1.926)	**−2.109** (−1.944,1.985)
	4	−0.889 (−2.001,1.958)	−0.794 (−2.013,1.975)	**−2.111** (−1.901,2.000)
	8	−1.327 (−1.996,1.900)	−1.250 (−2.021,1.910)	−1.818 (−1.895,1.989)
	16	−0.777 (−2.005,1.895)	−0.678 (−2.015,1.868)	−0.799 (−1.892,2.002)
IBEX35	2	0.446 (−1.995, 1.912)	0.908 (−2.026,1.909)	−0.831 (−1.954,1.954)
	4	−0.592 (−2.011,1.953)	−0.754 (−2.024,1.945)	−0.756 (−1.922,1.967)
	8	−1.105 (−2.005,1.942)	−1.459 (−2.005,1.910)	−0.179 (−1.904,1.964)
	16	−0.724 (−1.948,1.882)	−1.226 (−1.944,1.892)	0.674 (−1.894,1.975)

sults appear to be consistent with those obtained by using our neural network testing scheme.

The data analysis application clearly shows that when using variance-ratio tests, neglecting their multiple testing structure and/or neglecting peculiar aspects of the data (such as heteroschedasticity) might lead to spurious results with ambiguous conclusion on market efficiency. On the other hand, the usage of multiple testing schemes based on neural network modeling, which are able to correctly incorporate

Table 7 Chow-Denning test in the presence of homoscedasticity and of heteroscedasticity of the data. Statistically significant tests at the level 5% are indicated in bold

Index	homoscedasticity	heteroscedasticity
AEX	2.238	1.263
CAC40	**3.889**	2.278
DAX	2.118	1.355
FTSE100	**4.098**	2.140
FTSEMIB	1.763	1.023
IBEX35	2.462	1.435

peculiar characteristics of the data sets (heterogeneity, nonlinearity) appear to lead to more stable and trustable results.

5 Concluding Remarks

In this paper we analyzed the weak form efficiency of some European stock markets by using a nonparametric testing procedure based on neural networks. The test procedure has been structured as a multiple test and its advantages with respect to variance-ratio tests is the stability and robustness of the results. However some issues still remain open. It is well known that variance-ratio tests are not robust to structural breaks (see [8]) and some modification of the standard test procedure is needed to deal with it. Is the neural network based test able to accommodate structural breaks as well nonlinearities? There is a wide collection of variance-ratio tests with their advantages and drawbacks. What are their relative merits in terms of size accuracy and power properties in small samples? These aspects however are out of the scope of this paper and they will be part of further research studies.

Acknowledgements The authors would like to thank the anonymous reviewers for their valuable comments and suggestions, which led to a significant improvement of the paper. They also gratefully acknowledge funding from the Italian Ministry of Education, University and Research (MIUR) through PRIN project "Forecasting economic and financial time series: understanding the complexity and modelling structural change" (code 2010J3LZEN).

References

1. Antoniou, A., Ergul, N., Holmes, P.: Market Efficiency, Thin Trading and Non-linear Behaviour: Evidence from an Emerging Market. Eur. Financ. Manag. **3**(2), 175–190 (1997)
2. Charles, A., Darné, O.: Variance-Ratio tests of random walk: an overview. J. Econ. Surv. **23**(3), 503–527 (2009)

3. Choi, I.: Testing the random walk hypothesis for real exchange rates. J. Appl. Econom. **14**, 293–308 (1999)
4. Chow, K.V., Denning, K.C.: A Simple Multiple Variance Ratio Test. J. Econom. **58**, 385–401 (1993)
5. Diebold, F.X., Nason, J.A.: Nonparametric exchange rate prediction. J. Int. Econ. **28**, 315–332 (1990)
6. Fama, E.: Market efficiency, long-term returns, and behavioral finance. J. Financ. Econ. **49**, 283–306 (1998)
7. Kim, J.H.: Automatic Variance Ratio Test under Conditional Heteroskedascity. Financ. Res. Letters **6**(3), 179–185 (2009)
8. Kim, Y., Kim, T.-H.: Variance-ratio Tests Robust to a Break in Drift. Eur. J. Pure and Appl. Math. **3**(3), 502–518 (2010)
9. La Rocca, M., Perna, C.: Variable selection in neural network regression models with dependent data: a subsampling approach, Comput. Stat. Data An. **48**, 415–429 (2005a)
10. La Rocca, M., Perna, C.: Neural network modeling by subsampling. In: Cabestany, J., Prieto, A., Sandoval, F. (eds.) Computational Intelligence and Bioinspired Computational Intelligence and Bioinspired Systems. Lecture Notes in Computer Science, p. 3512. Springer (2005b)
11. La Rocca, M., Perna, C.: Neural Network Modelling with Applications to Euro Exchange Rates. In: Kontoghiorghes E. J., Rustem B., Winker P. (eds.) Computational Methods in Financial Engineering, pp. 163–189 (2009)
12. Lim, K.P., Brooks, R.D., Hinich M.J.: Nonlinear serial dependence and the weak-form efficiency of Asian emerging stock markets. J. Int. Financ. Mark., Inst. and Money **18**(5), 527–544 (2008)
13. Lo, A.W., MacKinlay, A.C.: Stock market prices do not follow random walks: evidence from a simple specification test. Rev. Financ. Stud. **1**(1), 41–66 (1988)
14. Politis, D.N., Romano, J.P.: The Stationary Bootstrap. J. Am. Stat. Assoc. **89**, 1303–1313 (1994)
15. Romano, J.P., Wolf, M.: Exact and approximate stepdown methods for multiple hypothesis testing. J. Am. Stat. Assoc. **100**, 94–108 (2005)
16. Romano, J.P., Wolf, M.: Stepwise multiple testing as formalized data snooping. Econometrica **73**, 1237–1282 (2005)
17. Worthington, A., Higgs, H.: Random walks and market efficiency in European equity markets. Glob. J. Financ. and Econ. **1**(1), 59–78 (2004)
18. Wright, J.H.: Alternative Variance-Ratio Tests Using Ranks and Signs. J. Bus. & Econ. Stat. **18**, 1–9 (2000)

An Empirical Comparison of Variable Selection Methods in Competing Risks Model

Alessandra Amendola, Marialuisa Restaino and Luca Sensini

Abstract The variable selection is a challenging task in statistical analysis. In many real situations, a large number of potential predictors are available and a selection among them is recommended. For dealing with this problem, the automated procedures are the most commonly used methods, without taking into account their drawbacks and disadvantages. To overcome them, the shrinkage methods are a good alternative. Our aim is to investigate the performance of some variable selection methods, focusing on a statistical procedure suitable for the competing risks model. In this theoretical setting, the same variables might have different degrees of influence on the risks due to multiple causes and this has to be taken into account in the choice of the "best" subset. The proposed procedure, based on shrinkage techniques, is evaluated by means of empirical analysis on a data-set of financial indicators computed from a sample of industrial firms annual reports.

1 Introduction

The evolution of a firm is a complex process with many interconnected elements. The firm's coming out of the market could be considered as the final step of the

A. Amendola
Department of Economics and Statistics, University of Salerno, Via Giovanni Paolo II, 132, 84084 Fisciano (SA), Italy
e-mail: alamendola@unisa.it

M. Restaino (✉)
Department of Economics and Statistics, University of Salerno, Via Giovanni Paolo II, 132, 84084 Fisciano (SA), Italy
e-mail: mlrestaino@unisa.it

L. Sensini
Department of Business Studies and Research, University of Salerno, Via Giovanni Paolo II, 132, 84084 Fisciano (SA), Italy
e-mail: lsensini@unisa.it

M. Corazza, C. Pizzi (eds.), *Mathematical and Statistical Methods for Actuarial Sciences and Finance*, DOI 10.1007/978-3-319-02499-8_2, © Springer International Publishing Switzerland 2014

process. The companies may exit the market for several reasons (e.g. bankruptcy, liquidation, merger and acquisition), and each exit may be influenced by different factors, having distinct effects for the market structure [34].

In the literature on corporate distress prediction, the main contributions have investigated the determinants of the general exit decision, without considering possible distinctions among different status.

Starting from the seminal paper of [3], the empirical works have either treated all exits as homogeneous events or focused on only one form of exit (mainly bankruptcy), neglecting others (see for example [10, 36]). Only recently the attention has been focused on the consequences of different types of financial distress [31, 32]. In order to take into account the effects related to each exit form, the most used methods are the advanced version of logistic regression (i.e. the mixed logit, multinomial error component logit and nested logit model) [14, 29] and the competing risks models [12, 16].

In this paper we refer to a competing risks approach because, unlike the traditional logistic framework, it enables to incorporate the time to event as dependent variable in determining the probability of a firm being in a distressed status. In addition, this model allows taking into account whether and when the exit occurs, monitoring the evolution of the risk of an exit type over time.

Since each exit status may be caused by different reasons, a relevant task is related to the identification of a subset of factors that are significantly correlated with a given exit and have to be included in the competing risks model. In other words, it is important to identify those financial indicators that may influence each reason of leaving the market at a specified time in order to accurately predict the exit of a new firm, thus allowing a better risk assessment and model interpretation.

Variable selection has been largely investigated in statistical literature, and different methods have been proposed throughout the years. The traditional approach, mainly based on automatic selection procedure, suffers from a number of drawbacks that lead to biased and unstable results [11]. Besides, they face significant challenges when dealing with high dimensionality [21].

The instability of these variable selection methods justifies the need of considering other approaches, based on the maximization of different forms of penalized likelihood. Methods of penalization include traditional approaches such as AIC [1] and BIC [35] as well as more recent developments including bridge regression [23], LASSO [38], SCAD [19], LARS [17], elastic net [40] and MM algorithms [26]. Most of them have been widely developed within the regression setting, and only recently some of them have been extended to the context of survival data analysis and, in particular, for Cox's proportional hazards (PH) model [6, 7, 20, 39]. However, since now there are no references about their application in competing risks model.

The aim of this paper is to investigate the determinants of the probability of different types of firms' market exit by the competing risks hazard model, focusing on the variable selection problem. The competing risks model is estimated by maximizing the marginal likelihood subject to a shrinkage-type penalty that encourages sparse solutions and hence facilitates the process of variable selection. The proposed approach is compared over traditional stepwise procedure and their performance is evaluated through the empirical analysis on a data-set of financial indicators drawn from a sample of industrial firms annual reports. The sample includes three mutually exclusive exit status, namely firms going bankrupt, being liquidated and becoming inactive. Numerical results seem to be in favour of the proposed method, indicating that the lasso technique produces accurate and interpretable models and it tends to provide better results in terms of predictive accuracy.

The rest of the paper is structured as follows. In the next section, the statistical method is briefly presented. The empirical results are discussed in Sect. 3, while Sect. 4 concludes.

2 Model Specification and Variable Selection

2.1 Competing Risks Model

In this section the methodology employed in the paper is described. The competing risks model is an extension of the mortality model for survival data and is based on one transient state (alive state) and a certain number of absorbing states, corresponding to death from different causes. Thus, all transitions are from the state alive (for details, see [8, 9, 25]).

Let T be the observed failure time and let D be the cause of failure (*event-causing failure*). Since the different risks are assumed to be independent [2], one and only one cause can be assigned to every failure.

In this framework the central quantity is the probability of failing due to a given cause k, after having reached the time point t, called *cause-specific hazard function*, given by:

$$\lambda_k(t) = \lim_{\Delta t \to 0} \frac{P[T \leq t + \Delta t, D = k | T \geq t]}{\Delta t}, \qquad k = 1, \ldots, K. \tag{1}$$

It may be influenced by covariates' vector Z_{ik}:

$$\lambda_{ik}(t | \mathbf{Z}_{ik}) = \lambda_{k,0}(t) \exp\{\boldsymbol{\beta}_k^T \mathbf{Z}_{ik}(t)\}, \tag{2}$$

where $\lambda_{k,0}(t)$ is the baseline cause-specific hazard of cause k which does not need to be explicitly specified, $\mathbf{Z}_{ik}(t)$ is a vector of covariates for firm i specific to k-type hazard at time t, and the vector $\boldsymbol{\beta}_k$ represents the covariate effects on cause k to be estimated. Since the same variables could have different effects on the distinct risks, it is reasonable to assume that, for each k, $\boldsymbol{\beta}_k$ is independent of each other. Consequently, the baseline cause-specific hazard $\lambda_{k,0}(t)$ for the cause k is not required to be proportional to the baseline cause-specific hazard for another cause k' ($k \neq k'; k = 1, \ldots, K$).

By using the univariate Cox Proportional Hazard approach [13], the estimate of the coefficients' vector is obtained through the partial likelihood function for each specific hazard k:

$$L_k(\boldsymbol{\beta}_k) = \prod_{i=1}^{n_k} \frac{\exp\{\boldsymbol{\beta}_k^T \mathbf{Z}_{ik}(t_{ik})\}}{\sum_{l \in R(t_{ik})} \exp\{\boldsymbol{\beta}_k^T \mathbf{Z}_{lk}(t_{lk})\}}, \tag{3}$$

where n_k is the number of units in specific hazard k, and $R(t_{ik})$ is the set of units at risk at time t_{ik}.

If one is interested in the overall hazard function, the sum of all cause specific hazard functions may be computed as:

$$\lambda(t) = \sum_{k=1}^{K} \lambda_k(t), \tag{4}$$

and consequently the overall partial likelihood function is given by:

$$L(\boldsymbol{\beta}_1, \ldots, \boldsymbol{\beta}_K) = \prod_{k=1}^{K} \prod_{i=1}^{n_k} \frac{\exp\{\boldsymbol{\beta}_k^T \mathbf{Z}_{ik}(t_{ik})\}}{\sum_{l \in R(t_{ik})} \exp\{\boldsymbol{\beta}_k^T \mathbf{Z}_{lk}(t_{lk})\}}. \tag{5}$$

2.2 Lasso in Competing Risks Model

Since not all the covariates may contribute to the prediction of survival outcomes, the problem of interest is to select a subset of variables significantly associated with a specific failure type. Some of the variable selection techniques proposed for linear regression models have been extended to the context of survival models. They include best-subset selection, stepwise selection, asymptotic procedures based on score tests, bootstrap procedures [33] and Bayesian variable selection [22, 27]. However, the theoretical properties of these methods are generally unknown and they can be computationally too expensive in case of high dimensionality [20, 21]. Later, a family of penalized partial likelihood methods, such as the lasso [39], has been developed for Cox's proportional hazards model in survival analysis setting.

In the competing risks framework, the stability of the variable selection method is a major concern due to the presence of different types of risk. Our aim is to extend the lasso technique to the case of competing risks model.

Following [39], the lasso for the failure k is given by:

$$\hat{\boldsymbol{\beta}}_k = \underset{\boldsymbol{\beta}_k}{\operatorname{argmax}} \ \log(L_k(\boldsymbol{\beta}_k))$$

$$= \underset{\boldsymbol{\beta}_k}{\operatorname{argmax}} \sum_{i=1}^{n_k} \left[\boldsymbol{\beta}_k^T \mathbf{Z}_{ik}(t_{ik}) - \log \sum_{l \in R(t_{ik})} \exp\{\boldsymbol{\beta}_k^T \mathbf{Z}_{lk}(t_{lk})\} \right], \tag{6}$$

subject to

$$||\boldsymbol{\beta}_k||_1 \leq s_k,$$

where $||\boldsymbol{\beta}_k||_1 = |\beta_k^1| + |\beta_k^2| + \cdots + |\beta_k^p|$ is the L_1 norm of the coefficients vector $\boldsymbol{\beta}_k$ for the failure cause k, s_k is the tuning parameter that quantifies the magnitude of the constraints on the L_1 norm and determines the number of coefficients estimated as zero in the model, and p is the number of covariates.

As this is an optimization problem with constraint, the Lagrange multipliers method is applied. Thus, the Eq. (6) is equivalent to:

$$\hat{\boldsymbol{\beta}}_k = \operatorname*{argmin}_{\boldsymbol{\beta}_k} \ [-\log(L_k(\boldsymbol{\beta}_k)) + \lambda_k ||\boldsymbol{\beta}_k||_1], \tag{7}$$

where λ_k is the tuning parameter[1] determining the magnitude of penalty on the log partial likelihood.

Considering all independent failure causes K, the lasso estimator is given by:

$$(\hat{\boldsymbol{\beta}}_1, \ldots, \hat{\boldsymbol{\beta}}_K) = \operatorname*{argmax}_{\boldsymbol{\beta}_k} l(\boldsymbol{\beta}_1, \ldots, \boldsymbol{\beta}_K)$$

$$= \operatorname*{argmax}_{\boldsymbol{\beta}_k} \sum_{k=1}^{K} \sum_{i}^{n_k} \left[\boldsymbol{\beta}_k^T \mathbf{Z}_{ik}(t_{ik}) - \log \sum_{l \in R(t_{ik})} \exp\{\boldsymbol{\beta}_k^T \mathbf{Z}_{lk}(t_{lk})\} \right], \tag{8}$$

subject to

$$||\boldsymbol{\beta}_1||_1 \le s_1$$
$$||\boldsymbol{\beta}_2||_1 \le s_2$$
$$\ldots$$
$$||\boldsymbol{\beta}_K||_1 \le s_K.$$

This is equivalent to:

$$(\hat{\boldsymbol{\beta}}_1, \ldots, \hat{\boldsymbol{\beta}}_K) = \operatorname*{argmin}_{\boldsymbol{\beta}_k} \left[-l(\boldsymbol{\beta}_1, \ldots, \boldsymbol{\beta}_K) + \sum_{k=1}^{K} \lambda_k ||\boldsymbol{\beta}_k||_1 \right]. \tag{9}$$

In this case, a different tuning parameter for each vector of coefficients and for each type of failure has been considered, because the influence of the same variables on different failure types could be different, and we are interested in studying the effect of variables on the single survival state and not on the overall survival function.

3 Empirical Results

3.1 The Data

The proposed procedure is validated on a financial data set drawn from Amadeus database of Bureau van Dijk. The reference population consists of building Italian firms and the period considered here is from 2004 to 2009. The main interest is to investigate the determinants of firms that end up in financial distress for different causes. Namely, our attention is focused on three mutually exclusive states of exit from the market [34]: firms that have gone bankrupt, firms that have been liquidated, and firms being inactive. The bankrupt status includes those firms that have been

---

[1] In order to estimate the tuning parameter λ, the leave-one-out cross-validation is used [24, 37].

legally declared unable to pay its creditors and are under a Court supervision. The second status includes those companies that no longer exist because they have ceased their activities and are in the process of liquidation. The last state includes those firms that exit the database for other unknown reason. The reference group is given by active firms. From the population of active and non-active firms, a cluster random sample of $n = 1462$ firms based on the geographical distribution of the industrial firms across the regions is extracted. The final sample consists of 221 companies that went bankrupt, 129 that had entered voluntary liquidation, 228 that were inactive and finally 884 companies being in the active state.

Starting from the financial statements of each firm included in the sample, $nv = 20$ potential predictors, chosen among the most relevant in financial distress literature [4, 5, 15], are computed.

The original data set suffers from the presence of some accounting data observations that are severe outliers. In order to reduce the effects of anomalous observations, a pre-processing procedure is performed and those firms showing values of the financial predictors outside the 3th and 97th percentiles are excluded from the analysis. In order to achieve stability, a modified logarithmic transformation, defined for non-positive argument [30] is also applied.

Finally, the sample is divided into two parts: *in-sample set*, in order to determine how accurately a model classifies businesses, and *out-of-sample*, in order to determine how accurately a model classifies new businesses.

3.2 The Variables Selected

The goal of the analysis is to investigate the determinants of multiple causes of financial distress in competing risks model (in which the three exits are considered as competing exit routes) and compare the results with single-risk model (where the three exits are pooled). Moreover, once implemented the proposed variable selection procedure based on the lasso, the effects of the more significant variables on each exit type are investigated, and the gain of using an innovative selection method is evaluated over the traditional stepwise procedure.

The results for the estimated models are shown in Tables 1–3. They display the total number of variables selected in the single-risk and competing risks models according to the variable selection methods used (Table 1), the sign of the coefficients' estimates (Table 2) and the hazard ratios obtained by computing the exponential of coefficients β (Table 3).

From the results of Table 1, it can be observed that the lasso procedure leads to a smaller number of selected variables for both the single-risk and the competing risks approaches. Moreover, in the second case there is a relevant saving in terms of number of covariates. Actually, for classifying the bankrupted firms, only 8 variables are needed if the lasso approach is used, while 12 variables are selected as relevant by using the stepwise procedure. Again, for the inactive state, 11 and 2 variables are chosen by stepwise and lasso, respectively. Finally, for classifying the liquidated firms the stepwise approach selects 7 variables as the most significant, while the lasso

Table 1 Total number of variables selected by stepwise and lasso in the competing risks and the single risk models

Stepwise				Lasso			
Bankruptcy	Inactive	Liquidation	Single-Risk	Bankruptcy	Inactive	Liquidation	Single-Risk
12	11	7	12	8	2	3	10

chooses only 3 variables. Therefore, it seems that there is an advantage in using a shrinkage method over the classical one, in terms of model parsimony.

In order to further analyse the role of the selected variables, the sign of the coefficients' estimates (Table 2) for the competing risks (columns 3-5) and the pooled models (column 6) is analyzed. It is important to underline the meaning of signs within this context. A positive (negative) sign of coefficients means higher (lower) likelihood of becoming distressed and consequently lower (higher) probability of surviving in the market.

The sign of variables selected by stepwise accords with the one chosen by lasso, except for the return on capital employed. However, this discrepancy comes true only for two different states and therefore it may be neglected. Comparing the competing risks and the pooled models in terms of sign of coefficients, some differences can be noticed. In fact, the ROA, which provides a measure of how efficient the management is in using assets to generate earnings, being an indicator of a company profitability relative to total assets, has a negative coefficient for the pooled model, the bankrupted and liquidated states, while it is positive for the inactive state. Furthermore, there are other two ratios having different sign between bankruptcy and inactivity. The first is the current ratio, measuring whether or not a firm has enough resources to pay its short-term debts. The second is the gearing, a measure of financial leverage indicating the degree to which a firm's activities are funded by owner's funds versus creditor's funds. These differences may be related to the characteristics of the ratios and the states. As mentioned in Sect. 3.1, since the inactive state includes those firms that exit the market for any unknown reason, there could be some firms that have decided to stop their activity, even though they have an efficient management, leading to different values of the ratios.

The hazard ratio, reported in Table 3, evaluates the effect of the covariates on the hazard. A hazard ratio equal to one means that the variable has no effect on survival, whereas a hazard ratio greater (less) than one indicates that the effect of the covariate is to increase (decrease) the hazard rate. Looking at the results, it seems that the variables selected by lasso have a similar effect on the probability of becoming distressed according to the financial area of the ratios, even though the magnitude of impact is slightly different.

Table 2 Sign of coefficients' estimates for the competing risks and the single risk models

Variables	Area[a]	Bankruptcy Stepwise	Bankruptcy Lasso	Inactive Stepwise	Inactive Lasso	Liquidation Stepwise	Liquidation Lasso	Single-Risk Stepwise	Single-Risk Lasso
Return on shareholder funds	1			Negative					
Return on capital employed	1	Positive	Positive	Negative					
Return on total assets	1	Negative							
Profit margin	1	Negative							
EBITDA	1								
EBIT	1			Positive		Positive		Positive	Positive
Cash flow/Operating revenue	1			Positive				Positive	
ROE	1	Negative				Negative		Negative	Negative
ROA	1					Negative		Negative	
ROCE	1	Positive	Positive	Positive				Positive	Positive
Net assets turnover	2	Positive	Positive	Positive		Positive	Positive	Positive	Positive
Interest cover	2	Negative	Negative					Negative	Negative
Stock turnover	2	Negative						Negative	Negative
Collection period	2			Negative	Negative	Negative		Negative	Negative
Credit period	2	Negative	Negative	Negative	Negative	Negative	Negative	Negative	Negative
Current ratio	3	Negative	Negative	Positive					
Liquidity ratio	3	Positive						Positive	Positive
Shareholders liquidity ratio	3								
Solvency ratio	3	Negative	Negative	Negative		Negative	Negative	Negative	Negative
Gearing	3	Positive	Positive	Negative					

[a] Following the Amadeus classification, the numbers from 1 to 3 refer to the profitability, operational and structure area, respectively.

Table 3 Hazard ratios for the competing risks and the single risk models

Variables	Area[a]	Bankruptcy		Inactive		Liquidation		Single-Risk	
		Stepwise	Lasso	Stepwise	Lasso	Stepwise	Lasso	Stepwise	Lasso
Return on shareholder funds	1			0.9624					
Return on capital employed	1		1.0058	0.9248					
Return on total assets	1	1.4717							
Profit margin	1	0.9204							
EBITDA	1								
EBIT	1			1.1408		1.1129		1.0806	1.0267
Cash flow/Operating revenue	1			1.1613				1.0824	
ROE	1					0.9754		0.9770	0.9863
ROA	1			1.0785		0.9134		0.9484	
ROCE	1	0.7374	1.0814					1.0632	1.0311
Net assets turnover	2	1.0660	1.0865	1.0754		1.2108	1.0290	1.1340	1.0901
Interest cover	2	0.7439	0.8958					0.9399	0.9665
Stock turnover	2	0.9486						0.9725	0.9967
Collection period	2			0.7921	0.8141	0.9142		0.8842	0.8995
Credit period	2	0.7524	0.8286	0.8931	0.9660	0.8045	0.8700	0.8122	0.8257
Current ratio	3	0.6889	0.9882	1.2283					
Liquidity ratio	3	1.3949						1.2251	1.0073
Shareholders liquidity ratio	3								
Solvency ratio	3	0.8346	0.8270	0.8466		0.7213	0.8495	0.7879	0.8124
Gearing	3	1.1260	1.0046	0.9193					

[a] Following the Amadeus classification, the numbers from 1 to 3 refer to the profitability, operational and structure area, respectively.

3.3 The Accuracy Measures

In this section, the performance of the two variable selection methods here considered, in both competing risks and single-risk models, is evaluated by computing some accuracy measures: the proportion of firms classified correctly (accuracy), the *false-positive rate* (FP), that is the proportion of distressed firms misclassified as a non-distressed firm, the *false-negative rate* (FN), i.e. the proportion of non-distressed firms wrongly assigned to the distressed group, and the AUC, that is equal to the probability that a firm will rank a randomly chosen non-distressed firm higher than a randomly chosen distressed one [6, 18].

Table 4 shows these measures for both variables selection methods in competing risks and single-risk frameworks. Looking at the ratios computed for the *in-sample set*, used for assessing the classification ability, it is discovered that the lasso has better performance than the stepwise in terms of accuracy, while it has lower per-

Table 4 The accuracy measures for in-sample and out-of-sample sets

	In-Sample			
	Stepwise method			
	Bankruptcy	Inactive	Liquidation	Single-Risk
Accuracy	0.76225	0.78819	0.78153	0.74881
FP rate	0.40310	0.58025	0.47647	0.61441
FN rate	0.22188	0.17134	0.19222	0.11071
AUC	0.76051	0.68797	0.73364	0.73459
	Lasso method			
	Bankruptcy	Inactive	Liquidation	Single-Risk
Accuracy	0.78384	0.79022	0.79606	0.75234
FP rate	0.51938	0.58848	0.52500	0.72103
FN rate	0.18705	0.16817	0.17128	0.06458
AUC	0.74209	0.65524	0.72540	0.72669
	Out-of-Sample			
	Stepwise method			
	Bankruptcy	Inactive	Liquidation	Single-Risk
Accuracy	0.83158	0.85263	0.84662	0.87218
FP rate	0.66667	0.56250	0.44068	0.52564
FN rate	0.16616	0.13713	0.12541	0.07496
AUC	0.52971	0.68957	0.77809	0.72044
	Lasso method			
	Bankruptcy	Inactive	Liquidation	Single-Risk
Accuracy	0.84211	0.86165	0.86466	0.88421
FP rate	0.66667	0.56250	0.45763	0.67949
FN rate	0.15559	0.12789	0.10396	0.04089
AUC	0.79809	0.67604	0.78254	0.72985

formance in terms of both FP rate and AUC. However, by analyzing the results for the *out-of-sample set*, used for evaluating the prediction ability of the techniques, it can be noted that the lasso has a higher accuracy rate in all settings; it has a higher AUC for bankruptcy and inactivity states; it is a lower FN rate for all states, while it has equal FP rate in two cases, i.e. for the bankruptcy and the inactivity states. Therefore, the lasso seems to be a good alternative to the stepwise not only in terms of number of variables selected, but also in terms of accuracy, producing a relatively higher performance in predicting the firms' distress.

4 Conclusions

The literature on firms' distress has mainly investigated the exit from the market as a bivariate event analyzing each form of exit in unified framework. Moreover, most of the studies have not paid much attention to how selecting the variables that influence the exit from the market. This paper provides an innovative contribution in explaining the role of variable selection methods in competing risks setting for evaluating the decision to leave the market. A competing risks model is developed for examining the probability of going bankrupt, being inactive and liquidated on some financial characteristics of firms. The use of the lasso technique has been proposed as an alternative variables selection method, extending the available references to the competing risks model and comparing its performance over the widely used stepwise procedure.

The reached results have shown that using the lasso in selecting the most significant predictors for each type of exit can lead an advantage not only in terms of number of selected covariates but also in terms of predictability of the risk.

In fact, the shrinkage method allows classifying the distressed firms according to the different risks of coming out from the market by means of a considerably lower number of variables. Moreover, the role and the sign of the ratios selected by lasso are coherent with those selected by stepwise in terms of their interpretability.

Finally, the classification ability and the predictive accuracy of the methods, evaluated by means of correct classification, FP rate, FN rate and AUC, further confirm the benefit of using the shrinkage method.

References

1. Akaike, H.: Information Theory and an Extension of the Maximum Likelihood Principle. In: Petrov, B.N., Csaki F. (eds.) Second International Symposium on Information Theory, pp. 267–281 (1973)
2. Allison, P.D.: Discrete-time Methods for the Analysis of Event Histories. In: Leinhardt S. (eds.) Sociological Methodology, pp. 61–98. Jossey-Bass, San Francisco (1982)
3. Altman, E.I.: Financial Ratios, Discriminant Analysis and the Prediction of Corporate Bankruptcy. J. of Finance **23**, 589–609 (1968)

4. Altman, E.I.: Predicting financial distress of companies: revisiting the Z-score and ZTM model. New York University. Working Paper (2000)
5. Altman, E.I., Hochkiss, E.: Corporate Financial Distress and Bankruptcy: Predict and Avoid Bankruptcy, Analyze and Invest in Distressed Debt. John Wiley & Sons, New York (2006)
6. Amendola, A., Restaino, M., Sensini, L.: Variable selection in default risk model. J. of Risk Model Valid. 5(1), 3–19 (2011)
7. Antoniadis, A., Fryzlewicz, P., Letué, F.: The Dantzig selector in Cox's proportional hazards model. Scand. J. of Stat. 37(4), 531–552 (2010)
8. Andersen, P.K., Abildstrøm, S.Z., Rosthøj, S.: Competing Risks as a Multi-State Model, Stat. Methods in Med. Res. 11, 203–215 (2002)
9. Andersen, P.K., Borgan, Ø., Gill, R.D., Keiding, N.: Statistical Models based on Counting Processes. Springer, Berlin (1993)
10. Brabazon, A., Keenan, P.B.: A Hybrid Genetic Model for the Prediction of Corporate Failure. Comput. Manag. Sci. 1(3-4), 293–310 (2004)
11. Breiman, L.: Heuristics of instability and stabilization in model selection. Ann. of Stat. 24(6), 2350–2383 (1996)
12. Chancharat, N., Tian, G., Davy, P., McCrae, M., Lodh, S.: Multiple States of Financially Distressed Companies: Tests using a Competing-Risks Model. Australas. Account. Bus. and Finance J. 4(4), 27–44 (2010)
13. Cox, D.R.: Partial likelihood. Biom. 62(2), 269–276 (1975)
14. Dakovic, R., Czado C., Berg D.: Bankruptcy prediction in Norway: a comparison study. Appl. Econ. Lett. 17, 1739–1746 (2010)
15. Dimitras, A., Zanakis, S., Zopudinis, C.: A survey of businesses failures with an emphasis on failure prediction methods and industrial applications. Eur. J. of Oper. Res. 27, 337–357 (1996)
16. Dyrberg, A.: Firms in Financial Distress: An Exploratory Analysis. Working Paper 17/2004, Financial Market, Danmarks Nationalbank and Centre for Applied Microeconometrics (CAM), Institute of Economics, University of Copenhagen (2004)
17. Efron, B., Hastie, T., Johnstone, T., Tibshirani, R.: Least angle regression. Ann. of Stat. 32, 407–499 (2004)
18. Engelmann, B., Hayden, E., Tasche, D.: Testing rating accuracy. Risk 16, 82–86 (2003)
19. Fan, J., Li, R.: Variable Selection via Nonconcave Penalized Likelihood and its Oracle Properties. J. of the Am. Stat. Assoc. 96(456), 1348–1360 (2001)
20. Fan, J., Li, R.: Variable selection for Cox's proportional hazards model and frailty model. Ann. of Stat. 30, 74–99 (2002)
21. Fan, J., Lv, J.: Selective overview of variable selection in high dimensional feature space. Stat. Sinica 20, 101–148 (2010)
22. Faraggi, D., Simon, R.: Bayesian variable selection method for censored survival data. Biom. 54(4), 1475–1485 (1998)
23. Frank, I.E., Friedman, J.H.: A Statistical View of Some Chemometrics Regression Tools. Technometrics 35, 109–148 (1993)
24. Friedman, J., Hastie, T., Tibshirani, R.: Regularization Paths for Generalized Linear Models via Coordinate Descent. J. of Stat. Softw. 33(1), 1–22 (2010)
25. Hougaard, P.: Analysis of Multivariate Survival Data, Statistics for Biology and Health. Springer, New York (2000)
26. Hunter, D.R., Li, R.: Variable selection using MM algorithm. Ann. of Stat. 44(5), 1617–1642 (2005)

27. Ibrahim, J.G., Chen, M.-H., McEachern, S.N.: Bayesian Variable Selection for Proportional Hazards Models. Can. J. of Stat. **27**, 701–717 (1999)
28. Johnsen, T., Melicher, R.W.: Predicting corporate bankruptcy and financial distress: Information value added by multinomial logit models. J. of Econ. and Bus. **46**(4), 269–286 (1994)
29. Jones, S., Hensher, D.A.: Modelling corporate failure: a multinomial nested logit analysis for unordered outcomes. The Brit. Acc. Rev. **39**, 89–107 (2007)
30. Perederiy, V.: Bankruptcy Prediction Revisited: Non-Traditional Ratios and Lasso Selection. (Available at SSRN, 2009) http://ssrn.com/abstract=1518084 or http://dx.doi.org/10.2139/ssrn.1518084
31. Prantl, S.: Bankruptcy and voluntary liquidation: Evidence for new firms in East and West Germany after unification. Discussion paper no.03-72, ZEW, Centre for European Economic Research, London, pp. 1–42 (2003)
32. Rommer, A.D.: Firms in financial distress: an exploratory analysis. Working paper no.17, Danmarks Nationalbank and Centre for Applied Microeconometrics (CAM), Institute of Economics, University of Copenhagen, Copenhagen, pp. 1–68 (2004)
33. Sauerbrei, W., Schumacher, M.: A bootstrap resampling procedure for model building: application to the Cox regression model. Stat. in Med. **11**(16), 2093–2109 (1992)
34. Schary, M.: The probability of exit. RAND J. of Econ. **22**, 339–353 (1991)
35. Schwarz, G.: Estimating the Dimension of a Model. Ann. of Stat. **6**(2), 461–464 (1978)
36. Shumway, T.: Forecasting bankruptcy more accurately: A simple hazard model. J. of Bus. **74**(1), 101–124 (2001)
37. Simon, N., Friedman, J., Hastie, T., Tibshirani, R.: Regularization Paths for Cox's Proportional Hazards Model via Coordinate Descent. J. of Stat. Softw. **39**(5), 1–13 (2011)
38. Tibshirani, R.: Regression Shrinkage and Selection via the Lasso. J. of R. Stat. Soc. Ser. B **58**, 267–288 (1996)
39. Tibshirani, R.: The lasso method for variable selection in the Cox model. Stat. in Med. **16**, 385–395 (1997)
40. Zou, H., Hastie, T.: Regularization and Variable Selection via the Elastic Net. J. of the R. Stat. Soc. Ser. B **67**(Part 2), 301–320 (2005)

A Comparison Between Different Numerical Schemes for the Valuation of Unit-Linked Contracts Embedding a Surrender Option

Anna Rita Bacinello, Pietro Millossovich and Alvaro Montealegre

Abstract In this paper we describe and compare different numerical schemes for the valuation of unit-linked contracts with and without surrender option. We implement two different algorithms based on the Least Squares Monte Carlo method (LSMC), an algorithm based on the Partial Differential Equation Approach (PDE) and another based on Binomial Trees. We introduce a unifying way to define and solve the valuation problem in order to include the case of contracts with premiums paid continuously over time, along with that of single premium contracts, usually considered in the literature. Finally, we analyse the impact on the fair premiums of the main parameters of the model.

1 Introduction

Life insurance contracts are often very complex products that embed several types of options, more or less implicitly defined. The most popular implicit options are undoubtedly those implied by the presence of minimum guarantees in unit-linked life insurance. These options do not require any assumption on the policyholder behaviour unless early exercise features are also involved. Hence, their exercise is preference-free and their valuation resemble that of European-style Exotic options. The

A.R. Bacinello (✉)
Department of Business, Economics, Mathematics and Statistics 'B. de Finetti', University of Trieste, Trieste, Italy
e-mail: bacinel@units.it

P. Millossovich
Faculty of Actuarial Science and Insurance, Cass Business School, London, UK
e-mail: pietro.millossovich.1@city.ac.uk

A. Montealegre
Risk Methodology Team at Banco Santander, Madrid, Spain
e-mail: almontealegre@gruposantander.com

M. Corazza, C. Pizzi (eds.), *Mathematical and Statistical Methods for Actuarial Sciences and Finance*, DOI 10.1007/978-3-319-02499-8_3, © Springer International Publishing Switzerland 2014

surrender option, the possibility for the policyholder to early terminate the contract and receive a cash amount (the surrender value), is instead deeply affected by the policyholder behaviour. This option has become a major concern for insurers, especially in recent years, as policyholders are becoming increasingly attentive to alternative investment opportunities available in the market.

The surrender option is a typical American-style contingent-claim that does not admit a closed-form valuation formula, so that a numerical approach is called for. The valuation approaches proposed in the literature are based on binomial trees (see e.g. [2]), partial differential equations with free boundaries (PDE, e.g. [14]), or least squares Monte Carlo simulation (LSMC, [3,4]). The aim of this study is to compare the prices of unit-linked policies obtained by means of these different approaches. In particular this allows us to test the goodness of the LSMC approach, that provides in absolute terms the most general and flexible method and is unaffected by the dimensionality of the problem. We take as benchmarks the other approaches, relying on well established convergence results but requiring customization to the specific model. We introduce and treat, in a unified way, both the case of single premium contracts, usually considered in the literature, and that of periodic premiums paid continuously over time.

The paper is structured as follows. In Sect. 2 we present our valuation framework defining, in particular, the valuation problem, that is an optimal stopping problem. In Sect. 3 we tackle this problem with the LSMC approach, by providing two alternative algorithms for solving it, as proposed by [4] for quite general contracts. In Sect. 4 we use the PDE approach to formulate the problem, which is then solved by finite differences. For the binomial approach we directly refer the reader to [2], while in Sect. 5 we present some numerical results.

2 Valuation Framework

We analyse an endowment life insurance contract issued to a policyholder aged x at time 0. The contract provides a benefit B_T^s at the maturity T upon survival, or a benefit B_t^d in case of death at time t, with $0 < t \leq T$. Both benefits depend on the current value of a reference asset whose price at t is denoted by S_t. Hence

$$B_T^s = f^s(S_T), \quad B_t^d = f^d(t, S_t). \tag{1}$$

We consider both the case of single premium and that of periodic premiums paid upon survival. For the sake of simplicity, we assume that periodic premiums are paid continuously over the entire contract life at a rate π, so that in the interval $[t, t + \Delta]$ the amount of premiums paid is equal to $\pi\Delta$. Besides death and survival benefits, the policyholder is allowed to exit the contract before maturity, provided the policy has been in force for at least t^w years, with $0 \leq t^w \leq T$. In this case he receives a benefit

$$B_t^w = f^w(t, S_t) \tag{2}$$

if surrender takes place at $t \geq t^w$. Indeed, it is common practice for periodic premium contracts to require that the policyholder remains in the contract at least a few years (e.g. 2 or 3) in order to recover the initial expenses.

The model is specified under a probability $\mathbb{Q}$, which is assumed to be an equivalent martingale measure. Hence, under $\mathbb{Q}$, the price of any traded security is given by its expected discounted cash-flows (see [8]). Discounting is performed at the risk-free rate r, assumed here to be constant.

We use a geometric Brownian motion to model the $\mathbb{Q}$-dynamics of the reference asset price, defined as

$$\frac{dS_t}{S_t} = r\,dt + \sigma\,dW_t,$$

with (W_t) a Wiener process, $\sigma > 0$ the volatility parameter and $S_0 > 0$ given.

We denote by τ the time of death of the policyholder and by $m(t)$ its deterministic force of mortality. More precisely, $m(t)dt$ denotes the instantaneous conditional probability of death at age $x+t$. Under $\mathbb{Q}$, it is assumed that τ and (W_t) are independent.

Let $\theta \geq t^w$ be the time at which the policyholder decides to terminate the contract. Early termination can clearly occur only if the individual is still alive and the policy is still in force. Hence, surrender happens only if $\theta < \tau \wedge T$. The time θ is in general a stopping time with respect to the filtration $\mathbb{G} = (\mathscr{G}_t)$ jointly generated by (W_t) and τ, and we call it an *exercise policy*.

For a given exercise policy $\theta \geq t^w$ and a fixed time $t \leq T$ such that the contract is still in force (that is $t < \tau \wedge \theta$), the stochastic discounted benefit is given by

$$
\begin{aligned}
G_t(\theta) =& B_T^s\,e^{-r(T-t)}\,1_{\tau>T,\theta\geq T} + B_\tau^d\,e^{-r(\tau-t)}\,1_{\tau\leq T\wedge\theta} \\
&+ B_\theta^w\,e^{-r(\theta-t)}\,1_{\theta<T\wedge\tau} - \pi\int_t^T e^{-r(u-t)}\,1_{u<\tau\wedge\theta}\,du,
\end{aligned}
\tag{3}
$$

where 1_A is the indicator of the set A and $\pi = 0$ in the single premium case. Note that in the periodic premium case the amount $G_t(\theta)$ is net of future premiums payable by the policyholder.

The time t value of the contract (net in the case of periodic premium) is given by the usual risk neutral formula:

$$V_t(\theta) = E^{\mathbb{Q}}[G_t(\theta)|\mathscr{G}_t], \quad t \leq T. \tag{4}$$

The independence of τ and (W_t) implies an alternative expression for the contract value (see [4] for a more general framework):

$$V_t(\theta) = 1_{\tau>t}E^{\mathbb{Q}}\left[\widehat{G}_t(\theta)|\mathscr{F}_t\right], \quad t \leq T, \tag{5}$$

where $\mathbb{F} = (\mathscr{F}_t)$ is the filtration generated by (W_t) and, for $\theta > t$,

$$
\begin{aligned}
\widehat{G}_t(\theta) =& B_T^s\,_{T-t}E_{x+t}^r\,1_{\theta\geq T} + \int_t^T B_u^d\,m(u)\,_{u-t}E_{x+t}^r\,1_{\theta\geq u}du \\
&+ B_\theta^w\,_{\theta-t}E_{x+t}^r\,1_{\theta<T} - \pi\int_t^T\,_{u-t}E_{x+t}^r\,1_{u<\theta}\,du
\end{aligned}
\tag{6}
$$

with $_qE_y^j = \mathrm{e}^{-jq}{}_qp_y = \mathrm{e}^{-jq-\int_0^q m(y-x+u)\mathrm{d}u}$ being the expected present value of a pure endowment with term q and interest rate j, interpretable as a *mortality risk adjusted discount factor*, and $_qp_y$ is the q-years survival probability for a life aged y.

Finally, the contract value at time 0, V_0^*, is obtained by solving the optimal stopping problem

$$V_0^* = \sup_\theta V_0(\theta), \tag{7}$$

where the supremum is taken over all $\mathbb{G}$-stopping times $\theta \geq t^{\mathrm{w}}$. As shown by [4], and with (5) in mind, the supremum can alternatively be taken over the corresponding set of $\mathbb{F}$-stopping times.

In the single premium case we define a fair contract when the premium, say U^*, coincides with its initial value, that is $U^* = V_0^*$. Since in the periodic premium case V_0^* is a net value, we say that the contract is fair if instead

$$V_0^* = 0. \tag{8}$$

Equation (8) implicitly defines a fair premium rate π^* that has to be found numerically through an iterative procedure.

In order to single out the fair premium for the surrender option, it is convenient to compute the fair premium for the corresponding European version of the contract, namely without the surrender option. To this end, we denote by V_0^{E} the time 0 value of the European contract, given by $V_0^{\mathrm{E}} = V_0(\infty)$ (see (4) and (5)). Then, the fair premium in the single premium case, say $U^{*\mathrm{E}}$, coincides with the initial value of the contract V_0^{E}. In the periodic premium case, instead, the fair premium $\pi^{*\mathrm{E}}$ solves the equation $V_0^{\mathrm{E}} = 0$. Finally, the fair premiums of the surrender option are given by $U^{*\mathrm{S}} = U^* - U^{*\mathrm{E}}$, $\pi^{*\mathrm{S}} = \pi^* - \pi^{*\mathrm{E}}$ respectively.

3 Least-squares Monte Carlo Approach

The LSMC method, proposed by [11] for the valuation of American-style contingent claims, is based on the joint use of Monte Carlo simulation and Least Squares regression. We divide the interval $[0, T]$ in n subintervals of equal length $\Delta = T/n$, and let $\mathbb{T} = \{t_0, t_1, \ldots, t_n\}$ with $t_i = i\Delta$, $i = 0, 1, \ldots, n$. The contract can be surrendered at $t_i \in \mathbb{T}$, $t_i \geq t^{\mathrm{w}}$. Problem (7) is then replaced by its discretized version with the supremum computed over all stopping times $\theta \geq t^{\mathrm{w}}$ taking values in $\mathbb{T}$.

The LSMC approach relies on the dynamic programming principle and on estimating continuation values by regressing them against a set of suitable basis functions of the relevant state variables. In our setting we consider as state variables the reference asset price S (and related functions, see Sect. 5), and denote by $e = (e_1, \ldots, e_H)$ the set of basis functions. As in [4], we describe in what follows two alternative LSMC algorithms for solving (7). More in detail, Algorithm 1 is based on definitions (3)–(4), and the supremum in (7) is taken over $\mathbb{G}$-stopping times. As shown in [4], this results in the backward procedure starting at each simulated time of death (or maturity, whichever comes first). Algorithm 2 is instead based on definitions (5)–(6) and the supremum in (7) is taken over $\mathbb{F}$-stopping times. In the description of the algorithms, we will add the superscript h to all simulated variables to denote their value in the h-th simulation.

Algorithm 1

- **Step 0** *(Simulation)*. Simulate M paths of S over the time grid $\mathbb{T}$. Simulate further M times of death valued in $\mathbb{T} \cup \{\infty\}$, where conventionally we set $\tau^h = \infty$ if the policyholder is alive at maturity.
- **Step 1** *(Initialization)*. For $h = 1, \ldots, M$, set $C_T^h = B_T^{s,h}$ if $\tau^h > T$ and $C_{\tau^h}^h = B_{\tau^h}^{d,h}$ if $\tau^h \leq T$.
- **Step 2** *(Backward iteration)*. For $j = n-1, n-2, \ldots, 0$,
  - I *(Continuation values)*. Set $I_j = \{1 \leq h \leq M : \tau^h > t_j\}$ and, for $h \in I_j$, set $C_{t_j}^h = C_{t_{j+1}}^h e^{-r\Delta} - \pi(1 - e^{-r\Delta})r^{-1}$.
  - II If $t_j \geq t^w$:
    (Regression and comparison). Regress the continuation values $(C_{t_j}^h)_{h \in I_j}$ against $(e(S_{t_j}^h)_{h \in I_j})$, to obtain $\widetilde{C}_{t_j}^h = \beta_j^* \cdot e(S_{t_j}^h)$ for $h \in I_j$, where

$$\beta_j^* = \arg\min_{\beta_j \in \mathbb{R}^H} \sum_{h \in I_j} \left(C_{t_j}^h - \beta_j \cdot e(S_{t_j}^h) \right)^2;$$

    if $B_{t_j}^{w,h} > \widetilde{C}_{t_j}^h$ then set $C_{t_j}^h = B_{t_j}^{w,h}$.
- **Step 3** *(Initial value)*. Compute the time 0 value of the contract:

$$V_0^* = M^{-1} \sum_{h=1}^M C_0^h.$$

In the following we need the expected present value of a continuous life annuity with term q and interest rate j, defined by $_q\bar{a}_y^j = \int_0^q {}_uE_y^j \, du$.

Algorithm 2

- **Step 0** *(Simulation)*. Simulate M paths of S over the time grid $\mathbb{T}$.
- **Step 1** *(Initialization)*. Set $C_T^h = B_T^{s,h}$ for $h = 1, \ldots, M$.
- **Step 2** *(Backward iteration)*. For $j = n-1, n-2, \ldots, 0$,
  - I *(Continuation values)*. For $h = 1, \ldots, M$ let

$$C_{t_j}^h = {}_\Delta E_{x+t_j}^r C_{t_{j+1}}^h + \int_{t_j}^{t_{j+1}} B_u^{d,h} m(u)_{u-t_j} E_{x+t_j}^r \, du - \pi {}_\Delta \bar{a}_{x+t_j}^r.$$

  - II If $t_j \geq t^w$:
    (Regression and comparison). Regress the continuation values $(C_{t_j}^h)_{h=1,\ldots,M}$ against $(e(S_{t_j}^h))_{h=1,\ldots,M}$ to obtain $\widetilde{C}_{t_j}^h = \beta_j^* \cdot e(S_{t_j}^h)$ for $h = 1, \ldots, M$, where

$$\beta_j^* = \arg\min_{\beta_j \in \mathbb{R}^H} \sum_{h=1}^M \left(C_{t_j}^h - \beta_j \cdot e(S_{t_j}^h) \right)^2;$$

    if $B_{t_j}^{w,h} > \widetilde{C}_{t_j}^h$ set $C_{t_j}^h = B_{t_j}^{w,h}$.

• **Step 3** *(Initial value).* The time 0 value of the contract is:

$$V_0^* = M^{-1} \sum_{h=1}^{M} C_0^h.$$

Algorithms 1 and 2 can be used to compute the value V_0^E of the European version of the contract by ordinary Monte Carlo. To this end, one has to skip the regression step 2 II in the above algorithms, or set $t^w = T$.

4 Partial Differential Equation Approach

In this section we outline the partial differential equation (PDE) associated with the contract value, by using definitions (5) and (6) for it.

For $(t, S) \in [0, T] \times (0, \infty)$ denote by $v^*(t, S)$ the value at $t \leq T$ of a contract still in force, when $S_t = S$, so that $V^*(0) = v^*(0, S_0)$. By arbitrage free principles, it can be proved that v^* solves the following linear complementary problem (See [9, 14, 17]):

$$\begin{cases} (\mathscr{L}v^* + f^d m - \pi)(v^* - f^w) = 0 \\ \mathscr{L}v^* + f^d m - \pi \leq 0 \\ v^* \geq f^w \\ v^*(T, S) = f^s(S) \end{cases} , \qquad (9)$$

where the operator $\mathscr{L}$ is given by

$$\mathscr{L}v = -(r+m)v + \frac{\partial v}{\partial t} + \frac{\partial v}{\partial S} S r + \frac{1}{2} \frac{\partial^2 v}{\partial S^2} S^2 \sigma^2$$

and f^d and f^s are the death and survival benefits defined in (1). In (9), f^w is defined by (2) for $t \geq t^w$ while we set $f^w(t, S) = -\infty$ for $0 \leq t < t^w$. Note that this convention forces v^* to satisfy the PDE $\mathscr{L}v^* + f^d m - \pi = 0$ for $0 \leq t < t^w$. Together with the conditions in (9), one may need to add boundary conditions $v^*(t, 0) = g^0(t)$ and $v^*(t, +\infty) = g^{+\infty}(t)$ for $0 \leq t < T$. The functions g^0, $g^{+\infty}$ depend on f^d, f^s, f^w and whether the premium is single or periodic.

To simplify (9), we first transform the (t, S) plane into the (u, y) plane setting

$$\widetilde{v}^*(u, y) = v^*(t, S) \left({}_{T-t}E_{x+t}^r \right)^{-1}, \quad y = \log S + (T-t)r - u, \quad u = (T-t)\frac{\sigma^2}{2}. \quad (10)$$

The original linear complementary problem (9) can then be rewritten as follows:

$$\begin{cases} \left(\widetilde{\mathscr{L}v}^* - \widetilde{f}^d \right) \left(\widetilde{v}^* - \widetilde{f}^w \right) = 0 \\ \widetilde{\mathscr{L}v}^* - \widetilde{f}^d \geq 0 \\ \widetilde{v}^* \geq \widetilde{f}^w \\ \widetilde{v}^*(0, y) = \widetilde{f}^s(0, y) \end{cases} , \qquad (11)$$

for $y \in \mathbb{R}$ and $0 < u \leq T\frac{\sigma^2}{2}$, together with the boundary conditions $\widetilde{v}^*(u, -\infty) = \widetilde{g}^{-\infty}(u)$ and $\widetilde{v}^*(u, +\infty) = \widetilde{g}^{+\infty}(u)$ for $0 < u \leq T\frac{\sigma^2}{2}$, where

$$\mathscr{L} = \frac{\partial}{\partial u} - \frac{\partial^2}{\partial y^2}, \quad \widetilde{f}^d(u, y) = (m(t)f^d(t, S) - \pi)\frac{2}{\sigma^2}(_{T-t}E^r_{x+t})^{-1}, \quad \widetilde{f}^s(u, y) = f^s(S),$$

$$\widetilde{f}^w(u, y) = f^w(t, S)(_{T-t}E^r_{x+t})^{-1}, \quad \widetilde{g}^{-\infty}(u) = g^0(t), \quad \widetilde{g}^{+\infty}(u) = g^{+\infty}(t)$$

with (t, S) on the right hand side of these equalities expressed through (u, y) according to the transformation defined in (10). Note that the price v^{*E} of the contract without surrender option can be similarly found by solving the ordinary PDE $\mathscr{L}v^E + f^d m = \pi$. Problem (11) can now be solved using a standard finite difference algorithm (see, e.g., [8]).

5 Numerical Examples

In this Sect. we present some numerical results for the fair (single and periodic) premium of a contract embedding a surrender option. We consider an insured aged $x = 40$ at time 0 and assume for m a Weibull force of mortality, that is $m(t) = c_1^{-c_2}c_2(x + t)^{c_2 - 1}$, with parameters $c_1 = 83.6904$ and $c_2 = 8.2966$ as in [4]. We apply both algorithms described in Sect. 3 to compute contract values by LSMC, and take as benchmark the values obtained by solving (9).

In the LSMC scheme we consider a vector of basis functions of the form $(S_t)^{n_1}(B_t^w)^{n_2}$ with $n_1 + n_2 \leq 2$. It is well known that the choice of the type and number of basis functions adopted in the regression step plays a relevant role. An extensive analysis of its implications on robustness and convergence and its interaction with the number of simulations can be found in [1, 13, 15, 16]. A further investigation in the context of the present work to assess the impact of the choice made above would be required, and we leave it to future research.

In the LSMC method the time grid consists of 2 intervals per year and we run 800 000 simulations (20 groups of 40 000 simulations each) using antithetic variables to reduce the variance (see [10]). In the finite difference scheme we adopt a grid with 200 steps in the time dimension and 400 in the reference asset dimension. In the binomial model we fix 250 steps per year (see [2]). In the numerical experiments we choose the following set of basic parameters: the maturity $T = 20$, the risk-free rate $r = 4\%$, the initial value of the reference asset $S_0 = 100$ and its volatility $\sigma = 20\%$.

Single Premium We assume here that $t^w = 0$ and that the contract provides guarantees given by

$$B_T^s = f^s(S_T) = F_0 \max\left\{\frac{S_T}{S_0}, e^{r^s T}\right\}, \quad B_t^i = f^i(t, S_t) = F_0 \max\left\{\frac{S_t}{S_0}, e^{r^i t}\right\}$$

for i=d,w, where F_0 is the principal of the contract. The European value of the contract admits a closed form expression (see [6,7]) given by

$$V_0^E = U^{*E} = \frac{F_0}{S_0}\left(S_0 + \int_0^T \text{put}(t, S_0 e^{r^d t})\,_t p_x\, m(t)\mathrm{d}t + \text{put}(T, S_0 e^{r^s T})\,_T p_x\right),$$

where $_z p_y = e^{-\int_0^z m(y-x+u)\mathrm{d}u} = e^{-c_1^{-c_2}((y+z)^{c_2}-y^{c_2})}$ and $\text{put}(t,K)$ is the time 0 value of the European put with maturity t, strike K, underlying S in the Black-Scholes model (see [5]). In all the numerical examples we set $r^s = r^d = r^w = 2\%$ and $F_0 = S_0$. To solve problem (9), the appropriate boundary conditions are: for $0 \le t < T$

$$g^0(t) = F_0 \max\Big\{ \sup_{t \vee t^w \le u < T}\left(e^{r^d t}\,_{u-t}\bar{A}_{x+t}^{r-r^d} + e^{r^w t}\,_{u-t}E_{x+t}^{r-r^w}\right),$$

$$e^{r^d t}\,_{T-t}\bar{A}_{x+t}^{r-r^d} + e^{r^s t}\,_{T-t}E_{x+t}^{r-r^s}\Big\}, \qquad g^{+\infty}(t) = F_0\frac{S_t}{S_0},$$

where $_q\bar{A}_y^j = \int_0^q {}_s E_y^j\, m(y-x+s)\mathrm{d}s$ is the expected present value of a term life assurance payable at the time of death for a policyholder aged y, with term q years and interest rate j.

In the case considered in the numerical examples ($r^s = r^d = r^w \doteq r^g \le r$ and $t^w = 0$) it is easy to see that

$$g^0(t) = F_0 e^{r^g t}\max_{t \le u \le T}\bar{A}_{x+t:\overline{u-t|}}^{r-r^g} = F_0 e^{r^g t},$$

where $\bar{A}_{y:\overline{q|}}^j = {}_q\bar{A}_y^j + {}_q E_y^j$ is the expected present value of an endowment.

Periodic Premium Here we take $t^w = 3$ years. The contract provides guarantees given by

$$B_T^s = f^s(S_T) = F_0 \max\left\{\frac{S_T}{S_0}, e^{r^s T}\right\}, \quad B_t^i = f^i(t, S_t) = \frac{t}{T}F_0 \max\left\{\frac{S_t}{S_0}, e^{r^i t}\right\}$$

for i=d,w, where now B^w is relevant only for $t \ge t^w$. Hence benefits in case of death or surrender follow a pro-rata rule. The fair premium rate for the European contract admits the following closed form expression, that can be obtained through an argument similar to the one used in the single premium case:

$$\pi^{*E} = \frac{F_0\left[\left(S_0 + \text{put}(T, S_0 e^{r^s T})\right)\,_T p_x + \int_0^T \frac{t}{T}\left(S_0 + \text{put}(t, S_0 e^{r^d t})\right)\,_t p_x\, m(t)\mathrm{d}t\right]}{S_0\, T\, \bar{a}_x^r}.$$

Again, we set $r^s = r^d = r^w \doteq r^g = 2\%$ and $F_0 = S_0$. To solve problem (9), the appropriate boundary conditions are now: for $0 \le t < T$

$$g^0(t) = \max \left\{ \frac{F_0}{T} \left(e^{r^d t} \left(t \, _{T-t}\bar{A}^{r-r^d}_{x+t} + _{T-t}(\bar{I}\bar{A})^{r-r^d}_{x+t} \right) + e^{r^s T} \, _{T-t}E^{r-r^s}_{x+t} \right) - \pi \, _{T-t}\bar{a}^r_{x+t}, \right.$$

$$\left. \sup_{t \vee t^w \le u < T} \left(\frac{F_0}{T} \left[e^{r^d t} \left(t \, _{u-t}\bar{A}^{r-r^d}_{x+t} + _{u-t}(\bar{I}\bar{A})^{r-r^d}_{x+t} \right) + e^{r^w t} \, _{u-t}E^{r-r^w}_{x+t} \right] - \pi \, _{u-t}\bar{a}^r_{x+t} \right) \right\},$$

$$g^{+\infty}(t) = \max_{t \vee t^w \le u \le T} \left(\frac{F_0}{T} \frac{S_t}{S_0} \left(t + (\bar{I}\bar{A})^0_{x+t:\overline{u-t}|} \right) - \pi \, _{u-t}\bar{a}^r_{x+t} \right),$$

where $_q(\bar{I}\bar{A})^j_y = \int_0^q {_s}E^j_y \, m(y-x+s) \, s \, ds$ and $(\bar{I}\bar{A})^j_{y:\overline{q}|} = {_q}(\bar{I}\bar{A})^j_y + {_q}{_q}E^j_y$ are the expected present values of an increasing term assurance payable at the time of death and an increasing endowment, for a policyholder aged y, term q years and with instantaneous interest rate j.

In the case considered in the numerical examples it can be shown that

$$g^0(t) = \max_{t \vee t^w \le u \le T} b^0_t(u), \quad g^{+\infty}(t) = \max \left\{ b^{+\infty}_t(t \vee t^w), b^{+\infty}_t(T) \right\}$$

where

$$b^0_t(u) = \frac{F_0}{T} e^{r^g t} \left(t \bar{A}^{r-r^g}_{x+t:\overline{u-t}|} + (\bar{I}\bar{A})^{r-r^g}_{x+t:\overline{u-t}|} \right) - \pi \, _{u-t}\bar{a}^r_{x+t},$$

$$b^{+\infty}_t(u) = \frac{F_0}{T} \frac{S_t}{S_0} \left(t + (\bar{I}\bar{A})^0_{x+t:\overline{u-t}|} \right) - \pi \, _{u-t}\bar{a}^r_{x+t}.$$

Results In Table 1 we report the results for the fair single premium U^* and the fair periodic premium rate π^* obtained by applying the LSMC Algorithm 1 (A1 column), Algorithm 2 (A2), the PDE approach (PDE) and the binomial model (B), as well as the fair premium for the European version of the contract U^{*E} and π^{*E} respectively (E), for different levels of the risk-free rate r. We notice that, as expected, the fair single premiums U^* and U^{*E} decrease with r. Recall, in fact, that these premiums are simply the initial value of the liabilities of the insurance company (with and without surrender option). As for the fair periodic premiums, we notice instead an increasing trend, apart from the case of the European contract when r is very close (or equal) to the guaranteed rate r^g. This pattern has a less intuitive explanation. The fair periodic premium π^* (π^{*E} respectively) is the (unique) zero of a decreasing function of π (π^E), that represents the initial American (European) contract net value. For a given level of π (π^E), this function is the difference between the fair values of the insurance company liabilities and the policyholder liabilities, both decreasing with r. The first component can be further split into the fund value and the value of the guarantee (American or, respectively, European, 'Titanic' put option, see [12]). While the fund value is independent of r, the value of the guarantee decreases with it. Under realistic assumptions for the contract parameters, in particular when the guaranteed rate r^g is (sufficiently) below r, the value of the guarantee is relatively small with respect to the fund value, that represents the predominant component of the insurer's liability. In this case the premium component prevails on the guarantee component,

Table 1 Fair premiums (standard errors in round brackets) for different risk-free rates

	Single premium					Periodic premium				
r%	A1	A2	PDE	B	E	A1	A2	PDE	B	E
2.0	134.117	134.340	134.207	134.179	134.179	8.958	8.967	8.967	8.956	8.136
	(0.221)	(0.223)				(0.001)	(0.001)			
2.5	129.306	129.508	129.442	129.422	128.083	9.074	9.070	9.101	9.088	8.128
	(0.238)	(0.238)				(0.002)	(0.002)			
3.0	125.681	125.827	125.821	125.801	122.916	9.212	9.213	9.252	9.237	8.157
	(0.227)	(0.217)				(0.002)	(0.002)			
3.5	122.681	122.848	122.898	122.879	118.569	9.353	9.353	9.423	9.403	8.222
	(0.230)	(0.222)				(0.002)	(0.002)			
4.0	120.297	120.471	120.483	120.466	114.937	9.513	9.509	9.610	9.584	8.321
	(0.235)	(0.244)				(0.001)	(0.001)			
4.5	118.278	118.494	118.463	118.443	111.927	9.704	9.700	9.808	9.780	8.455
	(0.228)	(0.241)				(0.001)	(0.001)			
5.0	116.573	116.743	116.748	116.729	109.450	9.913	9.911	10.021	9.989	8.619
	(0.220)	(0.228)				(0.001)	(0.001)			
5.5	115.106	115.269	115.288	115.263	107.430	10.142	10.139	10.248	10.212	8.814
	(0.228)	(0.226)				(0.001)	(0.001)			
6.0	113.809	114.037	114.026	114.001	105.796	10.389	10.388	10.488	10.446	9.035
	(0.207)	(0.221)				(0.001)	(0.001)			

unless the periodic premium is unreasonably low, making thus the contract value increasing with r. To compensate this, the fair periodic premium increases as well. For unrealistic contract parameters (in which the guaranteed rate is very close to, or even above, r), instead, the guarantee component can be important and prevail on the premium component not only for small levels of π (π^E). This can produce a contract value decreasing with r, hence a fair periodic premium decreasing as well. In Table 1 this happens only for European contracts, but in other experiments (not reported here) where we have kept $r \leq r^g$ this happens also for American contracts. Moreover, from Table 1 we observe that the sensitivity of all premiums with respect to r is rather strong, as expected, that the value of the surrender option is almost never negligible (apart from the single premium case when $r = 2\%$), and that the American premiums obtained with the different numerical approaches are very close to each other.

In Table 2 we show similar results as before (for single and periodic fair premiums), when the contract maturity changes. In the single premium case, of course, the initial value of the American contract (U^*) increases with its maturity. Recall in fact that the guarantee component is an American Titanic put, while the fund component is independent of T. This is not true, instead, for European contracts (in particular for the European Titanic put): in this case U^{*E} initially increases, reaches a maximum for $T = 15$ years and after decreases. The influence of T on the fair single premium is not so strong, while it becomes very important in the periodic premium case, in which

Table 2 Fair premiums (standard errors in round brackets) for different maturities

	Single premium					Periodic premium				
T	A1	A2	PDE	B	E	A1	A2	PDE	B	E
10.0	116.919	116.937	117.176	117.172	114.559	15.045	15.046	15.009	15.004	13.896
	(0.108)	(0.110)				(0.001)	(0.001)			
12.5	118.142	118.154	118.297	118.293	114.937	12.859	12.862	12.867	12.853	11.678
	(0.146)	(0.135)				(0.001)	(0.001)			
15.0	119.027	119.080	119.181	119.175	115.080	11.374	11.374	11.428	11.408	10.191
	(0.179)	(0.162)				(0.001)	(0.001)			
17.5	119.782	119.825	119.898	119.886	115.064	10.335	10.336	10.387	10.369	9.125
	(0.187)	(0.186)				(0.001)	(0.001)			
20.0	120.297	120.471	120.483	120.466	114.937	9.513	9.509	9.607	9.584	8.321
	(0.235)	(0.244)				(0.001)	(0.001)			
22.5	120.685	120.936	120.967	120.942	114.734	8.889	8.881	8.987	8.967	7.693
	(0.126)	(0.154)				(0.002)	(0.001)			
25.0	121.279	121.578	121.372	121.333	114.480	8.376	8.373	8.494	8.466	7.187
	(0.258)	(0.274)				(0.002)	(0.001)			
27.5	121.515	121.810	121.708	121.656	114.197	7.948	7.943	8.075	8.046	6.767
	(0.351)	(0.414)				(0.001)	(0.001)			
30.0	121.532	121.999	121.985	121.919	113.901	7.581	7.571	7.718	7.684	6.410
	(0.260)	(0.324)				(0.001)	(0.001)			

the premium is halved when the maturity increases from 10 to 30 years. The periodic premium is decreasing, because the longer is the contract duration, the longer is the period in which the payment for the insurance company liabilities is split. Also, in this table, we can see that the different numerical approaches for computing the American contract value do not lead to significant differences in the results. Recall, in particular, that when computing the fair periodic premium two approximations are involved: the first concerns the contract value, the second the zero search, and numerical errors could propagate.

In Table 3 we report the results for different levels of the reference fund volatility parameter σ. In this case, as expected, both single and periodic premiums strongly react to changes in the volatility parameter, increasing with it. The other results (comparison between different numerical approaches, value of the surrender option) are similar to those of the previous tables.

Finally, in Table 4 we report the mean relative errors of the LSMC results with respect to the benchmark, that is the value obtained with the PDE approach.

Table 3 Fair premiums (standard errors in round brackets) for different fund volatilities

	Single premium					Periodic premium				
σ%	A1	A2	PDE	B	E	A1	A2	PDE	B	E
10.0	107.355	107.544	107.495	107.477	103.685	7.986	7.985	7.995	7.973	7.505
	(0.083)	(0.092)				(0.000)	(0.000)			
12.5	110.442	110.657	110.602	110.586	106.248	8.341	8.344	8.360	8.339	7.691
	(0.112)	(0.111)				(0.001)	(0.001)			
15.0	113.714	113.928	113.847	113.829	109.050	8.711	8.714	8.755	8.733	7.894
	(0.142)	(0.144)				(0.001)	(0.001)			
17.5	117.031	117.214	117.157	117.137	111.970	9.102	9.106	9.172	9.149	8.106
	(0.191)	(0.192)				(0.001)	(0.001)			
20.0	120.297	120.471	120.483	120.466	114.937	9.513	9.509	9.609	9.584	8.321
	(0.235)	(0.244)				(0.001)	(0.001)			
22.5	123.600	123.760	123.804	123.782	117.906	9.931	9.921	10.061	10.036	8.537
	(0.293)	(0.278)				(0.002)	(0.002)			
25.0	126.762	126.862	127.092	127.066	120.844	10.351	10.348	10.528	10.503	8.750
	(0.321)	(0.301)				(0.002)	(0.002)			
27.5	129.546	129.752	130.328	130.298	123.730	10.804	10.791	11.010	10.984	8.960
	(0.463)	(0.453)				(0.002)	(0.002)			
30.0	132.521	132.550	133.500	133.464	126.547	11.267	11.237	11.504	11.477	9.164
	(0.553)	(0.433)				(0.002)	(0.002)			

Table 4 Relative root mean squared errors (in percentage)

	Single		Periodic	
	A1	A2	A1	A2
interest rate	0.141	0.013	0.743	0.752
maturity	0.174	0.027	0.846	0.881
volatility	0.268	0.129	1.058	1.107

Acknowledgements We acknowledge financial support from PRIN 2008 research project 'Retirement saving and private pension benefits: individual choices, risks borne by the providers' and from Ministerio de Economia y Competitividad grant ECO2011-28134, partially supported by FEDER funds. Earlier versions of the paper have been presented at the 15th IME conference in Trieste, the 'Italy-China Joint Workshop on Risk Evaluation and Insurance Techniques for Social Security Management' in Florence, the Meeting of the PRIN 2008 research group in Parma, the 5th MAF conference in Venice and the XIII Iberian-Italian Congress of Financial and Actuarial Mathematics in Cividale del Friuli.

References

1. Areal, N., Rodrigues, A., Armada, M.R.: On improving the least squares Monte Carlo option valuation method. Rev Deriv. Res. **11**(1–2), 119–151 (2008)
2. Bacinello, A.R.: Endogenous model of surrender conditions in equity-linked life insurance. Insur. Math. Econ. **37**(2), 270–296 (2005)
3. Bacinello, A.R., Biffis, E., Millossovich, P.: Pricing life insurance contracts with early exercise features. J. Comput. Appl. Math. **233**(1), 27–35 (2009)
4. Bacinello, A.R., Biffis, E., Millossovich, P.: Regression-based algorithms for life insurance contracts with surrender guarantees. Quant. Financ. P. (9), 1077–1090 (2010)
5. Black, F., Scholes, M.: The pricing of options and corporate liabilities. J. Polit. Econ. **81**(3), 637–654 (1973)
6. Boyle, P.P., Schwartz, E.S.: Equilibrium prices of guarantees under equity-linked contracts. J. Risk Insur. **44**(4), 639–660 (1977)
7. Brennan, M.J., Schwartz, E.S.: The pricing of equity-linked life insurance policies with an asset value guarantee. J. Financ. Econ. **3**(3), 195–213 (1976)
8. Duffie, D.: Dynamic asset pricing theory, 3rd edn. Princeton University Press, Princeton (2001)
9. Friedman, A., Shen, W.: A variational inequality approach to financial valuation of retirement benefits based on salary. Financ. Stoch. **6**(3), 273–302 (2002)
10. Glasserman, P.: Monte Carlo methods in financial engineering. Springer (2003)
11. Longstaff, F.A., Schwartz, E.S.: Valuing American options by simulation: a simple least-squares approach. Rev. Financ. Stud. **14**(1), 113–147 (2001)
12. Milevsky, M.A., Posner, S.E.: The Titanic option: valuation of the guaranteed minimum death benefit in variable annuities and mutual funds. J. Risk Insur. **68**(1), 93–128 (2001)
13. Moreno, M., Navas, J.F.: On the robustness of least-squares Monte Carlo (LSM) for pricing American derivatives. Rev. Deriv. Res. **6**(2), 107–128 (2003)
14. Shen, W., Xu, H.: The valuation of unit-linked policies with or without surrender options. Insur. Math. Econ. **36**(1), 79–92 (2005)
15. Stentoft, L.: Assessing the least squares Monte Carlo approach to American option valuation. Rev. Deriv. Res. **7**(2), 129–168 (2004)
16. Stentoft, L.: American option pricing using simulation and regression: numerical convergence results. In: Cummins, M., Murphy, F., Miller, J.J.H. (eds.) Topics in Numerical Methods for Finance. Springer Proceedings in Mathematics & Statistics **19**, pp. 57–94 (2012)
17. Wilmott, P., Dewynne, J., Howison, S.: Option pricing: mathematical models and computation. Oxford Financial Press (1993)

Dynamic Tracking Error with Shortfall Control Using Stochastic Programming

Diana Barro and Elio Canestrelli

Abstract In this contribution we tackle the issue of portfolio management combining benchmarking and risk control. We propose a dynamic tracking error problem and we consider the problem of monitoring at discrete points the shortfalls of the portfolio below a set of given reference levels of wealth. We formulate and solve the resulting dynamic optimization problem using stochastic programming. The proposed model allows for a great flexibility in the combination of the tracking goal and the downside risk protection. We provide the results of out-of-sample simulation experiments, on real data, for different portfolio configurations and different market conditions.

1 Introduction

Measuring risk is a crucial issue in financial modeling and it is relevant both for pricing purposes and for asset allocation problems. In particular, investors are concerned with the measurement and the management of risk in such a way that they can obtain a portfolio which is compliant with their risk attitude.

The majority of investors are more concerned with downside risk rather than upside risk, and there is experimental evidence that they treat differently losses from gains. This introduces the need for asymmetric risk measures and can account for the growing interest of investors in mean return-downside risk portfolio models. The key idea of these approaches is to separate the downside deviations from the upside

D. Barro (✉)
Department of Economics and SSAV, University Ca' Foscari Venice, Cannaregio 873, 30121 Venice, Italy
e-mail: d.barro@unive.it

E. Canestrelli
Department of Economics, University Ca' Foscari Venice, Cannaregio 873, 30121 Venice, Italy
e-mail: canestre@unive.it

M. Corazza, C. Pizzi (eds.), *Mathematical and Statistical Methods for Actuarial Sciences and Finance*, DOI 10.1007/978-3-319-02499-8_4, © Springer International Publishing Switzerland 2014

potential and control only the first part of risk. For a classification of different risk measure in portfolio selection see, for example, [7].

Risk measurement is strictly connected with the definition of a term of comparison with respect to which we can contrast and compare the risk/return profile of our portfolio. Moreover, it is common practice to monitor the performance of a portfolio, or a fund manager, with reference to an explicitly declared, or implicitly assumed, benchmark. This allows for a more objective assessment of the risk profile and the performance evaluation of the investment, linking it with current market conditions.

The tracking error and tracking error volatility are widely used measures of how closely the investments behave with respect to the reference portfolio. However, they are symmetric measures of distance and dispersion and cannot account for investor aversion for downside rather than upside deviations.

In this contribution we aim at jointly considering the presence of a benchmark and the issue of controlling downside risk. We introduce a set of barriers which accounts for loss/gain preferences of the investor, i.e. the shortfall can be computed with respect to a given level of acceptable losses or with respect to a given desired level of minimum return for the portfolio.

The risk is thus measured by two different components, the first reflects the risk profile of the benchmark; nevertheless, we consider that investors are more concerned with the downside risk and, in particular, with negative deviations from certain reference levels. The resulting portfolio accounts for these two aspect of risk. An indirect measure of risk through the choice of a benchmark and a more direct control on the values of the portfolio through the reference levels. The approach is flexible and allows to easily accounting for different investor preferences. To express risk aversion in portfolio management problems other approaches are possible, mainly based on the definition of a proper utility function and of risk aversion coefficients. They are particularly interesting from a theoretical point of view and have been explored in the literature.

However, in this contribution, we are interested in investigating the connection between the tracking error goal and the control of downside risk both from the point of view of an investor and of a fund manager. To this aim, the use of reference points and the management of risk through shortfalls from the set of specified threshold levels of wealth represent, in our opinion, an easily understandable way of measuring and communicating risk.

The structure of the paper is as follows. In Sect. 2, we briefly present the contributions in the literature which deal with benchmarking and shortfall control. In Sect. 3, we present and discuss our model for multiperiod tracking error with shortfall. In Sect. 4, we present an application of the proposed model and, in order to account for different market conditions, we consider out-of-sample simulation experiments for different periods. Section 5 concludes.

2 Literature Review

Different contributions in the literature tackled the issue of benchmarking and tracking error. In a static portfolio selection framework, see, for example, [3, 11, 24]. For a discussion on the reliability of tracking error as a measure of risk see [26]. While for the use of asymmetric tracking error see [16, 20, 24].

Among the contributions on dynamic tracking error problems, in continuous or discrete time setting, we refer to [2, 8, 12, 17].

The concern of investors for downside risk has led to the development of a huge stream of financial literature on the use of asymmetric and tail risk measures in portfolio selection problems. There are many contributions which propose the use of alternative risk measures, among them, see for example, [1, 7, 9, 10, 15, 18, 19, 25].

We are interested in considering a multiperiod tracking error problem and a discrete time monitoring of the shortfalls below given threshold levels of wealth, we consider both symmetric and asymmetric tracking error measures. To this aim we formulate and solve a multistage stochastic programming problem which provides us with enough flexibility in the formulation of the objective function and of the constraints. To deal with uncertainty in optimization problems other approaches are possible and, in particular, we mention Robust Optimization and its application also to financial optimization problems (see, for example, [4, 6, 23]).

Dempster et al., in [13, 14], tackled the problem of dynamic portfolio management for a pension fund in presence of minimum guarantees. They propose to consider, as objective function, for their multistage stochastic programming problem, the minimization of expected average shortfall and of expected maximum shortfall.

Different contributions in the literature consider the introduction of a shortfall constraint in portfolio management; in particular, for a discussion on the use of shortfall as a risk measure in asset allocation and in static tracking error problem, we refer to [5]. In our problem we consider a discrete monitoring of the portfolio level through the measurement of the shortfalls with respect to a set of reference levels of wealth. This goal is then combined with a tracking error objective and our model can be specified in different ways to account for symmetric and asymmetric distance measures both with respect to the risky benchmark and with respect to the wealth barriers. Moreover, we can allow for a trade-off between the two terms according to the investor's preferences.

3 Model Formulation

We consider a dynamic tracking error problem and we assume that the investor is interested in tracking the performance of a risky benchmark over time, where the benchmark itself is treated as a stochastic component.

We consider the arborescent formulation of the problem and a scenario tree from $t = 0$ (current state) to T; we denote with k_t a generic node in the event tree at time t and with π_{k_t} the associated probability. We denote with y_{k_t} the value of the managed portfolio in each node and with x_{k_t} the value of the stochastic benchmark. To control the value of the portfolio we introduce a set of reference levels of wealth, z_{jt} with $j = 1, \ldots, J$, which act as thresholds with respect to which we monitor the behavior of the portfolio. In particular, we are interested in monitoring the shortfalls of the portfolio value below each threshold level. The threshold levels are not stochastic but can be time dependent.

The investor can choose among a set of risky assets and a liquidity component. We denote with q_{ik_t}, $i = 1, \ldots, n$, and l_{k_t} the holdings in each asset, while, we use a_{ik_t} and v_{ik_t} to denote the amounts of risky asset purchased and sold in each node, respectively. The liquidity component absorbs the turnover in the portfolio and accounts also for proportional transaction costs (tc). Moreover, for each node k_t, we denote with $r_{k_t} = (r_{1k_t}, \ldots, r_{nk_t}, r_{lk_t})$ the vector of returns of the risky assets and for the liquidity component for period $[t-1;t]$.

We want our model to account both for symmetric and asymmetric distance measures from the risky benchmark and from the reference levels. To this aim we choose to use a mean absolute deviation model (MAD). The mean absolute deviation measure presents many advantages: it leads to a linear optimization problem (see, for example, [21, 22, 24]), and can be easily separated into positive and negative deviations, allowing for the required flexibility.

In more detail, we define the distance measure of the managed portfolio from the risky benchmark, in node k_t, as follows

$$|y_{k_t} - x_{k_t}| = \max[y_{k_t} - x_{k_t}; 0] + \max[-y_{k_t} + x_{k_t}; 0] = \theta_{k_t}^+ + \theta_{k_t}^-. \qquad (1)$$

With respect to the threshold levels of wealth z_{jt}, with $j = 1, \ldots, J$, we are interested in considering only negative deviations from the reference levels and thus we propose to use the following asymmetric distance measure

$$[y_{k_t} - z_{jt}]^- = \max[-y_{k_t} + z_{jt}; 0] = \gamma_{jk_t}^-. \qquad (2)$$

For a discussion on how the MAD model can be transformed into a linear optimization problem see, for example, [22, 24, 27].

The resulting multiperiod stochastic programming problem is

$$\min \sum_{t=0}^{T} \left[\sum_{k_t=K_{t-1}+1}^{K_t} \pi_{k_t} \left(c^+ \theta_{k_t}^+ + c^- \theta_{k_t}^- \right) + \sum_{k_t=K_{t-1}+1}^{K_t} \pi_{k_t} \sum_{j=1}^{J} d_j^- \gamma_{jk_t}^- \right] \qquad (3)$$

$$\theta_{k_t}^+ - \theta_{k_t}^- = y_{k_t} - x_{k_t} \qquad (4)$$

$$\gamma_{jk_t}^- \geq -y_{k_t} + z_{jt} \quad j = 1, \ldots, J \qquad (5)$$

$$y_{k_t} = l_{k_t} + \sum_{i=1}^{n} q_{ik_t} \qquad (6)$$

$$q_{ik_t} = (1 + r_{ik_t}) \left[q_{if(k_t)} + a_{if(k_t)} - v_{if(k_t)} \right] \qquad (7)$$

$$l_{k_t} = (1 + r_{lk_t}) \left[l_{f(k_t)} - \sum_{i=1}^{n} (1 + tc) a_{if(k_t)} + \sum_{i=1}^{n} (1 - tc) v_{if(k_t)} \right] \qquad (8)$$

$$a_{ik_t} \geq 0 \quad v_{ik_t} \geq 0 \tag{9}$$

$$q_{ik_t} \geq 0 \quad l_{k_t} \geq 0 \tag{10}$$

$$\theta_{k_t}^+ \geq 0 \quad \theta_{k_t}^- \geq 0 \tag{11}$$

$$\gamma_{k_t}^- \geq 0 \tag{12}$$

$$q_{i0} = \bar{q}_i \quad l_0 = \bar{l} \tag{13}$$

$$i = 1, \ldots, n \quad k_t = K_{t-1} + 1, \ldots, K_t \quad t = 1, \ldots, T$$

where c^+, c^- and d_j^- are positive weights, which may account for a progressive penalization of the deviations or for goal preferences with respect to different threshold levels.

Equation (6) represents the portfolio composition in each node, while Eqs. (7)–(8) describe the dynamics of the assets in the portfolio moving from an ancestor node $f(k_t)$ to a descendent node k_t. Finally, Eq. (10) provide the non-negativity conditions on the portfolio composition, ruling out the possibility for short-selling and borrowing, and (13) give the initial portfolio endowments. The objective function of our problem accounts for two different terms. The first is a tracking goal with respect to the risky benchmark while the second term accounts for the shortfalls below the reference levels. The investor is interested in minimizing the distance from the risky benchmark while at the same time limiting the downside risk measured through the shortfalls.

The risk profile of the resulting optimal portfolios takes into account both the connection with the benchmark and the risk aversion for gain or losses lower than a set of specified barriers which are settled by the investors according to their investment goals. The control through reference levels is flexible and presents different advantages in the formulation of the problem. First, barriers are allowed to be time dependent and the risk control can be tailored along the investment horizon. For example, it can be made tighter towards the end of the planning period to account for wealth conservation objectives. Second, the control is introduced in the objective function, rather than in the constraints, in this way we allow for a trade-off with the benchmarking goal and we are able to account for different risk profiles in the investment.

4 Computational Experiments

In the following we provide an application of the proposed model to real data through an out-of-sample exercise of portfolio management. We assume that our investor is interested in tracking the MSCI Europe Index using a subset of MSCI Style Indexes. To test our model we use a weekly dataset from June 6, 2007 to May 16, 2012. Summary statistics on the Indexes are provided in Table 1.

We apply our model simulating the management of the portfolio over a 10-week period using a rolling-horizon procedure. In our experiments we consider two periods to account for different market conditions. The first simulation period ranges

Table 1 Summary statistics MSCI Europe Index and MSCI Europe Style Indexes, weekly data from June 6, 2007 to May 16, 2012

	Mean	Variance	Skewness	Kurtosis
MSCI EUROPE	−0.0014	0.0009	−0.5579	1.4773
MSCI EUROPE LG	−0.0004	0.0007	−0.6569	1.9806
MSCI EUROPE LV	−0.0021	0.0012	−0.4599	1.2498
MSCI EUROPE MG	−0.0010	0.0011	−0.5418	1.6296
MSCI EUROPE MV	−0.0024	0.0013	−0.1217	0.8914
MSCI EUROPE SMG	−0.0008	0.0011	−0.5507	1.7773
MSCI EUROPE SMV	−0.0019	0.0013	−0.1505	0.8251
MSCI EUROPE SG	−0.0005	0.0012	−0.5613	1.9879
MSCI EUROPE SV	−0.0013	0.0013	−0.1420	0.7773

from December 14, 2011 to February 22, 2012, during this period the benchmark has a significant positive trend, as can be seen from Fig. 1. The second simulation period is from February 8, 2012 to April 18, 2012 and the market experiences both huge drops and raises, see Fig. 2.

For each simulation period, we test different configurations of the objective function in order to analyze the behavior of the optimized tracking portfolio with respect to the benchmark and the threshold levels. In more detail we consider the following settings (in Table 2 we summarize the coefficients of the objective function for the portfolio configurations considered in the experiments):

- Portfolio 1a – pure benchmark tracking;
- Portfolio 1b – benchmark tracking plus a shortfall control;
- Portfolio 2a – asymmetric benchmark tracking plus shortfall control;
- Portfolio 2b – asymmetric benchmark tracking plus enhanced shortfall control;
- Portfolio 2c – asymmetric benchmark tracking plus enhanced shortfall control;
- Portfolio 3a – pure shortfall control – one threshold – (no benchmark tracking);
- Portfolio 3b – pure shortfall control – two thresholds – (no benchmark tracking).

Table 2 Parameters settings for the different portfolio configurations considered in the computational experiments

	c^+	c^-	d_1^-	d_2^-
portfolio 1a	1	1	0	0
portfolio 1b	1	1	1	1
portfolio 2a	0	1	1	1
portfolio 2b	0	1	1	10
portfolio 2c	0	1	1	100
portfolio 3a	0	0	1	0
portfolio 3b	0	0	1	1

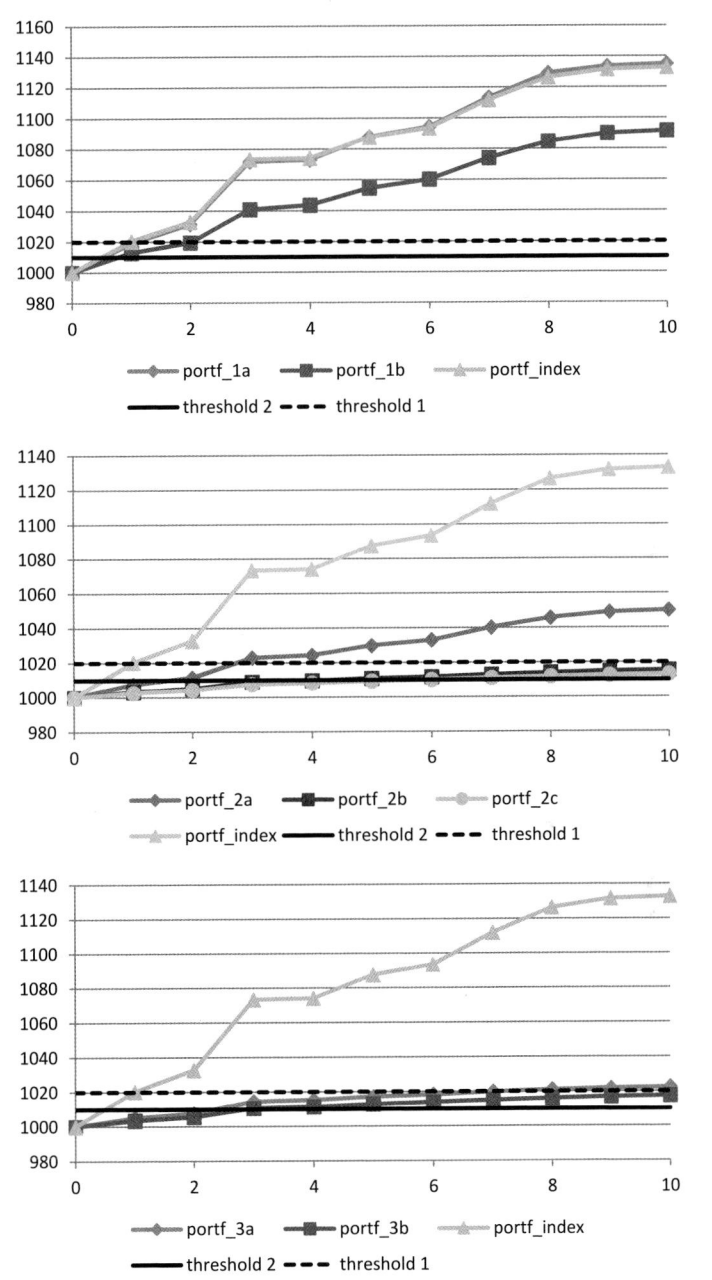

Fig. 1 Comparison between optimized tracking portfolios and benchmark, rolling simulation over 10-week period – December 14, 2011 to February 22, 2012 – two threshold levels $z_1 = 1020$ and $z_2 = 1010$

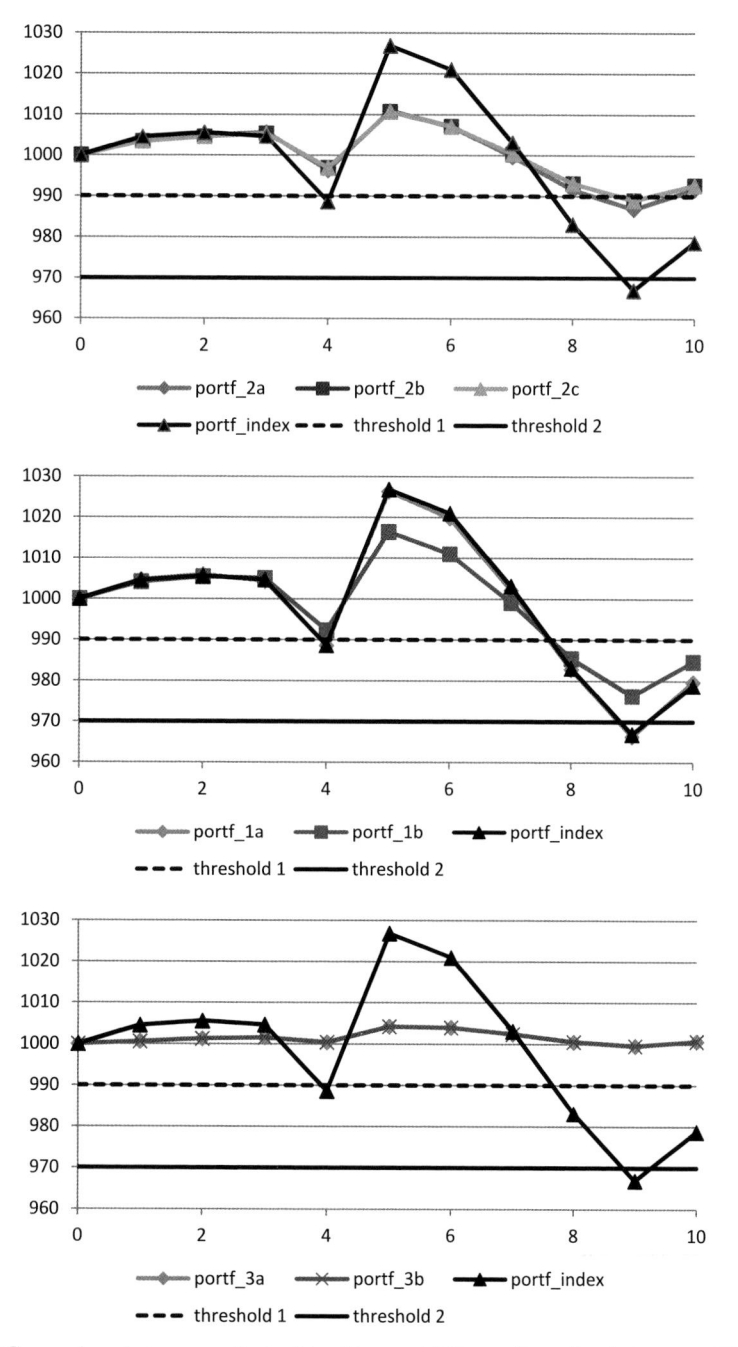

Fig. 2 Comparison between optimized tracking portfolios, rolling simulation over 10-week period - February 8, 2012 to April 18, 2012- two threshold levels $z_1 = 990$ and $z_2 = 970$

The threshold levels used in the simulation periods ($J = 2$) are as follows. For the first period we set $z_{1,t} = 1010$ and $z_{2,t} = 1020$ constant for all $t = 0, \ldots, T$; while we choose $z_{1,t} = 970$ and $z_{2,t} = 990$, for all $t = 0, \ldots, T$, for the second simulation period.

In more detail, the management experiments are carried out as follows. At each step of the simulation we generate a 2-stage scenario tree; we solve the optimization problem (3)–(13) and take the first period optimal decision. The portfolio composition is then evaluated using the true realized returns in the market and the resulting value represents the new endowment for the following period. Clearly, the results of the management experiments depend on the reliability of the generated scenario trees. Many different models can be used to estimate future expected returns and generate the event trees. In this contribution we do not tackle the issue of comparing different models; we propose to generate the scenarios using historical simulation, which assumes only that past returns are good predictors for future behavior without further hypothesis on the return distributions.

In a first analysis we compare a pure tracking error model (portfolio 1a), with portfolio 1b in which we add the controls on the downside deviations from the threshold levels.

A second set of experiments consider only asymmetric distance measures from the risky benchmark and the thresholds levels of wealth. We analyze three different configuration of the objective function where we progressively increase the penalty on the lower barrier; we refer to portfolio 2a, 2b and 2c in Table 2 for the choices of the parameters.

Finally, we consider an experiment in which we drop the tracking goal with respect to the benchmark and consider only the downside penalization of negative deviations from the barriers (see, portfolio 3a and 3b, in Table 2).

In Fig. 1 we present the results for the portfolio management experiments for the first period.

We consider, as tracking assets, the following MSCI Style Indexes: LG, LV, SG, SV. From the top graph we can see that the four Style Indexes guarantee a good tracking performance of the index (portfolio 1a), while, when we introduce a shortfall control for the threshold levels (portfolio 1b) we sacrifice a potential upside capture. In the middle graph we can observe the portfolio behavior when we consider asymmetric tracking for the risky benchmark and we introduce a progressively higher penalization for the shortfalls (portfolios 2a, 2b and 2c). Finally, the bottom graph displays the behavior of the portfolios which account only for shortfall penalization without any tracking component in the objective function.

The same set of experiments has been carried out for the second simulation period. This case is more interesting from the point of view of a downside protection since the index experiences a sharp drop even below the lower threshold.

The same considerations as in the previous set of experiments apply. In particular, the introduction of a shortfall control is effective but is done at the cost of a reduction in the gains when the market rises. Moreover, if we compare portfolios 2a and 1b, it is interesting to observe that when we allow for a higher tracking error we can improve the downside protection. From the bottom graph, which displays the behavior of non-tracking portfolios, we can observe that there is an improvement in the downside protection but at a higher cost in terms of upside capture.

The four Style Indexes, we used as tracking assets, allowed us to obtain a good tracking performance in the out-of-sample experiments as it can be seen if we compare portfolios 1*a* with the Index for both simulation periods. However, in order to analyze the possible improvement in the tracking performances we carried out a further experiment using all the eight Style Indexes.

We computed the Tracking Error (TE) and Tracking Error Volatility (TEV) for both simulation periods for tracking portfolios using as tracking assets the eight Style Indexes. The results, in the two periods, are comparable, and we decide to present only the statistics for the second period, which is more interesting from the point of view of the behavior of the benchmark.

In Tables 3 and 4 we report the percentage Tracking Error and percentage Tracking Error Volatility for the analyzed portfolios using four Style Indexes (LG, LV, SG, SV), while Tables 5 and 6 presents the same statistics using all the eight Style Indexes (LG, LV, MG, MV, SMG, SMV, SG, SV). Including all the Style Indexes improves the performance but, in our opinion, the tracking results with four Indexes are already satisfying.

The proposed model allows for a great flexibility in the formulation of the objective function and can thus accommodate for different combinations of the tracking

Table 3 Percentage Tracking Error with respect to the benchmark, over 10-week period from February 8, 2012 to April 18, 2012, using 4 Style Indexes (LG,LV,SG,SV)

	1	2	3	4	5	6	7	8	9	10
portf 1a	0.010	0.022	−0.063	0.026	−0.043	−0.063	0.016	0.066	−0.027	0.155
portf 1b	−0.040	0.024	0.055	0.325	−1.425	0.032	0.572	0.634	0.715	−0.363
portf 2a	−0.084	0.016	0.161	0.668	−2.392	0.188	1.013	1.188	1.175	−0.730
portf 2b	−0.113	0.014	0.170	0.747	−2.473	0.212	1.076	1.277	1.237	−0.852
portf 2c	−0.113	0.014	0.170	0.748	−2.472	0.213	1.076	1.277	1.237	-0.854
portf 3a	−0.394	−0.035	0.117	1.480	−3.476	0.544	1.594	1.799	1.555	−1.111
portf 3b	−0.395	−0.035	0.118	1.479	−3.477	0.544	1.595	1.805	1.554	−1.114

Table 4 Percentage Tracking Error Volatility with respect to the benchmark, over 10-week period from February 8, 2012 to April 18, 2012, using 4 Style Indexes (LG,LV,SG,SV)

	1	2	3	4	5	6	7	8	9	10
portf 1a	0.037	0.021	0.033	0.032	0.019	0.006	0.049	0.098	0.033	0.083
portf 1b	0.235	0.064	0.154	0.245	0.713	0.129	0.438	0.522	0.396	0.621
portf 2a	0.323	0.100	0.178	0.415	0.800	0.207	0.535	0.636	0.380	0.529
portf 2b	0.302	0.098	0.167	0.361	0.717	0.189	0.504	0.529	0.309	0.358
portf 2c	0.302	0.098	0.167	0.361	0.717	0.189	0.503	0.529	0.309	0.357
portf 3a	0.044	0.022	0.033	0.048	0.081	0.025	0.072	0.085	0.047	0.080
portf 3b	0.044	0.022	0.034	0.050	0.079	0.025	0.071	0.085	0.048	0.078

Table 5 Percentage Tracking Error with respect to the benchmark, over 10-week period from February 8, 2012 to April 18, 2012, using 8 Style Indexes (LG, LV, MG, MV, SMG, SMG, SG, SV)

	1	2	3	4	5	6	7	8	9	10
portf 1a	0.008	0.002	-0.002	0.014	0.004	-0.006	0.001	0.002	-0.001	-0.010
portf 1b	-0.078	0.028	0.075	0.300	-1.359	0.087	0.589	0.644	0.737	-0.536
portf 2a	-0.088	0.028	0.209	0.650	-2.439	0.260	1.019	1.268	1.185	-0.821
portf 2b	-0.098	0.024	0.203	0.706	-2.576	0.280	1.103	1.348	1.259	-0.893
portf 2c	-0.099	0.024	0.203	0.706	-2.576	0.280	1.103	1.348	1.261	-0.894
portf 3a	-0.399	-0.038	0.126	1.477	-3.458	0.545	1.598	1.780	1.558	-1.172
portf 3b	-0.405	-0.037	0.126	1.477	-3.462	0.546	1.603	1.774	1.559	-1.171

Table 6 Percentage Tracking Error Volatility with respect to the benchmark, over 10-week period from February 8, 2012 to April 18, 2012, using 8 Style Indexes (LG, LV, MG, MV, SMG, SMG, SG, SV)

	1	2	3	4	5	6	7	8	9	10
portf 1a	0.017	0.005	0.013	0.011	0.027	0.013	0.022	0.054	0.023	0.078
portf 1b	0.219	0.070	0.148	0.261	0.655	0.138	0.462	0.502	0.353	0.560
portf 2a	0.332	0.103	0.204	0.459	0.872	0.201	0.543	0.559	0.352	0.394
portf 2b	0.321	0.102	0.187	0.424	0.712	0.191	0.493	0.471	0.288	0.319
portf 2c	0.321	0.101	0.187	0.423	0.713	0.190	0.492	0.472	0.287	0.320
portf 3a	0.069	0.015	0.031	0.042	0.105	0.022	0.060	0.110	0.046	0.063
portf 3b	0.066	0.014	0.032	0.040	0.113	0.021	0.061	0.114	0.045	0.071

and protection goals. Different risk attitude of the investor can be considered in the form of combining a tracking goal with respect to a risky benchmark and introducing a set of desired barriers to control the behavior of the portfolio.

5 Concluding Remarks

In this contribution we propose a multiperiod tracking error problem which can account for shortfall control using a sequence of references levels for wealth. The use of thresholds to define the goals for the portfolio management problem is intuitive for the investor and avoids the choice of a proper utility function and the definition of risk attitude/tolerance parameters. The proposed model allows to consider asymmetric tracking measures and in particular to penalize only downside deviations from the reference wealth levels. The computational experiments discussed show the trade-off between the possibility of upside capture and the control on downside risk. The role of the number of thresholds and the choice of a progressive penalization could be further investigated.

Acknowledgements The authors thank dott. Fabio Lanza for the research assistance in the computational experiments.

References

1. Alexander, G.J., Baptista, A.M.: Portfolio selection with a drawdown constraint. Journal of Banking and Finance **30**, 3171–3189 (2006)
2. Barro, D., Canestrelli, E.: Tracking error: a multistage portfolio model. Annals of Operations Research. **165**(1), 44–66 (2009)
3. Basak, S., Shapiro, A., and Tepla, L.: Risk management with benchmarking. Working Paper Series Asset Management SC-AM-03-16. Salomon Center for the study of financial institutions, NYU (2003)
4. Ben'Tal, A., Nemirovski, A.: Robust optimization – methodology and applications. Mathematical programming, Ser. B **92**, 453–480 (2002)
5. Bertsimas, D., Lauprete, G.J., Samarov, A.: Shortfall as a risk measure: properties, optimization and applications. Journal of Economic Dynamics and control **28**, 1353–1381 (2004)
6. Bertsimas, D., Sim, M.: The price of robustness. Operations Research **52** 35–53 (2004)
7. Biglova, A., Ortobelli, S., Rachev, S., Stoyanov, S.: Different approaches to risk estimation in portfolio theory. The Journal of Portfolio Management, pp. 103–112 (2004)
8. Browne, S.: Beating a moving target: optimal portfolio strategies for outperforming a stochastic benchmark. Finance and Stochastics **3**, 255–271 (1999)
9. Chekhlov, A., Uryasev, S., Zabarankin, M.: Drawdown measure in portfolio optimization. International Journal of Pure and Applied Finance **8**(1), 13–58 (2005)
10. Cuoco, D., He, H., Isaenko, S.: Optimal dynamic trading strategies with risk limits. Operations Research **56**(2), 358–368 (2008)
11. Dembo, R., Rosen, D.: The practice of portfolio replication, a practical overview of forward and inverse probems. Annals of Operations Research **85**, 267–284 (1999)
12. Dempster, M.H.A., Thompson, G.W.P.: Dynamic portfolio replication using stochastic programming. In: Dempster, M.H.A. (ed.) Risk Management: value at risk and beyond, pp. 100–128. Cambridge University Press (2002)
13. Dempster, M.A.H., Germano, M., Medova, E.A., Rietbergen, M.I., Sandrini, F., Scrowston, M.: Managing guarantees. The Journal of Portfolio Management **32**(2) (2006)
14. Dempster, M.A.H., Germano, M., Medova, E.A., Rietbergen, M.I., Sandrini, F., Scrowston, M.: Designing minimum guaranteed return funds. Quantitative Finance **7**(2), 245–256 (2007)
15. Ebert, U.: Measures of downside risk. Economics Bulletin **4**(16), 1–9 (2005)
16. Franks, E.C.: Targeting excess-of-benchmark returns. The Journal of Portfolio Management **18**, 6–12 (1992)
17. Gaivoronski, A., van der Vijst, N.: Optimal portfolio selection and dynamic benchmark tracking. European Journal of Operational Research **163**, 115–131
18. Ibrahim, K., Kamil, A.A., Mustafa, A.: Portfolio selection problem with maximum deviation measure: a stochastic programming approach. International Journal of Mathematical Models and Methods in Applied Sciences **1**(2), 123–129 (2008)
19. Jansen, D.W., Koedijk, K.G., de Vries, C.G.: Portfolio selection with limited downside risk. Journal of Empirical Finance **7**, 247–269 (2000)
20. King, A.J.: Asymmetric risk measures and tracking models for portfolio optimization under uncertainty. Annals of Operations Research **45**, 165–177 (1993)

21. Konno, H., Yamazaki, H.: Mean absolute deviation portfolio optimization model and its applications to Tokyo stock market. Management Science **37**, 519–531 (1991)

22. Michalowski, W., Ogryczak, W.: Extending the MAD portfolio optimization model to incorporate downside risk aversion. Naval Research Logistics **48**, 185–200 (2001)

23. Quaranta, A.G., Zaffaroni, A.: Robust optimization of conditional value at risk and portfolio selection. Journal of Banking and Finance **32**, 2046–2056 (2008)

24. Rudolf, M., Wolter, H-J., Zimmermann, H.: A linear model for tracking error minimization. Journal of Banking and Finance **23**, 85–103 (1999)

25. Rockafellar, R.T., Uryasev, S.: Optimization of Conditional Value at risk. Journal of Risk **2**, 21–41 (2000)

26. Scowcroft, A., Sefton, J.: Do tracking errors reliably estimate portfolio risk? Journal of Asset Management **2**(3), 205–222 (2000)

27. Sharpe, D.F., Weil, R.L.: Linear Programming with Absolute-Value Functionals. Operations Research **19**(1), 120–124 (1971)

Firm's Volatility Risk Under Microstructure Noise

Flavia Barsotti[*] and Simona Sanfelici

Abstract Equity returns and firm's default probability are strictly interrelated financial measures capturing the credit risk profile of a firm. Following the idea proposed in [20] we use high-frequency equity prices in order to estimate the volatility risk component of a firm within a structural credit risk modeling approach. Differently from [20] we consider a more general framework by introducing market microstructure noise as a direct effect of using noisy high-frequency data and propose the use of non-parametric estimation techniques in order to estimate equity volatility. We conduct a simulation analysis to compare the performance of different non-parametric volatility estimators in their capability of i) filtering out the market microstructure noise, ii) extracting the (unobservable) true underlying asset volatility level, iii) predicting default probabilities deriving from calibrating Merton [17] structural model.

1 Introduction

Structural credit risk models, firstly formalized by [17], consider a firm's equity and debt as contingent claims partitioning the asset value of the firm. Empirical tests of structural credit risk models show poor predictions of default probabilities and credit spreads, especially for short maturities. Methods of strict estimation or calibration provide evidence that predicted credit spreads are far below observed ones [14], the

F. Barsotti (✉)
Risk Methodologies, Group Risk Methodologies & Architecture, Group Financial Risks, Group Risk Management, UniCredit S.p.A, Piazza Gae Aulenti, Tower A, Floor 19, 20154 Milan, Italy
e-mail: Flavia.Barsotti@unicredit.eu

S. Sanfelici
Department of Economics, University of Parma, Via J.F. Kennedy 6, 43125 Parma, Italy
e-mail: simona.sanfelici@unipr.it

[*] The views presented in this paper are solely those of the author and do not necessarily represent those of UniCredit Spa.

M. Corazza, C. Pizzi (eds.), *Mathematical and Statistical Methods for Actuarial Sciences and Finance*, DOI 10.1007/978-3-319-02499-8_5, © Springer International Publishing Switzerland 2014

structural variables explain little of the credit spread variation [12], pricing error is large for corporate bonds [18].

A critical issue about the implementation of these models is that firm's asset value and volatility are not directly observable. The idea is then to use the information content of equity prices and then back out the firm's asset volatility. Nevertheless, the market microstructure literature strongly suggests that trading noises can affect equity prices so that the estimation of equity volatility and other related quantities may become a difficult task. For instance, as well documented, observed equity prices can diverge from their equilibrium value due to illiquidity, asymmetric information, price discreteness and other measurement errors. The effects of trading noise on how frequently one should sample equity price are analyzed in [2,4]. In the specific context of structural credit risk models, the relationship between the unobservable asset volatility and the observed equity value predicted by the pricing model is masked by trading noise; ignoring microstructure effects could non-trivially inflate estimates for the "true" asset volatility, and this would produce misleading estimates for default probabilities and credit spreads. This issue has been analyzed in [9], where the authors extend [8] to explicitly account for trading noise contamination of equity prices: they devise a particle filter-based maximum likelihood method based solely on the time series of observed equity values, which is robust to market microstructure effects. The importance of using high frequency data to back out parameters involved in firm's value dynamics has been highlighted by [20], which propose a novel approach to identify the volatility and jump risks component of individual firms from high-frequency equity prices. Their analysis suggests that high-frequency-based volatility measures can help to better explain credit spreads, above and beyond what is already captured by the true leverage ratio. However, the highest frequency considered in this paper is the 5-minute conservative sampling frequency which allows to eliminate microstructure effects.

In this paper we propose a particular econometric approach to structural model calibration based on non-parametric estimation of equity volatility from high-frequency intra-day equity prices. Several non-parametric estimators of daily stock volatility have been proposed in the econometric literature allowing to exploit the information contained in intra-day high-frequency data neglecting microstructure effects [1, 2, 4, 11, 13, 15, 19]. We propose a Monte Carlo simulation study based on Merton [17] structural model, trying to compare the performance of different non-parametric volatility estimators in their capability of i) filtering out the market microstructure noise, ii) extracting the true underlying asset (unobservable) volatility level, iii) predicting default probabilities. We show that the choice of the volatility estimator can largely affect calibrated default probabilities and hence risk evaluation. In particular, the commonly used Realized Volatility estimator is unable to provide reliable estimates for the volatility risk leading to a significant underestimation of default probabilities.

2 Structural Model Under Market Microstructure Effects

We follow Merton [17] structural model by assuming a firm's value process described by a geometric Brownian motion, which evolves according to

$$dA_t = (\mu - \delta)A_t dt + \sigma A_t dW_t \tag{1}$$

where A_t is asset value at time t, μ the instantaneous asset return, δ the asset payout ratio and σ the asset volatility. The firm has two classes of outstanding claims: equity and a zero-coupon debt with promised payment B at maturity T. To price corporate debt as in [17] we assume that: (i) default occurs only at maturity with debt face value as default boundary; (ii) when default occurs, the absolute priority rule prevails. The payoffs to debt holders and equity holders at time T become, respectively

$$D_T = \min(A_T, B), \qquad S_T = \max(A_T - B, 0).$$

From now on, we focus our attention on equity value and default probabilities in order to develop our computational econometric analysis. Equity claim can be priced at each time $t < T$ through the standard Black-Scholes option pricing model as the price of a European call option given by

$$S_t = S(A_t; \sigma, B, r, \delta, T) = A_t \phi(d_t) - Be^{-r(T-t)} \phi(d_t - \sigma\sqrt{T-t}), \tag{2}$$

where

$$d_t := \frac{\log(\frac{A_t}{B}) + (r - \delta + \sigma^2/2)(T-t)}{\sigma\sqrt{T-t}}, \tag{3}$$

and $\phi(\cdot)$ is the standard normal distribution function. Therefore, by applying Itô's lemma, the instantaneous volatility Σ_t^s of the log equity price can be written as

$$\Sigma_t^s := \frac{A_t}{S_t} \frac{\partial S_t}{\partial A_t} \sigma. \tag{4}$$

Notice that the equity volatility is driven by the time-varying factor A_t, whereas the asset volatility σ is constant. The firm's probability of default at maturity T is the probability of A_T being below the constant barrier represented by the face value of debt B. Under the physical probability measure $\mathbb{P}$ we have

$$\mathbb{P}(A_T < B|A_t) = \phi(\sigma\sqrt{T-t} - d_t^{\mathbb{P}}), \tag{5}$$

with $d_t^{\mathbb{P}}$ given by Eq. (3) where we only replace the interest rate r with μ.

For a given firm, one can obtain a time series of equity prices $\{S_j, j = 0, \ldots, N\}$, with a given sampling frequency $h = t_j - t_{j-1}$ assumed to be constant, for ease of exposition. If one could observe the "true" equity price, than equity volatility could be easily estimated by the well known Realized Volatility estimator [7] at any desired accuracy level using high frequency data. However, as noted for instance by [9], the relationship between the unobserved asset and the observed equity value predicted by the pricing formula (2) may be masked by trading noise. We assume an additive error structure for the trading noise on the logarithmic equity value as follows

$$\log \widetilde{S}(t_j) = \log S(t_j) + \eta(t_j), \tag{6}$$

where the random shocks $\eta(t_j)$, for $0 \leq j \leq N$ are i.i.d. random variables with mean zero and bounded fourth moment and independent of the efficient log-return process. The assumption of independence can be relaxed by considering a particular form of dependent noise, given by [11], with market microstructure noise that is time-dependent in tick time and correlated with efficient returns

$$\widetilde{\eta}_j := \alpha[\log S(t_j) - \log S(t_{j-1})] + \eta_j, \tag{7}$$

where α is a real constant and $\widetilde{\eta}_j$ and η_j are the shorten notation for $\widetilde{\eta}(t_j)$ and $\eta(t_j)$. The case $\alpha = 0$ corresponds to the case of independent noise assumption.

3 Equity Volatility Estimation

The basic idea of our paper is that, using suitable volatility estimators, we can infer the true volatility process Σ_t^s of equity returns from noisy high-frequency data. Then, the equity volatility estimate can be used to back out the asset volatility σ such as to fit exactly, say, the 5-years probability of default by solving Equation (5) with respect to σ, as explained in details in the following.

For the reader's convenience, we now give some details about the implementation of the volatility measures employed in our analysis. We set $\tilde{p}_t := \log \widetilde{S}_t$, the noisy equity log-price. Time is measured in daily units. We build daily measure of volatility by considering daily windows of n intra-day equity data $\tilde{p}_{t,j}, j = 0, 1, \dots, n$. Besides the well known Realized Volatility estimator $\Sigma_t^{RV} := \sum_{j=1}^n \delta_j(\tilde{p})^2$, where $\delta_j(\tilde{p}) := \tilde{p}_{t,j} - \tilde{p}_{t,j-1}$ is the j-th within-day equity log-return on day t, we consider the following estimators of the volatility process Σ_t^s: the bias corrected estimator by Hansen and Lunde [11]

$$\Sigma_t^{HL} := \Sigma_t^{RV} + 2\frac{n}{n-1} \sum_{j=1}^{n-1} \delta_j(\tilde{p})\delta_{j+1}(\tilde{p})$$

the *flat-top realized kernels* by [4,6]

$$\Sigma_t^K := \sum_{h=-H}^{H} k\left(\frac{h}{H+1}\right) \sum_{j=|h|+1}^{n} \delta_j(\tilde{p})\delta_{j-|h|}(\tilde{p})$$

with kernels of TH_2 type $k(x) = \sin^2\left(\frac{\pi}{2}(1-x)^2\right)$. The realized kernels may be considered as unbiased corrections of the Realized Volatility by means of the first H autocovariances of the returns. In particular, when H is selected to be zero the realized kernels become the Realized Volatility. Our analysis includes also the *two-scale estimator* by [19]

$$\Sigma_t^{TS} := \frac{S}{S-1} \left(\frac{1}{S} \sum_{s=1}^{S} \Sigma_t^{G^{(s)}} - \frac{1}{S} \Sigma_t^{RV} \right).$$

The two-scale (subsampling) estimator is a bias-adjusted average of lower frequency realized volatilities computed on S non-overlapping observation subgrids $G^{(s)}$ containing n_S observations. Recently, [13] proposed a pre-averaging technique as an alternative to subsampling in order to reduce the microstructure effects. The idea is that if one averages a number of observed log-prices, one is closer to the latent process $p(t)$. This approach, when well implemented, gives rise to rate optimal estimators of power variations. In particular, a consistent estimator of the integrated volatility can be constructed as

$$\Sigma_t^{PA} = \frac{\sqrt{\Delta}}{\theta \psi_2} \sum_{s=0}^{n-k_n+1} \bar{\delta}_s(\tilde{p})^2 - \frac{\psi_1 \Delta}{2\theta^2 \psi_2} \sum_{s=1}^{n} \delta_s(\tilde{p})^2,$$

where the pre-averaged return process is given by

$$\bar{\delta}_s(\tilde{p}) := \sum_{r=1}^{k_n} g\left(\frac{r}{k_n}\right) \delta_{s+r}(\tilde{p}) = \frac{1}{k_n} \left(\sum_{j=k_n/2}^{k_n-1} \tilde{p}_{t,s+j} - \sum_{j=0}^{k_n/2-1} \tilde{p}_{t,s+j} \right),$$

$\theta = k_n \sqrt{\Delta}$, $\psi_1 = 1$ and $\psi_2 = 1/12$, corresponding to the "hat" weight function $g(x) = x \wedge (1-x)$. The Fourier estimator [15] is given by

$$\Sigma_t^{F} = \frac{(2\pi)^2}{2N+1} \sum_{s=-N}^{N} c_s(d\tilde{p}_n)c_{-s}(d\tilde{p}_n),$$

where $c_k(d\tilde{p}_n) = \frac{1}{2\pi} \sum_{i=1}^{n} \exp(-ikt_{i-1})\delta_i(\tilde{p})$. Finite sample MSE-based optimal rules for choosing the parameters employed by these estimators are discussed in [3, 5, 16, 19]. Here, we proceed according to the following rules: a simple approximation of the optimal sampling frequency for the Realized Volatility estimator is to choose the number of observations approximately equal to $n^* = (Q/4E[\eta^2]^2)^{1/3}$, where Q is the integrated quarticity estimated by means of low frequency returns. The optimal number of subgrids S is given by $c^* n^{2/3}$, where $c^* = (Q/48E[\eta^2]^2)^{-1/3}$. For the Kernel estimator, we apply the optimal mean square error bandwidth selection suggested by [5] and get $H = c^* \xi^{4/5} n^{3/5}$, where $c^* = (144/0.269)^{1/5}$, $\xi^2 = E[\eta^2]/\sqrt{Q}$. In the case of the Pre-averaging estimator, inspired by [5], we choose $k_n = c^* \xi^{4/5} n^{3/5}$. Finally, for the Fourier estimator, the optimal cutting frequency N can be easily obtained by direct minimization of the estimated MSE given by Theorem 3 in [16].

4 Volatility and Default Probability Computation

In this section we provide numerical results of our simulation study showing and quantifying the impact of different volatility measures on default probability estimation, based on high-frequency equity data affected by trading noise. We analyze the performance of the volatility estimators with respect to their ability of i) filtering the microstructure noise and correctly extract equity volatility, ii) backing out asset volatility and iii) predicting default probabilities.

We perform Monte Carlo simulations by generating the underlying asset dynamic according to model (1) for rating classes A, BBB, BB. Then, high-frequency equity prices are simulated through Equation (2). The sample contains 500 days and equispaced intra-day data are generated at a frequency $h = 4$ sec (i.e. $n = 21600$). Based on the calibration results by [20], we consider as model parameters $r = 0.05$, $\delta = 0.02$; rate A: $\sigma = 0.2128$, $\mu = 0.0643$, $B = 43.13$; rate BBB: $\sigma = 0.2296$, $\mu = 0.0655$, $B = 48.02$; rate BB: $\sigma = 0.2371$, $\mu = 0.057$, $B = 58.63$. Market microstructure noise is considered, alternatively, for both cases described by Eqs. (6)-(7). The random shocks η_j are i.i.d Gaussian random variables with zero mean and standard deviation equal to 2 times the log-equity return standard deviation. We set $\alpha = 0.5$ in the dependent noise case (7). Once we have noisy equity prices, we compare the performance of different volatility estimators in their ability of extracting the "true" equity volatility Σ_t^s and other related quantities for the underlying process describing firm's assets value dynamics.

Table 1 presents numerical evidence for each equity volatility estimator introduced in Sect. 3 and used in our comparison when (a) trading noise is independent of intra-day equity log-returns and (b) trading noise is correlated with intra-day equity log-returns, respectively. The table lists the average relative error (ARE), the mean squared error (MSE) and bias achieved by the different equity volatility estimators. Σ_t^{RV} represents the Realized Volatility estimator using all tick-by-tick equity data, while Σ_t^{RVSS} refers to the Realized Volatility estimator based on sparse sampling, where the sampling frequency is optimized in order to filter the microstructure effects, as explained in Sect. 3. Our results strongly confirm well known stylized facts documented by the econometric literature and highlighted by [16]: Σ_t^{RV} estimates are completely swamped by noise and sparse sampling can only moderately provide efficient estimates. The first order correction of Σ_t^{HL}, as an alternative to sparse sampling, can reduce the bias due to the spurious first order autocorrelation in equity

Table 1 Equity Volatility Estimators. The table shows average relative error (ARE), mean squared error (MSE) and BIAS for different equity volatility estimators for A-rated firms. Results are based on 500 daily Monte Carlo simulations. Panel (a) refers to the trading noise given in Equation (6), panel (b) to Equation (7), with $\alpha = 0.5$

	Σ_t^{RV}	Σ_t^{RVSS}	Σ_t^{TS}	Σ_t^{HL}	Σ_t^{K}	Σ_t^{PA}	Σ_t^{F}
(a) ARE	2.00e+0	3.28e-2	5.37e-4	9.89e-4	5.10e-4	−1.19e-3	1.33e-2
MSE	4.49e-1	4.13e-4	2.55e-5	2.21e-4	5.62e-5	6.59e-5	8.01e-5
BIAS	6.69e-1	1.10e-2	1.97e-4	3.73e-4	1.83e-4	−3.87e-4	4.46e-3
(b) ARE	2.24e+0	3.61e-2	6.31e-4	1.07e-3	6.29e-4	−1.43e-3	1.38e-2
MSE	5.65e-1	4.90e-4	2.79e-5	3.25e-4	5.78e-5	6.79e-5	8.58e-5
BIAS	7.50e-1	1.21e-2	1.96e-4	3.46e-4	2.03e-4	−4.81e-4	4.60e-3

returns introduced by the trading noise. The best results are provided by Σ_t^F and by the other estimators specifically designed to handle microstructure effects Σ_t^{TS}, Σ_t^K and Σ_t^{PA}. However, the rank of estimators is different when we consider absolute error measures such as MSE and bias versus percentage error measures such as ARE. In fact, in the latter case Σ_t^{HL} performs better than Σ_t^{PA} and Σ_t^F. This means that Σ_t^{HL} performs better under low volatility regimes. Finally, we notice that Σ_t^{PA} tends to underestimate volatility, differently from all the other estimators.

In order to study the influence of different equity volatility estimators on the default probability predicted by Merton [17] model, we proceed by developing the following calibration exercise. For each equity volatility estimator, generically denoted by E, and each day t in our sample, we find the corresponding asset volatility estimate $\hat{\sigma}^E$ by matching the 5-years default probability coming from our equity volatility estimate Σ_t^E with the one evaluated through the model using the given parameters values. In so doing, we act as if we did not know the asset values to mimic the real-life estimation situation and we conduct inference only based on observable quantities such as measurable equity volatility and the 5-years default probabilities. Default probabilities are computed, for any maturity, according to Eq. (5). In order to avoid arbitrage opportunities, following [20], we consider as key assumption that all securities written on the underlying firm value A_t must have the same Sharpe ratio, see [17], Equation (6). This consideration enables us to express the instantaneous asset return μ as a function of the unknown asset volatility, given each equity volatility estimate Σ_t^E, and then to solve Eq. (5) for the 5-years default probability with respect to the asset volatility, to obtain the corresponding asset volatility estimate $\hat{\sigma}^E$. Once asset volatility is known, we can compute default probabilities for any other maturity according to Eq. (5).

Before analyzing the impact of the different volatility estimators on the calibration procedure for default probabilities, we analyze their influence on the calibration of asset volatility σ. A visual insight on the variability of asset volatility estimates obtained with the procedures based on the different equity volatility estimators can be achieved from Fig. 1, plotting the ratio $\hat{\sigma}^E / \sigma$ in the case of independent noise for an A-rated firm. It is evident how the high-frequency Realized Volatility procedure (black line in the bottom) largely underestimate asset volatility. Similar results are obtained for the dependent noise case.

Table 2 shows descriptive statistics of calibrated asset volatility. We report results obtained by matching 5-years default probabilities for each equity volatility estimator E. Panel (a) of the table refers to a trading noise of the form (6); panel (b) refers to results obtained for a trading noise of the form (7). It is evident from the table that the Kernel and Two-scale estimators provide the most accurate estimation of the true value $\sigma = 0.2128$, with the smallest standard deviation. On the contrary, $\hat{\sigma}^{RV}$ is strongly biased due to microstructure effects, while the optimized $\hat{\sigma}^{RVSS}$ is less biased, at the price of a slightly larger variance. These results are confirmed by statistics for the ratio $\hat{\sigma}^E / \sigma$ as well.

The average (over all the 'daily' Monte Carlo replications) default probabilities for different maturities and calibration procedures are plotted in Fig. 2. For each day in the sample and for each volatility estimator E we use the calibrated asset volatility

Table 2 Calibrated Asset Volatility. The table shows descriptive statistics of calibrated asset volatility for different equity volatility estimators. Results are based on 500 daily Monte Carlo simulations for an A-rated firm. Panel (a) refers to the trading noise given in Eq. (6), panel (b) to Eq. (7), with $\alpha = 0.5$

	Mean	Median	10 perc.	90 perc.	Min	Max	Std Dev.
(a) $\hat{\sigma}^{RV}$	0.192652	0.192648	0.192569	0.192743	0.192419	0.192894	0.000069
$\hat{\sigma}^{RVSS}$	0.211811	0.211775	0.209762	0.213980	0.207321	0.217353	0.001575
$\hat{\sigma}^{TS}$	0.212759	0.212750	0.212129	0.213361	0.211354	0.214601	0.000503
$\hat{\sigma}^{HL}$	0.212812	0.212618	0.211089	0.214912	0.209575	0.219557	0.001515
$\hat{\sigma}^{K}$	0.212770	0.212734	0.211934	0.213688	0.210808	0.215141	0.000742
$\hat{\sigma}^{PA}$	0.212830	0.212790	0.211896	0.213878	0.210616	0.215468	0.000807
$\hat{\sigma}^{F}$	0.212355	0.212313	0.211390	0.213298	0.210435	0.214927	0.000749
$\hat{\sigma}^{RV}/\sigma$	0.905318	0.905299	0.904928	0.905745	0.904225	0.906457	0.000326
$\hat{\sigma}^{RVSS}/\sigma$	0.995352	0.995186	0.985726	1.005543	0.974253	1.021393	0.007401
$\hat{\sigma}^{TS}/\sigma$	0.999808	0.999767	0.996846	1.002634	0.993207	1.008465	0.002363
$\hat{\sigma}^{HL}/\sigma$	1.000056	0.999143	0.991960	1.009924	0.984843	1.031753	0.007120
$\hat{\sigma}^{K}/\sigma$	0.999860	0.999689	0.995929	1.004173	0.990637	1.011002	0.003488
$\hat{\sigma}^{PA}/\sigma$	1.000139	0.999951	0.995753	1.005067	0.989735	1.012535	0.003792
$\hat{\sigma}^{F}/\sigma$	0.997911	0.997711	0.993376	1.002341	0.988885	1.009996	0.003519
(b) $\hat{\sigma}^{RV}$	0.191966	0.191965	0.191886	0.192049	0.191798	0.192136	0.000062
$\hat{\sigma}^{RVSS}$	0.211722	0.211546	0.209694	0.213990	0.207310	0.218999	0.001698
$\hat{\sigma}^{TS}$	0.212756	0.212734	0.212077	0.213403	0.211356	0.214588	0.000524
$\hat{\sigma}^{HL}$	0.212841	0.212708	0.210635	0.215044	0.208440	0.220014	0.001801
$\hat{\sigma}^{K}$	0.212766	0.212746	0.211857	0.213716	0.210610	0.215369	0.000755
$\hat{\sigma}^{PA}$	0.212838	0.212813	0.211817	0.213845	0.210465	0.215823	0.000823
$\hat{\sigma}^{F}$	0.212340	0.212320	0.211347	0.213360	0.210210	0.214626	0.000774
$\hat{\sigma}^{RV}/\sigma$	0.902097	0.902090	0.901721	0.902485	0.901305	0.902894	0.000293
$\hat{\sigma}^{RVSS}/\sigma$	0.994936	0.994109	0.985405	1.005592	0.974203	1.029129	0.007982
$\hat{\sigma}^{TS}/\sigma$	0.999792	0.999689	0.996601	1.002834	0.993215	1.008401	0.002462
$\hat{\sigma}^{HL}/\sigma$	1.000193	0.999566	0.989825	1.010547	0.979512	1.033899	0.008463
$\hat{\sigma}^{K}/\sigma$	0.999840	0.999748	0.995566	1.004304	0.989710	1.012071	0.003550
$\hat{\sigma}^{PA}/\sigma$	1.000177	1.000061	0.995382	1.004913	0.989029	1.014208	0.003869
$\hat{\sigma}^{F}/\sigma$	0.997840	0.997745	0.993174	1.002631	0.987827	1.008582	0.003638

$\hat{\sigma}^{E}$ in order to compute the default probabilities for any maturity (from 1 to 5 years) by Eq. (5). From the figure, it appears evident how the Realized Volatility approach based on high frequency noisy data drastically underestimates default probabilities, so that sparse sampling becomes mandatory when equity data are affected by microstructure effects. On the whole, all the other procedures seem to provide sensible results and only a deeper analysis reveals differences among different calibration procedures.

The analysis of the effects of using alternative equity volatility estimators on default probabilities is conducted for different maturities (from 1 to 5 years) through

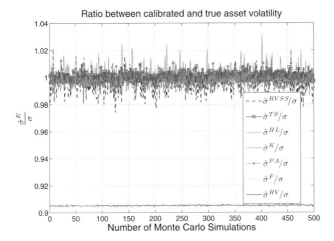

Fig. 1 Calibrated Asset Volatility. The plot shows the ratio between calibrated asset volatility $\hat{\sigma}^E$ and the true asset volatility σ for 500 daily Monte Carlo simulations. The plot refers to the trading noise given in Equation (6). Panel (a) of Table 2 gives descriptive statistics of these values

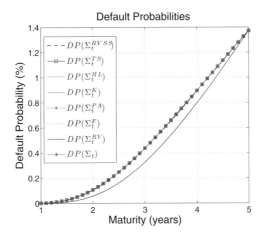

Fig. 2 Default Probability. The plot shows default probabilities given by (5). By matching 5-years default probability coming from our equity volatility estimate with the one provided by the model, we obtain the corresponding calibrated asset asset volatility and use it to compute default probabilities for all maturities

the comparison of the mean relative error between the estimated default probability and the theoretical one. For each maturity, we consider the following measure

$$DP_{Err}^{E} := \mathbb{E}\left[\frac{DP(\Sigma_t^E) - DP(\Sigma_t^s)}{DP(\Sigma_t^s)}\right] \cdot 100, \qquad (8)$$

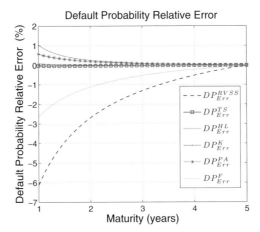

Fig. 3 Default Probability Relative Error. The plot shows default probability mean relative error given in Eq. (8) for maturities from 1 to 5 years. Results are based on 500 Monte Carlo simulations and refer to an A-rated firm with trading noise (7) and $\alpha = 5$

where $DP(\Sigma_t^E)$ is the default probability calibrated from Merton [17] structural model when equity and asset volatility are estimated through estimator E; $DP(\Sigma_t^s)$ is the corresponding theoretical default probability when equity and asset volatility are Σ_t^s and σ, respectively. Figure 3 shows the mean relative error on different calibration procedures for an A-rated firm with noise setting (7). Similar plots are obtained for the noise case (6) and for other credit qualities. The results for the Realized Volatility estimator using tick-by-tick data have been omitted, since the relative error in this case reaches 78% for the earliest maturities. When data are optimally sampled, the Realized Volatility calibration procedure is still affected by an error up to 6% for the shortest maturities. A negative (positive) error reveals that the calibration procedure underestimates (overestimates) default probabilities. Therefore, the classical Realized Volatility approach can severely underestimate risk. Except Fourier approach, all the other approaches slightly overestimate default probabilities, with Σ_t^{TS} and Σ_t^K providing the best estimation of risk.

Table 3 shows the corresponding numerical results for all the rating classes A, BBB, BB, suggesting that the choice of the volatility estimator largely affects the default probabilities estimation. The performance of each estimators in terms of default probability estimation is more strongly affected by the equity risk premium, i.e. ultimately by μ, than by the latent asset volatility σ. Therefore, generally, DP_{Err}^E does not worsen from A to BBB and BB classes as long as low-rated firms are associated with small values of μ, like the ones calibrated by [20]. Rather, the ranking among different estimators remains the same regardless of the firm credit rating. In particular, Σ_t^{TS}, Σ_t^{HL}, Σ_t^K and Σ_t^{PA} are the estimators providing the best risk evaluation in all the considered scenarios, immediately followed by Σ_t^F.

Table 3 Default Probability Relative Error. The table shows default probability mean relative error given in Eq. (8) for maturities from 1 to 4.5 years. Results are based on 500 daily Monte Carlo simulations using different equity volatility estimators. Relative errors are in percentage (%). Panel (a) refers to the trading noise given in Equation (6), panel (b) to Eq. (7)

		DP^{RV}_{Err}	DP^{RVSS}_{Err}	DP^{TS}_{Err}	DP^{HL}_{Err}	DP^{K}_{Err}	DP^{PA}_{Err}	DP^{F}_{Err}
(a)	1y	−77.552877	−5.653780	−0.016775	0.678380	0.105648	0.508828	−2.556536
	3y	−24.702511	−1.200495	−0.011160	0.057574	0.003154	0.077979	−0.518461
A	4.5y	−4.989094	−0.222674	−0.002300	0.008090	-0.000003	0.013540	−0.095374
	1y	−67.894485	−4.390339	−0.022799	0.416401	0.062076	0.363117	−1.955870
	3y	−19.816429	−0.933682	−0.009053	0.040953	0.002465	0.060442	−0.402094
BBB	4.5y	−3.944256	−0.174305	−0.001823	0.006151	0.000119	0.010747	−0.074603
	1y	−42.822431	−2.256778	−0.018665	0.136200	0.017413	0.163326	−0.983264
	3y	−10.750682	−0.483424	−0.004959	0.018367	0.001253	0.031100	−0.207333
BB	4.5y	−2.086650	−0.091030	−0.000966	0.003098	0.000153	0.005726	−0.038929
(b)	1y	−78.826465	−6.124004	−0.027360	1.037514	0.088215	0.572542	-2.640228
	3y	−25.516271	−1.309917	−0.014032	0.096727	−0.000922	0.089373	−0.536280
A	4.5y	−5.172970	−0.243269	−0.002852	0.014222	−0.000769	0.015589	−0.098680
	1y	−69.288313	-4.455845	0.101603	0.757382	0.313784	0.707459	-1.786171
	3y	−20.489877	−0.954666	0.014432	0.097111	0.050003	0.125589	−0.370176
BBB	4.5y	−4.089964	−0.178447	0.002465	0.016110	0.008800	0.022649	−0.068780
	1y	−44.056077	−2.416005	0.131715	0.721002	0.154349	0.328242	-0.891365
	3y	−11.137906	−0.518127	0.026070	0.135473	0.029731	0.065342	−0.188054
BB	4.5y	−2.164760	−0.097586	0.004837	0.024876	0.005486	0.012137	−0.035312

5 Conclusions

In this paper we consider Merton [17] structural model and use high-frequency equity prices in order to back out the unobservable asset volatility and calibrate default probabilities. We perform a Monte Carlo simulation study based on equispaced simulated equity prices and propose alternative equity volatility estimators, assuming data being contaminated by trading noise: the aim is to exploit the information content of high frequency intra-day data neglecting microstructure effects. While [9] propose a particle filter-based maximum likelihood method, we propose a different econometric approach. We consider alternative non-parametric (equity) volatility estimators and compare their performance in: i) filtering out the market microstructure noise, ii) extracting the (unobservable) true underlying asset volatility level, iii) predicting default probabilities. We consider, alternatively, trading noise being a) independent log-Gaussian distributed, b) correlated with intra-day equity log-returns. Non-observability of a firm's asset value does not actually impede the implementation of a structural credit risk model; nevertheless, the volatility estimator can largely affect calibrated default probabilities, thus risk evaluation as it happens for A, BBB and BB rating classes analyzed. The commonly used Realized Volatility estimator is unable to provide reliable estimates for the asset volatility under market microstructure, leading to a significant underestimation of asset volatility and default probabilities. Next step will be to develop the current analysis by extending the focus on credit spreads estimation and considering more sophisticated underlying asset dynamics, i.e. stochastic volatility and/or jump-diffusion models.

References

1. Aït-Sahalia, Y., Mykland, P., Zhang, L.: How often to sample a continuous-time process in the presence of market microstructure noise. Review of Financial Studies **18**, 351–416 (2005)
2. Bandi, F.M., Russel, J.R.: Separating market microstructure noise from volatility. Journal of Financial Economics. **79**, 655–692 (2006)
3. Bandi, F.M., Russell, J.R.: Market microstructure noise, integrated variance estimators, and the accuracy of asymptotic approximations. Working paper, Univ. of Chicago (2006). http://faculty.chicagogsb.edu/federicobandi
4. Barndorff-Nielsen, O.E., Hansen, P.R., Lunde, A., Shephard, N.: Designing realised kernels to measure the ex-post variation of equity prices in the presence of noise. Econometrica **76**(6), 1481–1536 (2008)
5. Barndorff-Nielsen, O.E., Hansen, P.R., Lunde, A., Shephard, N.: Multivariate realised kernels: consistent positive semi-definite estimators of the covariation of equity prices with noise and non-synchronous trading. Working paper (2008)
6. Barndorff-Nielsen, O.E., Hansen, P.R., Lunde, A., Shephard, N.: Subsampling realised kernels. Journal of Econometrics (2010)
7. Barndorff-Nielsen, O.E., Shephard, N.: Econometric analysis of realized volatility and its use in estimating stochastic volatility models. J. R. Statist. Soc., Ser. **B 64**, 253–280 (2002)
8. Duan, J.C.: Maximum likelihood estimation using price data of the derivative contract. Mathematical Finance **4**, 155–167 (1994)
9. Duan, J.C., Fulop, A.: Estimating the structural credit risk model when equity prices are contaminated by trading noises. Journal of Econometrics **150**, 288–296 (2009)
10. Ericsson, J., Reneby, J.: Estimating Structural Bond Pricing Models. Journal of Business **78**(2) 707–735 (2005)
11. Hansen, P.R., Lunde, A.: Realized variance and market microstructure noise (with discussions). Journal of Business and Economic Statistics **24**, 127–218 (2006)
12. Huang, J., Huang, M.: How much of the corporate-treasury yield spread is due to credit risk? Working Paper, Penn State University (2003)
13. Jacod, J., Li, Y., Mykland, P.A., Podolskij, M., Vetter, M.: Microstructure noise in the continuous case: the pre-averaging approach. Stochastic Processes and their Applications **119**, 2249–2276 (2009)
14. Jones, P.E., Scott, P.M., Rosenfeld, E.: Contingent claims analysis of corporate capital structures: an empirical investigation. Journal of Finance **39**, 611–625 (1984)
15. Malliavin, P., Mancino, M.E.: Fourier series method for measurement of multivariate volatilities. Finance and Stochastics **6**(1), 49–61. Springer (2009)
16. Mancino, M.E., Sanfelici, S.: Robustness of Fourier estimator of integrated volatility in the presence of microstructure noise. Computational Statistics & Data analysis **52**, 2966–2989. Elsevier (2008)

17. Merton, R.C.: On the Pricing of Corporate Debt: The Risk Structure of Interest Rates. The Journal of Finance **29**, 449–470 (1974)
18. Eom, Y.H., Helwege, J., Huang, J.: Structural models of corporate bond pricing: an empirical analysis. Review of Financial Studies **17**, 499–544 (2008)
19. Zhang, L., Mykland, P., Aït-Sahalia, Y.: A tale of two time scales: determining integrated volatility with noisy high frequency data. Journal of the American Statistical Association **100**, 1394–1411 (2005)
20. Zhang, B.Y., Zhou, H., Zhu, H.: Explaining Credit Default Swap Spreads with the Equity Volatility and Jump Risks of Individual Firms. Review of Financial Studies **22**(12), 5099–5131 (2009)

Socially Responsible Mutual Funds: An Efficiency Comparison Among the European Countries

Antonella Basso and Stefania Funari

Abstract The first objective of this contribution is to evaluate the performance of socially responsible investment (SRI) equity mutual funds in the main European countries with three different data envelopment analysis (DEA) models. Secondly, with a series of statistical tests we compare the performance of SRI and non SRI mutual funds in the various countries, to determine if SRI mutual funds have to sacrifice something in terms of financial performance; the results suggests that it is possible to invest in a socially responsible manner without giving up the financial reward. Thirdly, we compare the performance obtained by SRI mutual funds among different European countries.

1 Introduction

Socially responsible investment (SRI) funds have seen an increasing interest among investors.

Given the ethical considerations which drive socially responsible investments in mutual funds, investors might be willing to accept for SRI mutual funds lower financial returns. Actually, the literature on ethical investing has long investigated the issue of the eventual penalisation incurred by investments in SRI mutual funds, in search for an answer to the question whether it is possible "to do well while doing good"; see for example [9, 11] for a brief review. The answer which comes out from many empirical investigations are somewhat surprising, since most of the re-

A. Basso
Department of Economics, Ca' Foscari University of Venice, Cannaregio 873, 30121 Venice, Italy
e-mail: basso@unive.it

S. Funari (✉)
Department of Management, Ca' Foscari University of Venice, Cannaregio 873, 30121 Venice, Italy
e-mail: funari@unive.it

M. Corazza, C. Pizzi (eds.), *Mathematical and Statistical Methods for Actuarial Sciences and Finance*, DOI 10.1007/978-3-319-02499-8_6, © Springer International Publishing Switzerland 2014

sults suggest that it is not necessary to sacrifice returns in order to pursue the ethical objectives.

The main aims of this contribution are threefold. The first objective is to evaluate the performance of SRI equity mutual funds in the main European countries in which the socially responsible mutual funds play an important role. To this aim we apply three models designed in a DEA (data envelopment analysis) framework. DEA is an operational research technique widely used to assess the performance of a set of decision making units in many different fields (see for example [7]). In the field of socially responsible investing, the DEA methodology is especially useful because it enables to take into account both the financial objective to get an optimal reward–to–risk result and the ethical aim (see [5, 6]).

Secondly, we compare the performance indicators for SRI and non SRI mutual funds in the various countries, carrying out a series of statistical tests with the aim of determining if the socially responsible mutual funds really entail a sacrifice in terms of financial performance. It is interesting to note that the results of the statistical tests carried out on the DEA scores, applied for the first time to SRI funds, are in agreement with the conclusions of most of the empirical studies on the performance of SRI funds, suggesting that it is possible to invest in a socially responsible manner without giving up the financial reward.

A third original contribution is the comparison of the performance results obtained by SRI mutual funds in the different European countries.

The paper is organized as follows. Section 2 presents the main features of SRI mutual funds in Europe. Section 3 discusses the empirical results of the analysis carried out to evaluate the performance of SRI funds in the main European countries, while Sect. 4 presents the outcomes of the comparisons of the inefficiency scores carried out with a series of statistical tests.

2 SRI Mutual Funds in Europe

On 30/06/2006, at the beginning of the triennium considered in our analysis, the number of European SRI funds was equal to 388, spread over 15 countries (Austria, Belgium, Denmark, Finland, France, Germany, Ireland, Italy, Norway, Poland, Spain, Sweden, Switzerland, Netherlands, United Kingdom; see [13]). Three years later, on 30/06/2009, this number has increased to 683 (+76%). In the same period the total asset under management increased from 34009 to 53276 million euros, with a growth of +57%, showing the importance reached in Europe by the socially responsible investments.

For a more detailed presentation of the main features of socially responsible investing in Europe we refer to the Eurosif report [8] which analyses its presence in each European country. The analysis presented in this contribution considers the European socially responsible equity funds which use ethical, social and/or environmental screening to select the assets in their portfolios.

In the analysis carried out we have included all the SRI European equity funds for which the data in the 'SRI Funds Service' database were available for the period

Table 1 Average features of European SRI mutual funds and their non SRI counterparts by country

	Country	No. of funds	Ethical level	% Mean return	% St. Dev.	% Excess return	% Initial charges	% Exit charges
	SRI funds							
AT	Austria	10	2.69	−9.74	22.00	−13.23	4.45	0.00
BE	Belgium	10	2.95	−8.72	21.08	−12.19	3.20	0.00
CH	Switzerland	5	2.92	−4.06	19.98	−5.86	3.40	0.01
DE	Germany	4	1.76	−7.89	18.96	−11.40	3.63	0.00
ES	Spain	2	1.15	−11.40	18.47	−14.89	0.00	3.00
FR	France	36	1.29	−6.96	20.36	−10.50	2.64	0.15
IR	Irland	3	2.06	−8.09	20.54	−11.61	2.67	0.00
IT	Italy	3	1.62	−10.68	19.05	−14.15	1.00	0.00
LU	Luxembourg	38	2.10	−6.46	20.46	−10.05	4.34	0.24
NE	The Netherlands	7	2.17	−6.33	20.09	−9.83	0.33	0.26
NO	Norway	1	3.81	−12.00	22.13	−16.60	0.20	0.30
SE	Sweden	32	1.27	−1.36	21.41	−4.77	0.31	0.36
UK	United Kingdom	39	2.26	−3.58	19.64	−8.19	4.22	0.00
	Europe	190	1.92	−5.53	20.45	−9.23	2.93	0.18
	Non SRI funds							
AT	Austria	6	0.00	−11.66	20.63	−15.16	3.33	0.00
BE	Belgium	4	0.00	−9.27	18.64	−12.72	3.25	0.00
CH	Switzerland	3	0.00	−5.60	18.11	−7.44	4.33	0.33
DE	Germany	2	0.00	−5.82	17.09	−9.29	4.00	0.00
ES	Spain	2	0.00	−13.56	18.39	−16.97	0.00	1.00
FR	France	21	0.00	−7.38	19.77	−10.88	3.05	0.05
IT	Italy	3	0.00	−8.32	18.86	−11.82	2.33	0.00
LU	Luxembourg	16	0.00	−7.13	19.43	−10.69	3.45	0.00
NE	The Netherlands	4	0.00	−4.41	18.18	−7.93	0.35	0.35
SE	Sweden	10	0.00	−1.87	21.51	−5.29	0.10	0.00
UK	United Kingdom	20	0.00	−1.48	20.36	−6.10	4.15	0.00
	Europe	91	0.00	−5.74	19.79	−9.43	2.92	0.06

30/06/2006 to 30/06/2009. The overall number of SRI equity funds is equal to 190; their distribution in the various European countries is reported in Table 1, where they are grouped by country of domicile. As we can see, in the period considered the SRI funds are mainly concentrated in few countries, namely France, Luxembourg, Sweden and United Kingdom, and the analysis presented in this paper is focused on these countries.

In order to compare the performance of SRI mutual funds with that of traditional non SRI funds, besides the overall population of SRI funds, we have also analysed a set of non socially responsible funds. More precisely, we have included some non SRI equity funds with features analogous to those of the European SRI funds: for each SRI fund considered, a non SRI fund with similar features and a similar investment style was selected among those offered by the same fund company, whenever one such fund was available in the Morningstar Europe database (notice that one such fund do not always exists). For instance, the French SRI fund *AXA Euro Valeurs Responsables (C)* has features analogous to the non SRI fund *AXA Valeurs Euro A Acc*: they share the same country of domicile, the same management company (*Axa Investment Managers Paris*), the same category and investment style (Eurozone Large-Cap Equity); so the latter fund is included in the comparison data set.

The main features of the mutual funds considered are summarised in Table 1, which exhibits the average values by country. The return data taken into account are the monthly returns achieved by the mutual funds in the triennium 30/06/2006–30/06/2009 (source: Morningstar Europe), and the mean return of each mutual fund is the mean rate of return computed with the compound interest regime, measured on an annual basis (computed as the geometric mean of the capitalization factors minus 1). The average values of the various features shown in Table 1 are the arithmetic means of the values observed for the mutual funds domiciled in each country.

We may observe that the average values for the SRI and non SRI funds are fairly close, although the SRI funds exhibit a slightly higher mean return as well as a slightly higher standard deviation. The Welch's t test for equality of the means, however, indicate that the differences in the mean returns between SRI and non SRI mutual funds are not statistically significant (the p-value of the test, carried out on the entire data set considered in the analysis, is 0.7567), thus confirming the conclusions of most empirical studies that SRI funds do not give a lower financial return than non SRI funds; for a review of the empirical results on the comparison of the returns between SRI and non SRI mutual funds see e.g. [11].

With regard to the returns obtained by the mutual funds in the period considered, they are negative for most funds, due to the financial crisis, as are the excess returns. Of course, their average value differs among the various countries; in particular, the mutual funds of Sweden and UK seem to have better faced the crisis in this slump period.

The fourth column of Table 1 reports the mean ethical level of SRI mutual funds by country; the ethical level of all funds $1 \leq j \leq n$ has been computed with the ethical measure e_j proposed in [6], which takes into account both the positive and negative screening features and the eventual presence of an ethical committee and takes values in the interval $[0,5]$. As it can be seen, the mean ethical level varies substantially among the countries, meaning that in some countries the social responsibility of SRI mutual funds tends to be higher than in others.

Table 2 exhibits the frequency distribution for France, Luxembourg, Sweden and UK, the four countries with the highest numbers of SRI equity funds. We may observe that the rating distributions of France and Sweden are concentrated in the classes with lower values, while those of Luxembourg and UK show a somewhat more symmetric behaviour.

Table 2 Frequency distribution of the ethical level e_j of SRI funds of France, Luxembourg, Sweden and UK; the last column reports the comparison with Europe

Ethical rating class	FR	LU	SE	UK	Europe
$0 < e_j \leq 1$	18.2%	9.4%	24.5%	2.9%	21.6%
$1 < e_j \leq 2$	57.6%	20.8%	64.2%	25.0%	37.4%
$2 < e_j \leq 3$	18.2%	31.3%	11.3%	49.0%	24.7%
$3 < e_j \leq 4$	6.1%	33.3%	0.0%	23.1%	15.3%
$4 < e_j \leq 5$	0.0%	5.2%	0.0%	0.0%	1.1%

3 Empirical Results: Analysis of the Performance of SRI Funds of the Main European Countries

In this section we present the results of the empirical analysis carried out to assess the performance of SRI equity mutual funds in France, Luxembourg, Sweden and UK, i.e. the four European countries with the highest number of domiciled SRI mutual funds.

In this empirical analysis we have evaluated the performance of SRI and non SRI mutual funds by using three DEA models with constant returns to scale which can be applied even in slump periods and have been proposed in [6]; for the sake of brevity, we refer to [6] for the details of these models.

Adopting the same terminology and notation used in [6], we denote by I_{DEA-S} the efficiency score obtained with the DEA model for slump periods *DEA-S* by solving the relative optimisation problem; this model has non negative values of all the variables even when the mean returns are negative and does not take the ethical level into consideration. Its inputs are the initial capital invested (assumed equal to 1), the standard deviation of the returns of the mutual funds and the initial and exit charges, while the only output is the mean annual capitalization factor, i.e. 1 plus the mean return (see Table 1 for the average values for each country).

Analogously, we denote by I_{DEA-SE} the efficiency score computed with the *DEA-SE* model which inserts also the ethical level among the outputs. Finally, we indicate with $I_{DEA-SEef}$ the efficiency score computed with the *DEA-SEef* model, which assumes that the ethical level is exogenously fixed.

It can be proved that the values of the three performance indexes computed coincide for the non SRI funds, while for the socially responsible funds we have

$$I_{DEA-SE} \geq I_{DEA-SEef} \geq I_{DEA-S}. \tag{1}$$

Hence, the funds which are efficient with the *DEA-S* model (that have $I_{DEA-S} = 1$) remain efficient also with the other two models. Moreover, let us observe that the fact that the two DEA models devised for the socially responsible behaviour raise the value of the performance index of the SRI funds, while keeping it constant for the non SRI funds, does change the overall ranking, even for the non SRI funds.

In accordance with the fundamental idea of the DEA technique, it can be seen that a fund which excels with respect to one of the input or output variables is generally efficient: therefore it is efficient the fund with the highest mean return, but also the fund with the lowest standard deviation and the fund with the highest ethical level.

We may also notice that, for all the countries, the value of the I_{DEA-SE} and $I_{DEA-SEef}$ indexes and the relative ranking of the SRI funds are often very closed, while they differ more notably with respect to the value of I_{DEA-S}. This seems to indicate that considering the ethical level as fixed a priori does not affect the performance results significantly, while the inclusion of the ethical level in the analysis does raise the results of the SRI funds considerably.

On the other hand, when the ethical level is considered, the number of efficient funds among the SRI mutual funds roughly double. This can be seen from Table 3, which reports some statistics on the results of the analysis carried out on the single countries, useful to compare the performance results of the socially responsible and non socially responsible mutual funds computed with the three DEA models considered. From this table we may also observe that the rate of SRI funds above the median of the performance score of a country increases markedly for the two DEA models which takes the ethical level into consideration. Both effects show that the premium given to the socially responsible behaviour by the two DEA models that take the ethical level into account (*DEA-SE* and *DEA-SEef*) is sensible.

As for the differences among the various countries, we may observe that the SRI mutual funds on average exhibit a slightly better performance than the non SRI funds in France and Sweden, even considering the results of I_{DEA-S} which do not take the ethical level into account, while the opposite occurs in Luxembourg and UK. It remains to be seen if these differences are statistically significant, and this issue will be considered in next section. On the other hand, the results obtained using I_{DEA-SE} and $I_{DEA-SEef}$, which explicitly consider the socially responsible behaviour, considerably improve the performance of SRI funds for all the countries. In next section we will also test whether the results among the different countries are statistically significant.

4 Empirical Results: Efficiency Comparisons

As we have outlined in the introduction, the literature is not in complete accord on the connection between social responsibility and the financial performance of SRI mutual funds; for a discussion on this issue see for example [9, 11]. It is therefore interesting to see which indications come out from the results of our analysis concerning the European funds in the period 30/06/2006–30/06/2009.

We have seen in the previous section that the average value of the mean returns of SRI mutual funds is not statistically different from that of non SRI funds. Now let us compare the performance results of SRI and non SRI mutual funds and test whether their differences are statistically significant. To this aim we apply some statistical tests, specially designed for the DEA performance scores and presented in [3], to the performance indexes I_{DEA-S} and I_{DEA-SE}.

Table 3 Summary statistics of the empirical results of the analysis of the performance in the single countries (France, Luxembourg, Sweden and UK) considered; the results are compared for the three DEA model applied

	FR	LU	SE	UK
DEA-S				
Percentage of efficient funds	7.0%	13.0%	9.5%	10.2%
Percentage of SRI efficient funds	5.6%	10.5%	9.4%	7.7%
Percentage of non SRI efficient funds	9.5%	18.8%	10.0%	15.0%
Average performance	0.955	0.933	0.957	0.872
Average performance of SRI funds	0.957	0.930	0.957	0.864
Average performance of non SRI funds	0.951	0.940	0.955	0.887
Median of the performance score	0.952	0.923	0.959	0.865
Percentage of SRI funds above the median	52.8%	47.4%	50.0%	48.7%
Percentage of non SRI funds above the median	47.6%	56.3%	50.0%	55.0%
DEA-SE				
Percentage of efficient funds	10.5%	25.9%	19.0%	15.3%
Percentage of SRI efficient funds	11.1%	28.9%	21.9%	15.4%
Percentage of non SRI efficient funds	9.5%	18.8%	10.0%	15.0%
Average performance	0.962	0.952	0.968	0.904
Average performance of SRI funds	0.968	0.957	0.972	0.913
Average performance of non SRI funds	0.951	0.940	0.955	0.887
Median of the performance score	0.963	0.956	0.975	0.898
Percentage of SRI funds above the median	61.1%	55.3%	59.4%	64.1%
Percentage of non SRI funds above the median	33.3%	37.5%	20.0%	25.0%
DEA-SEef				
Percentage of efficient funds	10.5%	25.9%	19.0%	15.3%
Percentage of SRI efficient funds	11.1%	28.9%	21.9%	15.4%
Percentage of non SRI efficient funds	9.5%	18.8%	10.0%	15.0%
Average performance	0.962	0.951	0.968	0.898
Average performance of SRI funds	0.968	0.955	0.972	0.903
Average performance of non SRI funds	0.951	0.940	0.955	0.887
Median of the performance score	0.962	0.955	0.974	0.886
Percentage of SRI funds above the median	58.3%	52.6%	59.4%	56.4%
Percentage of non SRI funds above the median	38.1%	43.8%	20.0%	40.0%

Indeed, an advantage of the DEA methodology is that it gives the possibility to test the (eventual) presence of differences in the performance score between two groups of decision making units. The statistical tests proposed in the literature to verify the presence of these differences come from two different approaches which date back to Banker [1] and Simar and Wilson [12], respectively, and are based on the characterisation of the DEA efficiency scores as stochastic variables, with different hypothesis on the underlying data-generating process. There is discussion on which approach is to be preferred, and we can find empirical applications of both; in this paper we apply several tests reported in [3], which are based on different assumptions on the distribution of the "true" inefficiency measure, where the inefficiency is defined as the reciprocal of the DEA performance score.

More precisely, we have computed the three tests which assume that the deviations of the actual output from the production frontier arise only from a stochastic inefficiency term and exploit the asymptotic properties of the DEA inefficiency estimator studied by Banker in [1] (see also [3, Par. 11.2.2], and [2]):

A1. a test based on the assumption that the logarithm of the true inefficiency is exponentially distributed; in this case, under the null hypothesis H_0 the test statistics is asymptotically distributed as an F distribution;
A2. a test based on the assumption that the logarithm of the true inefficiency is distributed as half-normal; under H_0 the test statistics is again asymptotically distributed as an F distribution;
A3. a test with no assumptions on the distribution of the true inefficiency: the Kolmogorov-Smirnov's test statistics for the equality of the distributions of the logarithm of the true inefficiency between the two groups.

In addition, we have computed five tests suitable when the data generating process involves both an inefficiency term and a noise term independent of the inefficiency (see [2] and [3, Par. 11.4.1]):

B1. a test based on the statistical significance of the slope parameter of a regression of the DEA inefficiency scores on a dummy variable;
B2. a test designed to evaluate the null hypothesis that there is no difference in the mean inefficiency between the two groups, under the assumption that the inefficiency score is lognormally distributed;
B3. a test designed to evaluate the equality of the median of the inefficiencies between the two groups;
B4. the Mann-Whitney test to compare the DEA efficiency scores of the two groups;
B5. a Kolmogorov-Smirnov's test to compare the distributions of the DEA inefficiencies between the two groups.

Let us observe that tests A1, A2 and B2 make specific assumptions on the probability distribution of the inefficiency scores, while the other tests do not; on the other hand, in our opinion the tests denoted by the B letter are more suitable to deal with mutual fund performance measures, since it would be unrealistic to think that the mutual fund inefficiencies do not suffer from a noise term (suffice it to remember that the fund returns are heavily affected by the prices of the stocks included in the fund portfolio).

We have checked the assumptions made on the probability distribution of the inefficiency scores by tests A1, A2 and B2; the assumptions can be rejected for all countries, and also for the overall set of European mutual funds, with the only exception of France, where we can accept the assumption that the inefficiency scores are lognormally distributed. Hence, we will present in detail the results of tests A3, B3, B4 and B5 (test B1 turned out to have very little power for the mutual fund inefficiency scores).

For each country (France, Luxembourg, Sweden, UK) we have first computed the inefficiency scores and then have applied the tests to compare the DEA performance of the SRI and non SRI mutual funds (these tests have been carried out separately, since they refer to disjoint computations of the DEA scores for the different countries). As for the DEA model used in these comparisons, we were specially interested in testing the differences for the *DEA-S* model that does not give any reward to the SRI funds. In agreement with most of the empirical results reported in the literature, with a 0.05 significance level all the tests carried out lead to accept the null hypothesis of no differences.

We have also replicated the tests with the *DEA-SE* model, and we expected results more favourable to the SRI funds. This actually happens; in particular, with the two Kolmogorov-Smirnov's tests (A3 and B5), which seem to reject the null hypothesis more frequently, for France, Sweden and UK we can accept the alternative hypothesis H_1 of different distributions of the DEA inefficiencies.

Moreover, we have carried out a second series of tests with the aim to compare the DEA efficiency of the SRI mutual funds across the countries. In order to do so, first we have computed the DEA efficiency scores for all the European funds considered all together and then we have tested the differences between pairs of countries; the tests have been carried out both with the *DEA-S* model that considers only the financial inputs and outputs and the *DEA-SE* model that takes into account also the ethical level; the null hypothesis tested is the equality of the inefficiency scores.

For all tests, we have used a significance level $\alpha = 0.05$ and we have applied the Holm-Bonferroni method [10] to take into account the problem of multiple comparisons. So the test results have been considered in ascending order of p-values, and the p-values (say $p_1, p_2, \ldots, p_m$, where m is the number of comparisons carried out and $p_1 \leq p_2 \leq \ldots \leq p_m$) have been compared with the values

$$\frac{\alpha}{m}, \frac{\alpha}{m-1}, \ldots, \frac{\alpha}{1}, \tag{2}$$

respectively. For $k = 1, 2, \ldots, m$, the null hypothesis H_0 is rejected at level α if $p_k \leq \frac{\alpha}{m-k+1}$, until the minimal index such that $p_{\bar{k}} > \frac{\alpha}{m-\bar{k}+1}$, while it is accepted for all $k \geq \bar{k}$.

First of all, test B1 seems to have very little power, since in our investigation it never leads to reject the null hypothesis. Secondly, tests A3, B4 and B5 lead to the same acceptance/rejection decisions, while test B3 seems to have a slightly more limited power, since it leads to accept the null hypothesis in one additional case.

The main results obtained with tests A3, B4 and B5 are summarized in Table 4, which shows which hypothesis, H_0 or H_1, is accepted using a 0.05 significance level with the Holm-Bonferroni method; for the sake of completeness we also report the p-values of one of these tests (B5). As for the SRI mutual funds, the results of the tests suggest that the inefficiency scores of Swedish mutual funds are different from those of the other countries (the differences in the performance scores are statistically significant for all the comparisons involving Sweden). On the contrary, using the

Table 4 Hypothesis accepted (using the Holm-Bonferroni correction for multiplicity) with tests A3, B4 and B5 carried out to compare the DEA inefficiency scores of the mutual funds across the countries (significance level 0.05); the p-values of the tests are also reported

	FR-LU	FR-SE	FR-UK	LU-SE	LU-UK	SE-UK
DEA-S						
SRI mutual funds	H_0	H_1	H_0	H_1	H_0	H_1
p-value	0.024	0.000	0.017	0.000	0.025	0.000
All mutual funds	H_0	H_1	H_1	H_1	H_1	H_1
p-value	0.071	0.000	0.000	0.000	0.003	0.000
DEA-SE						
SRI mutual funds	H_0	H_1	H_1	H_1	H_0	H_1
p-value	0.098	0.000	0.001	0.000	0.096	0.000
All mutual funds	H_0	H_1	H_1	H_1	H_0	H_1
p-value	0.047	0.000	0.000	0.000	0.106	0.000

Table 5 Winners of the pairwise comparisons of the DEA efficiency scores carried out with tests A3, B4 and B5. "*Country 1 $\succ$ country 2*" means that the winner (with the highest values of the scores) is country 1, while "*country 1 $\prec$ country 2*" denotes the pairs in which the winner is country 2 (significance level 0.05); the p-value of the tests are also reported

	FR-LU	FR-SE	FR-UK	LU-SE	LU-UK	SE-UK
DEA-S						
SRI funds	–	$FR \prec SE$	–	$LU \prec SE$	–	$SE \succ UK$
p-value		0		0		0
All funds	–	$FR \prec SE$	$FR \prec UK$	$LU \prec SE$	$LU \prec UK$	$SE \succ UK$
p-value		0	0.001	0	0.002	0
DEA-SE						
SRI funds	–	$FR \prec SE$	$FR \prec UK$	$LU \prec SE$	–	$SE \succ UK$
p-value		0	0	0		0
All funds	–	$FR \prec SE$	$FR \prec UK$	$LU \prec SE$	–	$SE \succ UK$
p-value		0	0	0		0

DEA-S model the differences in the DEA scores are not statistically significant for the other countries, while if we take the ethical level into consideration, using the *DEA-SE* model, we reject the null hypothesis of equality of the inefficiency scores also for the pair FR-UK.

We have also tested if the differences remain valid also for the non SRI funds, by considering all the funds (both SRI and non SRI ones) of the pair of countries compared; from Table 4 we may see that using the *DEA-S* model the differences in the DEA scores are statistically significant for all comparisons but FR-LU, while using the *DEA-SE* model we accept the null hypothesis of equality of the inefficiency scores also for the pair LU-UK.

Table 5 shows the winner of each pairwise comparison, when the differences in the performance scores are statistically significant. We denote by "*country 1* ≻ *country 2*" the pairs in which the "winner" (the country with the highest values of the performance scores) is country 1, and by "*country 1* ≺ *country 2*" the pairs in which the winner is country 2. We may observe that the winner among all the countries considered is undoubtedly Sweden, followed by UK.

References

1. Banker, R.D.: Maximum likelihood, consistency and data envelopment analysis: a statistical foundation. Management Science **39**, 1265–1273 (1993)
2. Banker, R.D., Zheng, Z.E., Natarajan, R.: DEA-based hypothesis tests for comparing two groups of decision making units. European Journal of Operational Research **206**, 231–238 (2010)
3. Banker, R.D., Natarajan, R.: Statistical tests based on DEA efficiency scores. In: Cooper, W.W., Seiford, L.M., Zhu, J. (eds) Handbook on Data Envelopment Analysis, 2nd edn., pp. 273–295. Springer, New York (2011)
4. Basso, A., Funari, S.: A data envelopment analysis approach to measure the mutual fund performance. European Journal of Operational Research **135**, 477–492 (2001)
5. Basso, A., Funari, S.: Measuring the performance of ethical mutual funds: A DEA approach. Journal of the Operational Research Society **54**, 521–531 (2003)
6. Basso, A., Funari, S.: DEA models for ethical and non ethical mutual funds. Mathematical Methods in Economics and Finance **2**, 21–40 (2007)
7. Cooper, W.W., Seiford, L.M., Tone, K.: Data envelopment analysis: A comprehensive text with models, applications, references and DEA-Colver Software. Kluwer Academic Publishers, Boston (2000)
8. European Sustainable and Responsible Investment Forum (Eurosif), 2008. European SRI Study 2008. Eurosif report, 1–55. http://www.eurosif.org
9. Hamilton, S., Jo, H., Statman, M.: Doing well while doing good? The investment performance of socially responsible mutual funds. Financial Analysts Journal **49**, 62–66 (1993)
10. Holm, S.: A simple sequentially rejective multiple test procedure. Scandinavian Journal of Statistics **6**, 65–70 (1979)
11. Renneboog, L., Ter Horst, J., Zhang, C.: Socially responsible investments: Institutional aspects, performance, and investor behavior. Journal of Banking & Finance **32**, 1723–1742 (2008)
12. Simar, L., Wilson, P.W.: Sensitivity analysis of efficiency scores: how to bootstrap in nonparametric frontier models. Management Science **44**, 49–61 (1998)
13. Vigeo SRI Research, 2009: Green, social and ethical funds in Europe – 2009 Review. Report (October 2009). http://www.vigeo.com

Fitting Financial Returns Distributions: A Mixture Normality Approach

Riccardo Bramante and Diego Zappa

Abstract An important research field in finance is the identification of probability distribution model that fits at the best the empirical distribution of time series returns. In this paper we propose the use of mixtures of truncated normal distributions in modelling returns. An optimization algorithm has been developed to obtain the best fit by using the minimum distance approach. Empirical results show evidence of the capability of the method to fit return distributions at a satisfactory level, completely maintaining local normality properties in the model. Moreover, the model provides a good tail fit thus improving the accuracy of Value at Risk estimates.

1 Introduction

In applied financial literature a relevant issue is the modelling of the empirical distribution of returns since many decision-making and asset pricing models depend on the assumptions related to the stochastic model generating the data (see e.g. [13]). Nevertheless many questions are still open:

1. Are the models supported by financial markets data?
2. How are the parameters in these models estimated?
3. Can the models be simplified?

Quoting the very famous fact by George Box "All the models are wrong but any are useful", it can generally be agreed that complex models may be closer to reality

R. Bramante (✉)
Department of Statistical Sciences, University Cattolica del Sacro Cuore, Largo Gemelli 1, Milan, Italy
e-mail: riccardo.bramante@unicatt.it

D. Zappa
Department of Statistical Sciences, University Cattolica del Sacro Cuore, Largo Gemelli 1, Milan, Italy
e-mail: diego.zappa@unicatt.it

M. Corazza, C. Pizzi (eds.), *Mathematical and Statistical Methods for Actuarial Sciences and Finance*, DOI 10.1007/978-3-319-02499-8_7, © Springer International Publishing Switzerland 2014

but often either involve many parameters or are not easy to be interpreted; on the other hand, too simple models may not capture important features of the data and can lead to serious bias.

Starting from these preliminaries, we made some considerations on how and whether the normal distribution might be still exploited in financial modelling [7].

It is obvious to say that the normal distribution is too simple in fitting returns distribution, but it is far to be excluded in practice. In fact, many practitioners still make extensive use of the normal distribution for returns even if different approaches are described in a very rich literature ([11, 13] for an extensive review).

Extending encouraging results obtained in the univariate approach, in this paper we propose a bivariate optimization algorithm for fitting returns of an investment conditional to a benchmark by using the minimum distance approach. This framework permits to decompose the global beta coefficient into local estimates, shading light to a different interpretation of dependence among returns.

2 Methodology

2.1 Fitting Distributions: Notation and Preliminary Exploratory Results

Let us fix some preliminary ideas and notation. Let $\mathbf{Z} = (X, Y)$ be a bivariate random variable and let $\hat{F}_n(z)$, $F_Z(z; \boldsymbol{\theta})$ be the empirical and the cumulative distribution function (*ECDF* and *CDF*) of (X, Y). Let $\{z_1, z_2, \ldots, z_n\}$ be a sample drawn from $\mathbf{Z}$. If $F_Z(z; \boldsymbol{\theta})$ describes the random nature of $\mathbf{Z}$, we expect that the bivariate *QQ*-plot of $\{z_i, F_Z^{-1}(\hat{F}_n(z_i); \boldsymbol{\theta})\}$, for $i = 1, 2, \ldots, n$, or equivalently the norm

$$\left\| z_i - F_Z^{-1}(\hat{F}_n(z_i); \boldsymbol{\theta}) \right\|_p \quad \text{for } i = 1, 2, \ldots, n \quad (1)$$

for some $p > 0$ is closed to zero. If $\{z_1, z_2, \ldots, z_n\}$ comes from a distribution $\mathbf{W}$ different from the one we have chosen we expect that the locus of points is locally different from zero. In an analogous manner, the analysis of the *PP*-plot over the domain $[0, 1]^2$ may be considered, i.e. the plot of $\{\hat{F}_n(z_i), F_Z(z_i; \boldsymbol{\theta})\}$, for $i = 1, 2, \ldots, n$.

Parameter estimation is typically based on standard maximum likelihood (*ML*) or by robust estimation procedures, such as the median and the median absolute deviation from median (*MAD*) [13]. Differently from *ML*, in order to let the fitting process as flexible and maximally data dependent as possible, the *MDA* approach (see [3]) is becoming popular in applications because of both its theoretical implications and the avilability of computational tools. It consists in solving the general unconstrained problem

$$\min_{\boldsymbol{\theta}} d\left(\hat{F}_n(z), F_Z(z; \boldsymbol{\theta})\right) \quad \mathbf{Z} \in \mathbb{R}^2 \quad (2)$$

where $d(\cdot)$ is an appropriate measure of discrepancy (i.e. loss function) and it can be interpreted as a transformation of the quantities used in the *PP*-plot. If $F_Z(z; \boldsymbol{\theta})$ is the "true" CDF then the unconstrained estimator $\widehat{\theta}$ minimizing (2) has been shown to be consistent.

Depending on which $d(\cdot)$ is used, further properties, e.g. robustness to extreme influence values, may be defined in addition. Let $A(z)$ and $B_\theta(z)$ be continuous functions. Examples of $d(\cdot)$ are

$$d(A(z),B_\theta(z)) := \begin{cases} KS: & sup \ |A(z) - B_\theta(z)| \\ MH: & E\left[|A(z) - B_\theta(z)|\right] \\ MQ: & E\left[(A(z) - B_\theta(z))^2\right] \end{cases} \tag{3}$$

i.e. the Kolgomorov, Manhattan, Euclidean (Cramer–von Mises) distances, respectively. A generalization of (2) is

$$\min_\theta \sum_i d\left(\left[\widehat{F}_n(z_i)\right]^q, \ [F_Z(z_i;\boldsymbol{\theta})]^q\right)^{1/q} \|z_i\|_p \quad \mathbf{Z} \in \mathbb{R}^2 \tag{4}$$

where $z_i = [x_i \ y_i]'$, $\|z_i\|_p = (\sum_i |z_i|^p)^{1/p}$ with $q > 0$, $p > 0$.

That means we may solve (4) by searching also for that parameters (q,p) that at the best guarantee a good fit between the two. In this paper, for the sake of simplicity and without loss of generality, we will use $q = 1$ and $p = 2$ to reconcile notation to the definitions in (3) and the norm of the weighs to the standard Euclidean distance of the vector z_i from the origin.

2.2 Fitting Bivariate Truncated Normal Distributions

Suppose to subdivide the domain of $\mathbf{Z}$ into a partition; assume that $F_Z(z_i;\boldsymbol{\theta})$ is represented by a sequence of truncated distributions, also called spliced distribution. Within each subset of the partition, we may interpret the distribution to be truncated. If by applying (4) the local solution does not improve with respect to the one obtained by using the whole domain, this should be an evidence that locally the process is not different from the unconditional one. By contrast if, constraining (4) over $\mathbb{R}^2_c \subseteq \mathbb{R}^2$, we obtain a solution better than the unconstrained one, there is an evidence that the underlying process may be considered as a mixture of distributions (see also [9]). A solution can be found by applying (4) in a two stage process: in the first step we estimate the parameters of the bivariate distribution $\mathbf{Z}$ and then, by keeping fixed these estimates, we look for a partition, if it exists, such that the optimum obtained in the first step has improved.

Let $\mathbf{Z} \sim N_2(\boldsymbol{\mu},\boldsymbol{\Sigma})$ be a bivariate normal random variable and indicate with $F_Z(z_i;\boldsymbol{\mu};\boldsymbol{\Sigma})$ its cumulative distribution function, where $\boldsymbol{\mu}$, $\boldsymbol{\Sigma}$ are the mean vector and the variance-covariance matrix, respectively. Let $\widehat{\boldsymbol{\mu}}$ and $\widehat{\boldsymbol{\Sigma}}$ be the estimates of $\boldsymbol{\mu}$, $\boldsymbol{\Sigma}$ obtained by solving (4). Let xtr_j be a generic threshold for the marginal X such that, for any $xtr_{j-1} < xtr_j$ and for $j = 1, 2, \ldots, k$

$$\bigcup_{j=1}^k \left(xtr_{j-1} \dashv xtr_j\right) = \mathbb{R}^2$$

$$\left(xtr_{j-1} \dashv xtr_j\right) \cap \left(xtr_{h-1} \dashv xtr_h\right) = \emptyset \quad for \ h \neq j \quad h,j = 1,\ldots,k \tag{5}$$

with $xtr_0 = -\infty$ and $xtr_k = +\infty$. Let

$$_Tfz\left(z;\widehat{\boldsymbol{\mu}};\widehat{\boldsymbol{\Sigma}};\, xtr_{j-1};\, xtr_j\right) =$$

$$
\begin{cases}
\dfrac{\exp\left(-\frac{1}{2}(z-\widehat{\boldsymbol{\mu}})'\widehat{\boldsymbol{\Sigma}}(z-\widehat{\boldsymbol{\mu}})\right)}{\int\left(\int_{xtr_{j-1}}^{xtr_j}\exp\left(-\frac{1}{2}(z-\widehat{\boldsymbol{\mu}})'\widehat{\boldsymbol{\Sigma}}(z-\widehat{\boldsymbol{\mu}})\right)dx\right)dy} & \text{for} \quad \begin{matrix}(x \in_x tr_{j-1}\dashv xtr_j) \\ \text{and } (y \in \mathbb{R})\end{matrix} \\[20pt]
0 & \text{otherwise}
\end{cases}
\tag{6}
$$

be the truncated normal pdf and let the *CDF* in z be

$$_TF_Z\left(z;\widehat{\boldsymbol{\mu}};\widehat{\boldsymbol{\Sigma}}\right) = \Sigma_{j=1}^i\left(\int\int_{-\infty}^z {}_Tfz\left(z;\widehat{\boldsymbol{\mu}};\widehat{\boldsymbol{\Sigma}};_x tr_{j-1};_x tr_j\right)dz\right)w_j \tag{7}$$

$$\text{with } w_j > 0 \ \forall j \quad \Sigma_{j=1}^{k-1}w_j = 1 - w_k.$$

From (7), $_TF_Z\left(z;\widehat{\boldsymbol{\mu}};\widehat{\boldsymbol{\Sigma}}\right)$ is a weighted sum of disjoint truncated distributions. The estimate of the weights and thresholds in (7) is obtained by solving

Find $w_1, \ldots, w_{k-1}$ and $tr_1, \ldots, tr_{k-1}$:

$$\sum_i d\left(\widehat{F}_n(z_i),\, F_Z\left(z_i;\widehat{\boldsymbol{\mu}};\widehat{\boldsymbol{\Sigma}}\right)\right)\|z_i\|_2 - \tag{8}$$

$$\min_{w,k}\sum_i d\left(\widehat{F}_n(z_i),\, {}_TF_Z\left(z_i;\widehat{\boldsymbol{\mu}};\widehat{\boldsymbol{\Sigma}}\right)\right)\|z_i\|_2 \geq \varepsilon$$

with $\varepsilon > 0$. It is the same problem stated in (4) but where the unknowns are not the parameters of the bivariate Gaussian distribution but the weights and the thresholds. Observe that we look for the smallest partition such that (8) is fulfilled, since for $k \to \infty$ the truncated distribution degenerates on the single observation.

3 Empirical Results

To exemplify of the aspects captured by the proposed model, we show results by using the returns of 300 Morgan Stanley Capital International (*MSCI*) indices, provided in the country and sector sub set, along with the World Index which is assumed to be the market benchmark . All the indices are denominated in US dollars and cover the period from January 1996 to March 2012. The sample period was divided into pre specified consecutive intervals with a fixed length of 500 observations[1] using

---

[1] In general, the choice of the window length involves balancing two opposite factors: on the one hand a larger window could embrace changing data generating processes, whereas on the other hand a shorter period implies a smaller data set available for estimation. In our opinion, the selection of the window size depends on the specific application. For the purposes of the empirical investigation

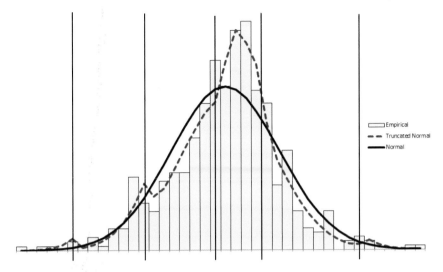

Fig. 1 World index marginal truncated normal distribution

a 20 trading session rolling window procedure: totally, approximately 20,000 data periods were tested.

In the univariate case, the optimization tool was initially used to estimate the location and variance of the distribution (with no thresholds). Results, in terms of degree of fit, were encouraging since a significant increment (76% on average) was provided with no relevant differences due to the index type. In the second step, a solution to (8) was found, obtaining the estimates of the unknowns and the distance from the solution found in (4) (see Fig. 1 for an example regarding the World Index).

The relative gain in the discrepancy measure by using the *MQ* distance with respect to the *ML* estimators for both the normal and the Skew-t distribution was satisfactory. The introduction of a mixture of truncated normal distributions has significantly increased the accuracy, capturing at an adequate degree the most relevant aspects of daily returns empirical distribution, by regulating both kurtosis and fat tails: starting from the case with 3 thresholds (i.e. in the 92.42% of the simulations) the average gain is higher than 80% with respect to both the normal and Skew-t distributions [7].

For the bivariate case, the optimization tool was used in a similar way to estimate, in the first step, the vector of the unknown parameters where the criterion function measures the weighted squared distance between the empirical and the bivariate normal distribution and to select, in the second step, the optimal truncation thresholds conditional to the benchmark (see Fig. 2 for an example regarding the *MSCI* Italy

proposed in this paper, the choice of equal sized sections of 500 data points seems to be a good compromise between the two factors.

Table 1 Optimization results (bivariate case)

	% Gain in degree of fit	
Type of Index	First Opt. Step	Second Opt. Step
Developed Markets Index	68.64	10.03
Emerging Markets Index	66.46	18.52
Emerging Sector Index	71.93	22.46
Europe Sector Index	76.93	17.55
World Sector Index	81.88	15.01

Truncated Multivariate Normal Density   **Contours of Truncated Multivariate Normal Density**

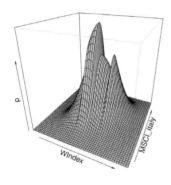
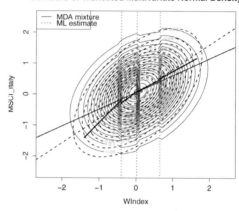

Fig. 2 MSCI Italy index bivariate truncated normal

Index). In Table 1 results regarding the relative gain in the discrepancy measure, for the two steps, are reported.

If compared to univariate outcomes, fitting results are less impressive. Nevertheless, the analysis of where thresholds are placed has provided useful information about the behavior of each index with respect to the benchmark. The partitioning algorithm returned roughly 90% of the cases with less than 3 cut off points (Table 2); moreover, if we focus on the first threshold location, analyzing separately the downside and the upside returns partition, a breakdown in the negative domain has been predominant (58% of cases): these results give evidence of clusters correlation occurring mainly in the negative domain, where index and benchmark series tend to be comonotone in their extreme low returns values.

Another interesting result was the inversion from positive to negative in the sign of the bivariate normal location parameters observed in the first optimization step (Table 3). Above all, reverse in sign in the correlation coefficient (12% on average) was obtained: generally this has occurred when extreme values were present or when the market model was not statistically significant.

Table 2 Threshold distribution (bivariate case)

| | | Type of First Threshold | |
Number of Threshold	% of Cases	% Positive	% Negative
No threshold	5.49		
1	69.34	51,77	48.23
2	15.22	24.80	75.20
3	8.17	3.86	96.14
4	1.55	0.62	99.38
5	0,22	0.00	100.00
6	0.00	0.00	100.00

Table 3 Bivariate Normal parameters

| | % Parameter Sign Inversion | |
Number of Threshold	Means	Correlation Coefficient
1	50.83	7.16
2	30.43	22.38
3	48.54	12.95
4	39.25	12.50
5	13.04	-

Analogously to relative risk analysis – typically measured by the classical beta coefficient – the domain partitioning provides a set of disjoint conditional regions where the local relationship between the index and the benchmark can be slightly different with respect to the one on the domain as a whole. In particular, it is interesting to analyze reverse in sign in the index – benchmark relationship (from a global positive beta to at least one negative local beta and the inverse case) which occurs, on average, in 46% of the total cases (see Table 4 where results referred to a partitions with more than 4 thresholds are omitted): this provides strong evidences that relative risk varies conditionally to the benchmark returns regions coherently with the well known clustering effect of returns.

Table 4 Beta parameters sign inversion

	% Sign Inversion Local Beta	
Number of Threshold	% negative	% positive
1	58.33	22.17
2	77.72	42.22
3	89.02	28.62
4	90.00	61.34

References

1. Aas, K., Hobæk Haff, I.: The Generalized Hyperbolic Skew Student's t-distribution. Journal of Financial Econometrics **4**, 275–309 (2006)
2. Azzalini, A., Capitanio, A.: Distributions Generated by Perturbation of Symmetry with Emphasis on a Multivariate Skew t Distribution. Journal of the Royal Statistical Society B **65**, 579–602 (2003)
3. Basu, A., Shiova, H., Park, S.: Statistical Inference: The Minimum Distance Approach. CRC Press, London (2011)
4. Black, F., Scholes, M.: The pricing of options and corporate liabilities. Journal of Political Economy **81**, 637–659 (1973)
5. Bates, D.M., Watts, D.G.: Nonlinear Regression Analysis and its application. Wiley, New York (2007)
6. Box, G.E.P., Draper, N.R.: Empirical Model-Building and Response Surfaces. Wiley, New York (1987)
7. Bramante R., Zappa D.: Value at Risk Estimation in a Mixture Normality Framework, Working paper (2013)
8. Chen, Y., Hardle, W., Jeong, S.: Nonparametric Risk Management with Generalized Hyperbolic Distributions. Journal of the American Statistical Association **103**, 910–923 (2008)
9. Cont, R.: Empirical properties of asset returns: stylized facts and statistical issues. Quantitative Finance **1**, 223–236 (2001)
10. Kon, S.J.: Models of stock returns – a Comparison. The Journal of Finance **39**, 147–165 (1984)
11. McNeil, A., Rudiger Frey, J., Embrechts, P.: Quantitative Risk Management. Princeton University Press, Princeton (2005)
12. Parr, W.C.: Minimum Distance Estimation, Encyclopedia of Statistical Science **5** (1985)
13. Ruppert, D.: Statistical and data analysis for financial engineering. Springer, Berlin (2011)
14. Wand, M.P., Jones, M.C.: Kernel smoothing. Chapman & Hall, London (1995)

Single-Name Concentration Risk Measurements in Credit Portfolios

Raffaella Calabrese and Francesco Porro

Abstract For assessing the effect of undiversified idiosyncratic risk, Basel II has established that banks should measure and control their credit concentration risk. Concentration risk in credit portfolios comes into being through an uneven distribution of bank loans to individual borrowers (single-name concentration) or through an unbalanced allocation of loans in productive sectors and geographical regions (sectoral concentration). In this paper six properties that ensure a coherent measure of single-name concentration are identified. To evaluate single-name concentration risk in the literature, Herfindahl–Hirschman index has been used. This index represents a particular case of Hannah–Kay index proposed in monopoly theory. In this work the proof that Hannah–Kay index satisfies all the six properties is given. Finally, the impact of the elasticity parameter in Hannah-Kay index on the single-name concentration measure is analysed by numerical applications.

1 Introduction

The Asymptotic Single-Risk Factor (ASRF) model [7] that underpins the Internal Rating Based (IRB) approach in the Basel II Accord (Basel Committee on Banking Supervision, BCBS 2004) assumes that idiosyncratic risk has been diversified away fully in the portfolio, so that economic capital depends only on systematic risk contributions. Systematic risk represents the effect of unexpected changes in macroeconomic and financial market conditions on the performance of borrowers. On the

R. Calabrese
Essex Business School, University of Essex, Colchester, UK
e-mail: rcalab@essex.ac.uk

F. Porro (✉)
Dipartimento di Statistica e Metodi Quantitativi, Università degli Studi di Milano-Bicocca, Milan, Italy
e-mail: francesco.porro1@unimib.it

M. Corazza, C. Pizzi (eds.), *Mathematical and Statistical Methods for Actuarial Sciences and Finance*, DOI 10.1007/978-3-319-02499-8_8, © Springer International Publishing Switzerland 2014

other hand, idiosyncratic risk represents the effects of risks that are particular to individual borrowers. In order to include idiosyncratic risk in economic capital, Basel II (BCBS, 2004) requires that banks estimate concentration risk. Concentration risks in credit portfolios arise from an unequal distribution of loans to single borrowers (*single-name concentration*) or industrial or regional sectors (*sector concentration*). This paper is focused only on the single-name concentration, in particular in the context of loan portfolios.

Since single-name concentration risk is relatively unexplored in the literature, the first aim of this paper is to identify the properties that a coherent measurement of single-name concentration risk should satisfy. In particular, six properties are identified: some of them were suggested for inequality measures (e.g. transfer principle, Lorenz-criterion, uniform distribution principle, superadditivity) and for monopoly theory (e.g. independence of loan quantity). The first theoretical result of this paper is the analysis of the relationship among the six properties.

A widely used index of single-name concentration risk is the Herfindahl–Hirschman index, proposed in monopoly theory [11–13]. Such an index represents a particular case of the Hannah–Kay index, proposed also in monopoly theory [10]. A second theoretical result of this work is that the Hannah–Kay index satisfies all the six properties for measurements of single-name concentration risk. This result justifies the wide use of the Herfindahl-Hirschman index in credit concentration risk analysis. The Herfindahl-Hirschman index is obtained as the Hannah-Kay index with an elasticity parameter equal to 2.

The last important result of this paper is the analysis of the impact of the elasticity parameter on the single-name concentration measure. The elasticity parameter determines the weight attached to the upper and the lower portions of the distribution. High elasticity parameter gives greater weight to the highest credit amounts and low elasticity emphasizes the small exposure. In order to highlight this characteristic, six portfolios with different levels of single-name concentration risk are analysed. The portfolio with the highest concentration risk is considered compliant to the regulation of the Bank of Italy [1].

The paper is organized as follows. Section 2 defines the six properties of a single-name concentration index and the relationships among them. Section 3 analyses the Hannah–Kay index and proves that it satisfies all the mentioned six properties. The Herfindahl–Hirschman index and the granularity adjustment are also examined. In the last section a numerical application on portfolios with different concentration risk is described.

2 Properties of a Single-Name Concentration Index

Consider a portfolio of n loans. The exposure of the loan i is represented by $x_i \geq 0$ and the total exposure of the portfolio is $\sum_{i=1}^{n} x_i = T$. In the following, a portfolio is denoted by the vector of the shares of the amounts of the loans $\mathbf{s} = (s_1, s_2, \ldots, s_n)$: the share $s_i \geq 0$ of i-th loan is defined as $s_i = x_i/T$. It follows that $\sum_{i=1}^{n} s_i = 1$. Whenever the shares of the portfolio $\mathbf{s}$ need to be ordered, the corresponding portfolio obtained by the increasing ranking of the shares will be denoted by $\mathbf{s}_{(.)} = (s_{(1)}, \ldots, s_{(n)})$. It is clear that any reasonable concentration measure C must satisfy $C(\mathbf{s}) = C(\mathbf{s}_{(.)})$. Whenever it is necessary, in order to remark the number n of the loans in the portfolio, the single-name concentration measure will be denoted with C_n.

The following six properties are the desirable ones that a single-name concentration measure C should satisfy. Indeed they were born in a different framework, nevertheless their translation to credit analysis can be considered successful (cf. [4,7]).

1. **Transfer principle** *The reduction of a loan exposure and an equal increase of a bigger loan that preserves the order must not decrease the concentration measure.*
   Let $\mathbf{s} = (s_1, s_2, \ldots, s_n)$ and $\mathbf{s}^* = (s_1^*, s_2^*, \ldots, s_n^*)$ be two portfolios such that

$$
s_{(k)}^* = \begin{cases} s_{(j)} - h & k = j \\ s_{(j+1)} + h & k = j+1 \\ s_{(k)} & otherwise, \end{cases} \tag{1}
$$

   where

$$
s_j < s_{j+1}, \qquad 0 < h < s_{(j+1)} - s_{(j)}, \qquad h < s_{(j+2)} - s_{(j+1)}.
$$

   Then $C(\mathbf{s}) \leq C(\mathbf{s}^*)$.

2. **Uniform distribution principle** *The measure of concentration attains its minimum value, when all loans are of equal size.*
   Let $\mathbf{s} = (s_1, s_2, \cdots, s_n)$ be a portfolio of n loans. Then $C(\mathbf{s}) \geq C(\mathbf{s_e})$, where $\mathbf{s_e}$ is the portfolio with equal-size loans, that is $\mathbf{s_e} = (1/n, \ldots, 1/n)$.

3. **Lorenz-criterion** *If two portfolios, which are composed of the same number of loans, satisfy that the aggregate size of the k biggest loans of the first portfolio is greater or equal to the size of the k biggest loans in the second portfolio for $1 \leq k \leq n$, then the same inequality must hold between the measures of concentration in the two portfolios.*
   Let $\mathbf{s} = (s_1, s_2, \ldots, s_n)$ and $\mathbf{s}^* = (s_1^*, s_2^*, \ldots, s_n^*)$ be two portfolios with n loans. If $\sum_{i=k}^{n} s_{(i)}^* \geq \sum_{i=k}^{n} s_{(i)}$ for all $k = 1, \ldots, n$, then $C(\mathbf{s}) \leq C(\mathbf{s}^*)$.

4. **Superadditivity** *If two (or more) loans are merged, the measure of concentration must not decrease.*
   Let $\mathbf{s} = (s_1, \ldots, s_i, \ldots, s_j, \ldots, s_n)$ be a portfolio of n loans, and $\mathbf{s}^* = (s_1, \ldots, s_{i-1}, s_{i+1}, \ldots, s_{j-1}, s_m, s_{j+1}, \ldots, s_n)$ a portfolio of $n-1$ loans such that $s_m = s_i + s_j$. Then $C_n(\mathbf{s}) \leq C_{n-1}(\mathbf{s}^*)$.

5. **Independence of loan quantity** *Consider a portfolio consisting of loans of equal size. The measure of concentration must not increase with an increase in the number of loans.*
   Let $\mathbf{s}_{e,n} = (1/n, \ldots, 1/n)$ and $\mathbf{s}_{e,m} = (1/m, \ldots, 1/m)$ be two portfolios with equal-size loans and $n \geq m$, then $C_n(\mathbf{s}_{e,n}) \leq C_m(\mathbf{s}_{e,m})$.

6. **Irrelevance of small exposures** *Granting an additional loan of a relatively low amount must not increase the concentration measure. More formally, if s' denotes a share of a loan and a new loan with a share $\tilde{s} \leq s'$ is granted, then the concentration measure must not increase.*
   Let $\mathbf{s} = (s_1, s_2 \ldots, s_n)$ be a portfolio of n loans with total exposure T. Then, there exists a share s' such that for all $\tilde{s} = \tilde{x}/(T + \tilde{x}) \leq s'$ the portfolio of $n + 1$ loans $\mathbf{s}^* = (s_1^*, s_2^*, \ldots, s_{n+1}^*)$ with shares

$$s_i^* = \begin{cases} x_i/(T + \tilde{x}) & i = 1, 2, \ldots, n \\ \tilde{x}/(T + \tilde{x}) & i = n + 1 \end{cases}$$

   is considered. It holds that $C(\mathbf{s}) \geq C(\mathbf{s}^*)$.

A few remarks on the aforementioned properties can be useful. The first three properties have been proposed for the concentration of income distribution. In the first three properties the number n of loans of the portfolio is fixed, while in the others n changes. This means that the properties 4, 5 and 6 point out the influence of the number of the loans on the concentration measure. The principle of transfers and the Lorenz-criterion have been proposed at the beginning of the last century: the former has been introduced by Pigou [16] and Dalton [5], the latter is related to the Lorenz curve proposed by Lorenz (see [14]). The property 4 can be applied more than one time by setting up the merge of three or more loans. Finally, the properties 4 and 5 have been suggested in the field of the industrial concentration where the issue of monopoly is very important.

Theorem 1 (Link among the properties) *If a concentration measure satisfies the properties 1 and 6, then it satisfies all the aforementioned six properties.*

Proof The outline of the proof is the following. It can be proved that a concentration index satisfying property 1 fulfills also properties 2 and 3. Further, if a concentration measure satisfies the properties 1 and 6, then it meets the property 4. Finally, properties 2 and 4 imply the property 5.

1. *Property 1 $\Rightarrow$ property 3 and property 2*
   A detailed proof can be found in [8].

2. *Properties 1 and 6 $\Rightarrow$ property 4*
   Let $\mathbf{s} = (s_1, \ldots, s_i, \ldots, s_j, \ldots, s_n)$ be a portfolio of n loans, and $\mathbf{s}^* = (s_1, \ldots, s_{i-1}, s_{i+1}, \ldots, s_{j-1}, s_m, s_{j+1}, \ldots, s_n)$ a portfolio obtained by the merge $s_m = s_i + s_j$. Let $\mathbf{s}^1 = (s_1, \ldots, s_{i-1}, 0, s_{i+1}, \ldots, s_{j-1}, s_m, s_{j+1}, \ldots, s_n)$ be the portfolio obtained from $\mathbf{s}^*$ by the adding of a null share. Since the property 6 holds, then $C_n(\mathbf{s}^1) \leq C_{n-1}(\mathbf{s}^*)$. Now, if the property 1 is satisfied then also property 3 is satisfied (see the previous point 1.). From the comparison of $\mathbf{s}$ with $\mathbf{s}^1$, through the property 3 it results that $C_n(\mathbf{s}) \leq C_n(\mathbf{s}^1)$. The conclusion is that $C_n(\mathbf{s}) \leq C_n(\mathbf{s}^1) \leq C_{n-1}(\mathbf{s}^*)$, and therefore the property 4 is true.

3. *Properties 2 and 4 $\Rightarrow$ property 5*
   Since the property 2 is satisfied, $C_{n+1}(1/n, \ldots, 1/n, 0) \geq C_{n+1}(1/(n+1), \ldots, 1/(n+1))$. After a merge, by property 4, it follows that

$C_n(1/n, \ldots, 1/n) \geq C_{n+1}(1/n, \ldots, 1/n, 0)$, and therefore
$C_n(1/n, \ldots, 1/n) \geq C_{n+1}(1/(n+1), \ldots, 1/(n+1))$. This means that $C_n(\mathbf{s_{e,n}}) \geq C_{n+1}(\mathbf{s_{e,n+1}})$. By iteration, the property 5 holds true.

3 Hannah–Kay Index

For industrial concentration Hannah and Kay [10] have proposed the following index (HK)

$$HK = \left(\sum_{i=1}^{n} s_i^\alpha \right)^{\frac{1}{1-\alpha}} \quad \text{with } \alpha > 0 \text{ and } \alpha \neq 1.$$

The HK index is inversely proportional to the level of concentration: if the concentration increases, the HK index decreases. For this reason in this paper, the Reciprocal of Hannah-Kay (RHK) index is considered:

$$RHK = \left(\sum_{i=1}^{n} s_i^\alpha \right)^{\frac{1}{\alpha-1}} \quad \alpha > 0 \text{ and } \alpha \neq 1, \tag{2}$$

so that the RHK index is proportional to the level of concentration. From a statistical point of view, the RHK index can be considered as a powered weighted mean

$$RHK = \left(\sum_{i=1}^{n} s_i^{\alpha-1} s_i \right)^{\frac{1}{\alpha-1}} \quad \alpha > 0 \text{ and } \alpha \neq 1$$

with exponent $\alpha - 1$ and weights s_i.

For a portfolio with equal-size loans the RHK index is

$$RHK = \left[\sum_{i=1}^{n} \left(\frac{1}{n} \right)^\alpha \right]^{\frac{1}{\alpha-1}} = \frac{1}{n}.$$

If the portfolio consists of only one non-null share, the RHK index is equal to 1. The role of the elasticity parameter α is to decide how much weight to attach to the upper portion of the distribution relative to the lower. High α gives greater weight to the role of the highest credit exposures in the distribution and low α emphasizes the presence or the absence of the small exposures.

Theorem 2 *The RHK index satisfies all the six properties considered in Sect. 2.*

Proof From Theorem 1 if the properties 1 and 6 are satisfied, all the six properties of a concentration measure are satisfied.

Property 1
Let **s** and **s*** be two portfolios that satisfy the condition (1). The following difference is computed:

$$f(h) = RHK(\mathbf{s}^*) - RHK(\mathbf{s}) = \left(\sum_{k \neq i,j} s_k^\alpha + (s_j + h)^\alpha + (s_j - h)^\alpha \right)^{\frac{1}{\alpha-1}} - \left(\sum s_k^\alpha \right)^{\frac{1}{\alpha-1}}.$$

The function f is continuous for $h > 0$ and $\lim_{h \to 0} f(h) = 0$. The derivative of $f(h)$ is

$$\frac{\partial f(h)}{\partial h} = \frac{\alpha}{\alpha - 1} \left(\sum_{k \neq i,j} s_k^\alpha + (s_j + h)^\alpha + (s_j - h)^\alpha \right)^{\frac{2-\alpha}{\alpha-1}} \left[(s_j + h)^{\alpha-1} - (s_i - h)^{\alpha-1} \right].$$

(3)

In order to determine the sign of this derivative, two cases are considered:

1. $0 < \alpha < 1$
   In Eq. (3) the first and the third factors of the product are negative and the second factor is positive, hence the derivative is positive.

2. $\alpha \geq 1$
   In Eq. (3) all the factors are positive, hence the derivative is positive.

It follows that the function f is increasing, hence $RHK(\mathbf{s}^*) > RHK(\mathbf{s})$.

Property 6
Let **s** and **s*** be two portfolios that satisfy the conditions given in the property 6. The following difference is computed:

$$g(\tilde{x}) = RHK(\mathbf{s}^*) - RHK(\mathbf{s}) = \left[\sum_{i=1}^{n} \left(\frac{x_i}{T + \tilde{x}} \right)^\alpha + \left(\frac{\tilde{x}}{T + \tilde{x}} \right)^\alpha \right]^{\frac{1}{\alpha-1}} - \left[\sum_{i=1}^{n} s_i^\alpha \right]^{\frac{1}{\alpha-1}}.$$

The function $g(\tilde{x})$ is continuous for $\tilde{x} > 0$ and $\lim_{\tilde{x} \to 0} g(\tilde{x}) = 0$, so the entry of a new loan with insignificant exposure $\tilde{x}$ in the portfolio has insignificant impact on the RHK index.

The derivative of $g(\tilde{x})$ is computed by obtaining

$$\frac{\partial g(\tilde{x})}{\partial \tilde{x}} = \frac{\alpha}{\alpha - 1} \left[\sum_{i=1}^{n} \left(\frac{x_i}{T + \tilde{x}} \right)^\alpha + \frac{\tilde{x}}{T + \tilde{x}} \right]^{\frac{2-\alpha}{\alpha-1}} \frac{\frac{\tilde{x}^{\alpha-1}}{T^{\alpha-1}} - \frac{\sum x_i^\alpha}{T^\alpha}}{(T + \tilde{x})^{\alpha+1}}.$$

(4)

Set Eq. (4) equal to zero, the value $\tilde{s}$ that does not change the concentration level coincides with the superior limit s' described in property 6:

$$s' = \left[\sum_{i=1}^{n} s_i^\alpha \right]^{\frac{1}{\alpha-1}} = RHK.$$

In order to determine the sign of the derivative (4) for $\tilde{s} < s'$, the three factors in Eq. (4) are analysed. The second factor is positive and the signs of the remaining factors depend on the value of α:

1. $0 < \alpha < 1$

   In the Eq. (4) the first factor of the product is negative and the third factor is positive, hence the derivative is negative.

2. $\alpha > 1$

   In the Eq. (4) the first factor is positive and the third factor is negative, hence the derivative is negative.

This means that by introducing a new loan with share $\tilde{s}$ lower than the RHK index, the RHK slightly decreases. On the contrary, if a new loan has a share $\tilde{s}$ higher than the RHK index, the effect of the new loan in reducing the share of the existing large exposures is offset to some extent by the fact that its exposure is large.

The next index represents a particular case of the RHK index for a given value of the elasticity parameter α.

3.1 Herfindahl–Hirschman Index

By considering $\alpha = 2$, the RHK index (2) becomes

$$HH = \sum_{i=1}^{n} s_i^2$$

the Herfindahl–Hirschman index (HH) proposed by Herfindahl [11] as an industrial concentration index, whose root has been proposed by Hirschman [12, 13]. For this reason, this index is known as Herfindahl–Hirschman index. It is defined as the sum of squared portfolio shares of all borrowers.

By considering the square of the portfolio share s_i in the HH index, small exposures affect the level of concentration less than a proportional relationship.
The main advantage of the HH index is that it satisfies all the six properties of an index of credit concentration, because it is a particular case of the RHK index. The HH index can be misleading as a risk measure because it does not consider the borrower's credit quality. An exposure to a Aaa-rated borrower, for example, is treated in the same way as an exposure to a B-rated borrower. This limitation is addressed by the granularity adjustment.

3.2 Granularity Adjustment

A granularity adjustment that incorporates name concentration in the Internal Rating model was already included in the Second Consultative Paper of Basel II [2] and was later significantly refined by the work of Gordy and Lutkebohmert [7]. It arises basically because the credit risk model of Basel II International Rating-Based (IRB) approach assumes that the bank's portfolios are perfectly fine-grained, meaning that each loan accounts only for a very small share of the total exposure. Real bank portfolios are, of course, not perfectly fine-grained, therefore an adjustment to the economical capital is needed.

Gordy and Lutkebohmert developed the following simplified formula for an add-on to the capital for unexpected loss in a single-factor model [7]

$$GA_n = \frac{1}{2K_n^*} \sum_{i=1}^{n} s_i^2 LGD_i [\delta(K_i + LGD_i PD_i) - K_i]$$ (5)

where K_i denotes the unexpected loss for the i-th exposure and it is defined as the difference between the Value-at-Risk (VaR) and the expected loss, LGD_i the expected Loss Given Default, PD_i the probability of default, $K_n^* = \sum_{i=1}^{n} s_i K_i$ and δ is a constant parameter [1].

From Eq. (5) it follows immediately that the granularity adjustment is linear in the HH index if the portfolio is homogeneous in terms of PD and LGD. In such case the granularity adjustment (5) measures only the single-name concentration risk, and it becomes a function of the HH index. For this reason, in the following section the granularity adjustment is taken into account through the analysis of the HH index.

4 Numerical Applications

In this section six portfolios of loans are considered and some RHK indices with different α are calculated on them.

For the construction of the most concentrated portfolio, the large exposure limits of the Bank of Italy [1] is considered. In this analysis the total exposure T of this portfolio is 1,000 euros. Therefore, the minimum regulatory capital charge of 8% is 80 euros and this is considered the capital requirement of the bank. The Bank of Italy establishes that an exposure is defined as *large* if it amounts to 10% or more of the bank's regulatory capital, in this case an exposure is large if it is greater than or equal to 8 euros. According to the Bank of Italy's regulation, a large exposure must not exceed 25% of the regulatory capital, in this case 20 euros. The sum of all large exposures is limited to eight times the regulatory capital, which corresponds to 640 euros in this case. By considering this regulation, the portfolio with the highest concentration risk P1 consists of 32 exposures equal to 20 euros, 51 equal to 7 euros and one equal to 3 euros. Hence, the total exposure of the portfolio P1 is 1,000 euros and its number of loans is 84.

In order to obtain the portfolio P2, each exposure of 20 euros in the portfolio P2 is divided into two exposures of 10 euros. It follows that the total exposure of the portfolio P2 remains constant ($T = 1,000$) and the number of the loans of the portfolio increases ($n = 116$). Moreover, the portfolio P3 is obtained from the portfolio P2 by merging two exposures of 10 euros in one of 20 euros. From the portfolio P3, by neglecting the exposure of 20 euros the portfolio P4 is defined. Finally, the last

---

[1] The constant δ depends on the VaR confidence level and also on the variance of the systematic factor. Gordy and Lutkebohmert (2007) suggest a value of 5 as a meaningful and parsimonious number.

two portfolios are obtained by introducing in P4 a medium exposure of 7 euros for the portfolio P5 and a low exposure of 3 euros for the portfolio P6. It is important to highlight that both the total exposure T and the number of loans n can change in these six portfolios.

The portfolios therefore are:

P1: $\underbrace{20\ldots20}_{32}\ \underbrace{7\ldots7}_{51}\ 3$ $T = 1,000$ $n = 84$

P2: $\underbrace{10\ldots10}_{64}\ \underbrace{7\ldots7}_{51}\ 3$ $T = 1,000$ $n = 116$

P3: $20\ \underbrace{10\ldots10}_{62}\ \underbrace{7\ldots7}_{51}\ 3$ $T = 1,000$ $n = 115$

P4: $\underbrace{10\ldots10}_{62}\ \underbrace{7\ldots7}_{51}\ 3$ $T = 980$ $n = 114$

P5: $\underbrace{10\ldots10}_{62}\ \underbrace{7\ldots7}_{52}\ 3$ $T = 987$ $n = 115$

P6: $\underbrace{10\ldots10}_{62}\ \underbrace{7\ldots7}_{51}\ 3\,3$ $T = 983$ $n = 115$

Table 1 summarizes the values of nine RHK indices for the six portfolios. Each RHK index corresponds to a given parameter α, ranging from 0.1 to 10. The value $\alpha = 2$ denotes the HH index.

We highlight that it is not easy to give a direct interpretation of the values of the RHK index in Table 1. Instead, banking managers or central bank supervisors can use this index for ranking portfolios on the basis of their single-name concentration risk.

Under this consideration, the first result is that, as aspected, all the indices take the highest value in P1, which is the portfolio as concentrated as possible. The comparisons between the portfolios P1 and P2 and the portfolios P2 and P3 fall in the subadditivity property 4. Since the portfolio P5 is obtained by adding a medium exposure to the portfolio P4, all the indices point out a reduction of the single-name concentration: this means that the effect of increasing the number of loans dominates the effect of introducing a medium exposure. The same outcome derives from the comparison of the portfolios P4 and P6, which is obtained from P4 by adding a small exposure. Furthermore, in this case it is worth to note that the difference between the concentration levels decreases as α increases.

Table 1 The values of the concentration indices are computed for the six portfolios and multiplied by 1000

	$\alpha = 0.1$	$\alpha = 0.5$	$\alpha = 0.99$	$\alpha = 1.01$	$\alpha = 2$	$\alpha = 3$	$\alpha = 4$	$\alpha = 5$	$\alpha = 10$
P1	12.07	12.77	13.65	13.84	15.31	16.54	17.18	17.93	19.07
P2	8.64	8.69	8.78	8.79	8.91	9.03	9.14	9.23	9.54
P3	8.72	8.79	8.89	8.92	9.11	9.35	9.67	10.07	13.04
P4	8.79	8.85	8.93	8.94	9.07	9.19	9.31	9.4	9.72
P5	8.71	8.77	8.85	8.87	8.99	9.11	9.25	9.32	9.65
P6	8.71	8.78	8.87	8.89	9.02	9.15	9.27	9.36	9.69

All the indices agree that the concentration risk decreases from P4 to P2. It follows that the impact of the increase of the number of loans in the portfolio is higher than the impact of larger loans. So far, the orderings of the indices for all the considered values of α are basically equivalent. An important difference arises by considering the portfolios P3 and P4. The former can be obtained by adding a loan with a large amount to the latter. The indices with α greater or equal to 2 show an increase of the single-name concentration, the others a decrease. This result can be explained since high values of α give greater weight to large exposures, while low values of α give more importance to the small ones.

As final remark, the numerical application shows that a suitable elasticity parameter α for RHK index is a value slightly higher than 2, in order to stress the impact of large loans.

References

1. Banca d'Italia: New regulations for the prudential supervision of banks. Circular 263 (27 December 2006)
2. Basel Committee on Banking Supervision: The New Basel Capital Accord, Consultative document. Basel (January 2001)
3. Basel Committee on Banking Supervision: International Convergence of Capital Measurement and Capital Standards: A Revised Framework. BIS, Basel (June 2004)
4. Becker, S., Dullmann, K., Pisarek, V.: Measurement of concentration risk – A theoretical comparison of selected concentration indices. unpublished Working Paper, Deutsche Bundesbank. Social Sciences. John Wiley, New York (2004)
5. Dalton, H.: The measurement of the inequality of incomes. Economics Journal **30**, 348–361 (1920)
6. Gordy, M.B.: A risk-factor model foundation for ratings-based bank capital rules. Journal of Financial Intermediation **12**(3), 199–232 (2003)
7. Gordy, M.B., Lütkebohmert, E.: Granularity adjustment for Basel II. Deutsche Bundesbank Discussion Paper (series 2), n. 1. (2007)
8. Encaoua, D., Jaquemin, A.: Degree of monopoly, indices of concentration and threat of entry. International economic review **21** (1980)
9. Gini, C.: Measurement of Inequality of Incomes. Economic journal **31**, 124–126 (1921)
10. Hannah L., Kay, J.A.: Concentration in modern industry. MacMillan Press, London (1977)
11. Herfindahl, O.: Concentration in the U.S. Steel Industry. Dissertion, Columbia University (1950)
12. Hirschmann, A.: National power and the structure of foreign trade. University of California Press, Berkeley (1945)
13. Hirschmann, A.: The paternity of an index. American Economic Review **54**(5), 761 (1964)
14. Lorenz, M.O.: Methods of measuring the concentration of wealth. Publications of the American Statistical Association **9**(70), 209–219 (1905)
15. Lütkebohmert, E.: Concentration risk in credit portfolios. Springer-Verlag (2009)
16. Pigou, A.C.: Wealth and Welfare. Macmillan Co., London (1912)

Bifactorial Pricing Models: Light and Shadows in Correlation Role

Rosa Cocozza and Antonio De Simone[*]

Abstract In modern option pricing theory many attempts have been accomplished in order to release some of the traditional assumptions of the Black and Scholes [5] model. Distinguished in this field are models allowing for stochastic interest rates, as suggested for the first time by Merton [20]. Afterwards, many stochastic interest rate models to evaluate the price of hybrid securities have been proposed in literature. Most of these are equilibrium pricing models whose parameters are estimated by means of statistical procedure, requiring a considerable computational burden. The recent financial crisis and the resulting instability of relevant time series may sensibly reduce the reliability of estimated parameters necessary to such models and, consequently, the calibration of the models. In this paper we discuss an original numerical procedure that can efficiently be adopted to the aim of pricing and the question of the correlation contribution in pricing framework. The procedure accounts for two sources of risk (the stock price and the spot interest rate) and, by means of an empirical evaluation tries to asses the relative contribution of the correlation component. The final target is to evaluate the "optimal" computation burden in pricing framework, given scarce dataset We show that the procedure proposed is a valuable compromise between computational burden and calibration efficiency, mainly because it overcomes difficulties and arbitrary choices in the estimation of the parameters.

R. Cocozza (✉)
Department of Business Administration, University of Napoli Federico II, Via Cinthia Monte S. Angelo, 80126 Naples, Italy
e-mail: rosa.cocozza@unina.it

A. De Simone
Department of Mathematics and Statistics, University of Napoli Federico II, Via Cinthia Monte S. Angelo, 80126 Naples, Italy
e-mail: a.desimone@unina.it

[*] Although the research is the result of a joint effort, Section 1 is due to Rosa Cocozza while Sections 2, 3, 4, 5 and 6 are due to Antonio De Simone.

M. Corazza, C. Pizzi (eds.), *Mathematical and Statistical Methods for Actuarial Sciences and Finance*, DOI 10.1007/978-3-319-02499-8_9, © Springer International Publishing Switzerland 2014

1 Introduction

In recent option pricing theory, many attempts have been carried out in order to release the assumption, typical of the Black and Scholes [5] model, of constant interest rate and, more specifically, to develop a pricing framework allowing for stochastic interest rates dynamics. Starting from the seminal paper by Merton [20], where the Gaussian process was adopted to describe the continuous-time short rate dynamics, several models for stock option pricing under stochastic interest rates have been proposed. Within this field, distinguished contributions are due to Ho and Lee [16], Rabinovitch [23], Miltersen and Schwartz [21], where the Gaussian process is adopted to describe the continuous-time short rate dynamics. Moreover, Amin and Jarrow [1], generalize the pricing framework proposed by Heat, Jarrow and Morton [15], by adding an arbitrary number of additional risky asset to their stochastic interest rate economy. Finally, Scott [24] and Kim and Kunimoto [18] proposed option pricing models where the short interest rate dynamics is described according to the Cox, Ingersoll and Ross (CIR) [10] model.

The success of the Gaussian specification of the term structure dynamics relies on the mathematical tractability and thus on the possibility of obtaining closed formulas and solutions for the price of stock and bond options, while the main drawbacks are related to the possibility for the interest rate trajectories to assume negative values. Moreover, if on the one hand the square root mean reverting process of the CIR model allows contemporaneously to prevent from negative interest rates and to preserve the mathematical tractability, on the other it is in general not able to ensure an acceptable fitting of the model prices to the observed market prices.

In an attempt to mitigate the above mentioned calibration drawbacks, an innovative two-factors numerical procedure is proposed to the aim of pricing different kind of financial contracts, such as stock option [7], participating policies [9] and convertible bonds [8]. In particular, this pricing framework assumes that the stock price dynamics is described according to the Cox, Ross and Rubinstein [11] binomial model (CRR) under a stochastic risk free rate, whose dynamics evolves over time accordingly to the Black, Derman and Toy [4] one-factor model (BDT). The BDT model avoids some drawbacks that typically affect equilibrium models of the term structure and offers, at the same time, an efficient calibration of the risk factor trajectories to the observed market prices.

In this article we discuss some issues related to the implementation and calibration of such a two factors numerical procedure. In particular, we discuss how it can be possible to calibrate the dynamics of each risk factor to the observed market prices. We also offer different possible ways for calibrating the correlation parameters. The main idea is that, as for the volatility of each risk factor, also the correlation can be such that the model can efficiently reproduce the observed market prices of the securities adopted as benchmark for the calibration. We then discuss how it is possible to determine an implied correlation coefficient from the observed market prices of stock options, showing whether and to what extent, such correlation measure can predict the future realized correlation.

The paper is organized as follows. Section 2 sketches the mechanics two-factor numerical procedure and how to apply it for pricing a stock option. Section 3 proposes some ways to calibrate the model, discussing in particular the opportunity of choosing a value for the correlation such that the observed option prices can efficiently be reproduced. Section 4 reports data and methodology adopted to test whether and to what extent the implied correlation can help to predict the future values of the realized correlation. Section 5 summarizes main results and Sect. 6 concludes the paper with final remarks.

2 The Two-Factors Numerical Procedure

In this section, we explain how to determine the arbitrage free price of an American stock option by means of a numerical procedure that accounts for two sources of risk: the stock price and the interest rate. In doing this, an important aspect to take into account is that the risk free rate dynamics influences the price of an equity derivative in two ways. First, the path followed by the interest rate can influence the discount factor adopted to determine the present value of the payoff of the derivative. Secondly, under the risk neutral probability measure, the expected stock (and derivative) price depends on the risk free rate level. Therefore, as the interest rate changes, both the final payoff and the discount factor change too while, on the contrary, we assume that the stock price dynamics do not influence the interest rate values. For this reason, the stock price dynamics cannot be specified until the interest rate is known. We assume that the interest rate considered as risk factor is the spot interbank offered rate, such as the m-month Libor or Euribor rate, and that its dynamics is described according to the BDT binomial model. More specifically, we assume that the m-month Libor rate at time t_j in the state of the world $j = u, d, L(t_i, t_{i+1})^j$ - with $\delta = t_{i+1} - t_i$ representing the tenor of the Libor rate expressed as fraction of year according to the market conventions - the successive time step can go up to $L(t_{i+1}, t_{i+2})^{ju} = L(t_i, t_{i+1})^j u_L$ or down $L(t_{i+1}, t_{i+2})^{jd} = L(t_i, t_{i+1})^j d_L$ with equal risk neutral marginal probability, where $u_L = \exp(\sigma_L, \delta)$, $d_L = \exp(-\sigma_L, \delta)$ and σ_L is the yearly volatility of the Libor rate that, for simplicity, is set constant over time. This means that the interest rate is *piecewise* constant over time so that, once the Libor rate is known, it can be adopted as risk free rate to determine the probability of the up movement of the stock price at the successive time step and the corresponding discount factor. Notice also that, at time t_0 the Libor rate $L(t_0, t_1)$ is not a random variable and is equal to the observed fixing. The second risk factor is the stock price and we assume that it evolves over time according to the CRR model. Assuming for simplicity that the time interval between two observations of the stock price (and of the interest rate) is equal to the tenor of the Libor rate δ, the risk neutral marginal probability of an up movement of the stock price between time t_i and time t_{i+1}, given the Libor rate at time t_i, p_t^j is equal to

$$p_t^j = \frac{B_t - d_s}{u_s - d_s} \tag{1}$$

where $B_t = 1 + L(t_i, t_{i+1})^j \delta$, $u_S = \exp(\sigma_S, \delta)$, $d_S = \exp(-\sigma_S, \delta)$ and σ_S is the yearly stock price volatility and, for simplicity, it is assumed to be constant over time. This implies that, at each time step, the up probability changes according to the changes of the Libor rate. However, to compute the probability associated to each couple of possible levels of stock price and interest rate, we need to specify the joint probability distribution function of the two risk factors. A naïve way to do this, very common especially in the practice of pricing convertible bonds (see for instance [17]), is by assuming that stock return and interest rate are independent, so that the joint probability is equal to the product of the marginal probabilities. However, considering the significant evidence of a negative correlation between the return on stocks and bonds (see among others [14]), the assumption of zero correlation cannot in general be satisfactory. Such a restrictive assumption can be released by redistributing in a different way the joint probabilities calculated in the case of independence among the possible states of the world. For instance, if we want to set a perfectly negative correlation, it is sufficient to equally distribute the probabilities, calculated in the case of independence, of contemporaneous up and down movements of both stock price and interest rate to the other two states of the world, according to Fig. 1.

In general, at each time t, to set a correlation equal to ρ_t, with $\rho_t < 0$, it will be sufficient to equally redistribute to the other two states of the world a percentage ρ_t of the probabilities of contemporaneous up and down movements of interest rate and stock price. On the contrary, to set a correlation equal to ρ_t, with $\rho_t > 0$, it will be necessary to equally redistribute to the other two states of the world a percentage ρ_t of the probability of opposite up and down movements (i.e. rate goes up/stock goes down and vice versa), of the two risk factors. Figure 2 shows an example of the tree representing the joint dynamics of the Libor rate and of the stock price.

To the aim of pricing of a stock option, it is necessary to calculate the terminal payoff of the option and then roll back the tree till time t_0. The terminal nodes are intuitive. Take node E, I and J as an example. The value of the stock option at node E,

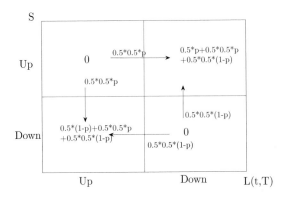

Fig. 1 Perfectly negative correlation between stock price S and Libor rate $L(t, T)$

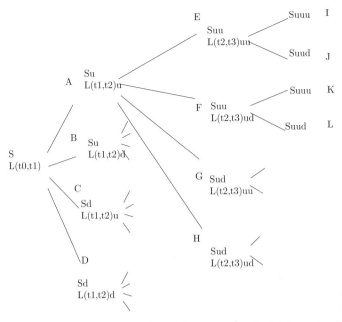

Fig. 2 Example of the joint dynamics of the stock price S and Libor rate $L(t,T)$

ξ_E, is therefore the present value[1], calculated at the rate $L(t_2,t_3)^u u$, of the weighted average of the payoff of the call at nodes I and J using as weights the probabilities p_{t+2}^{uu} and $1 - p_{t+2}^{uu}$ respectively. It is worth noting that at nodes I and K the stock price is the same, being equal to Su_S^3, and thus, the payoff of the call option is $(Su_S^3 - X)^+$, where X is the strike price. The difference between the two cases is in the discount rate that, at node I is equal to $L(t_2,t_3)^{uuu}$ while at node K is equal to $L(t_2,t_3)^{ud}$. Once the payoffs of the call at node E, F, G and H are calculated, we can use the joint probability function to determine the price of the call at node A, ξ_A. Such value is therefore equal to

$$\xi_A = \frac{\xi_E 0.5(p_t^* + 0.5\rho_t)}{1 + L(t_1,t_2)^u \delta} + \frac{\xi_F 0.5 p_t^* (1 - \rho_t)}{1 + L(t_1,t_2)^d \delta} + \frac{\xi_G (1 - p_t^*) 0.5(1 - \rho_t)}{1 + L(t_1,t_2)^u \delta}$$

$$+ \frac{\xi_H 0.5(1 - p_t^* + 0.5\rho_t)}{1 + L(t_1,t_2)^d \delta}.$$

Similarly, it is possible to determine the time t_0 value of the stock option by rolling back the tree.

---

[1] Notice that computing the present value of the average is possible only at the terminal nodes. In all the other cases, it is necessary to compute the average only after the present value is calculated.

3 Calibrating the Model

The adoption of the BDT and of the CRR models respectively for interest rate and stock price modeling facilitates the calibration of the tree, being most of the required input data directly observable on the financial markets. The only parameter that needs to be estimated is the correlation between Libor rate and stock price. To this aim, four different measures of correlation are concerned: i) historical correlation; ii) exponentially-weighted moving average (EWMA) correlation; iii) correlation forecasted by means of a bivariate GARCH(1,1) and iv) implied correlation. The first measure of correlation is the most simple and intuitive. The n-days correlation between Libor rate and stock return at time t, $\rho_{t,n}^{hist}$, can be easily calculated, using the past observations of the interbank offered rate and of the stock price. Notice that this measure of correlation attributes an equal weight to each observation from $t - n$ to t. On the contrary the EWMA correlation, whose diffusion amongst practitioners is mainly due to its application by RiskMetricsZ, gives more weight to recent data with respect to older data and, for this reason, it reacts faster to sudden changes of the risk factors. The n-days EWMA correlation between Libor rate and stock return at time t id denoted as $\rho_{t,n}^{EMWA}$ with a decay factor set equal to .94. Following Lopez and Walter [19], the third measure of correlation considered here is the forecast obtained by means of a bivariate GARCH(1,1), in the diagonal VECH specification proposed by Bollerslev et al. [6]. This means that the forecasts of the daily variance/covariance are obtained by means of the GARCH parameters estimated using 256 observations prior of the day t. Finally, the n-days correlation forecast at time t is denoted as $\rho_{t,n}^{GARCH}$. The major drawbacks related to the measures of correlation proposed above are that: i) they are "exogenous" to the model, meaning that the correlation is estimated regardless the mechanics of the model itself and thus, their use do not assure that the model can reproduce the observed prices of financial securities (e.g. of the options); ii) the correlation is always estimated using past data that, if the markets are efficient enough (see [13]), cannot significantly improve the information set contained in the current prices. For this reason, we propose a different way to calibrate the correlation, based only on the current market prices and volatilities of the risk factors. As a matter of fact, the BDT model can be efficiently calibrated by using the implied volatilities from caps, floors and swap options market obtained by means of the Black [3] formula (see [12]). At the same time, the CRR model can efficiently be calibrated by means of the implied volatility from stock option markets, calculated by mean of the Black and Sholes formula. Therefore, the price at time t of a stock option (e.g. an American call), calculated using the numerical procedure explained in Sect. 2, can be represented as a function of the current values of the stock price, of the Libor rate, of the corresponding volatilities and of the correlation parameter. Assuming for simplicity that the correlation and the volatility of both the interest rate and the stock price are constant over time, we can set the market price of a call option at time t, ξ_t^M, equal to the price ξ_t^{NP} of the same call calculated by means of the numerical procedure depicted in Sect. 2:

$$\xi_t^M = \xi_t^{NP}(S_t, L(t, T_k), \sigma_{S,t,n}^{BLS}, \sigma_{\delta_L,t,n}^{BLK}, \rho_{t,n}^I | X, n, y_t) \tag{2}$$

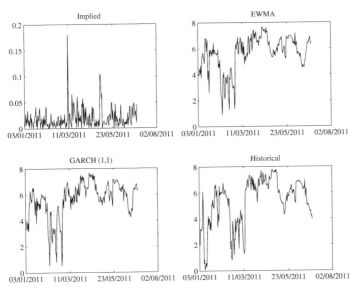

Fig. 3 Differences between NP and BAW price in percentage of the BAW price

where $n = T_k - t$ is the time to maturity of the option; $\sigma^{BLS}_{S,t,n}$ and $\sigma^{BLK}_{\delta_L,t,n}$ are respectively the yearly volatilities of the stock price and Libor rate at time t, calculated by means of the Black and Scholes and of the Black formulas with reference to options having a time to maturity equal to n; y_t is the dividend yield of the stock at time t and $rho^I_{t,n}$ is the n-days implied correlation measure that can be computed by using an appropriate algorithm of calculus. We therefore define as implied correlation that value of $rho^I_{t,n}$ that equals (at least approximately) the observed market price of a stock option (i.e. an American call) to its theoretical price as represented by the right-hand side of Eq. 2. Figure 3 reports the absolute value of differences, over the period 03/01/2011 - 30/12/2011, between the daily price of a 1 year constant maturity ATM call option calculated by means of the numerical procedure (NP price) using each of the above mentioned correlation measures, and the price of the same option calculated by using the the Barone Adesi and Whaley (1987) formula (BAW price). The difference is expressed as percentage of the BAW price that is one of the most used formulas for pricing American Options. As underlying, we select a security listed on the Italian stock market (Intesa San Paolo) while as risk free rate we use the 12 month Euribor rate. The time series of prices, rates and their volatilities are provided by Bloomberg^{TM}.

We notice that the best performance is due to the implied correlation, since only occasionally the price difference is higher than 0.05%. Moreover, we notice that also the other correlation measures allow for an appreciable fitting of the NP price to the BAW price and thus to the observed market prices, being the difference between them never higher than 7.7%. If we consider that the price of the option ranges from 0.2 to 0.3 Euros, it means that the maximum absolute difference between NP price and BAW price ranges from 1.54 cents to 2.31 cents. There are two main drawbacks related to the estimation of the proposed measure of implied correlation: i) the absence of a closed formula imply the necessity to state an algorithm of calculus to find the value of $rho^I_{t,n}$ allowing equation 2 to hold. The higher the precision of the calculus algorithm the higher the computational burden of computing the correlation parameter; ii) as for the volatility of the interest rate, it should be necessary to specify a term structure also for the implied correlation. To simplify the calibration procedure, we set the hypothesis that the correlation is constant over time even if it can be remarked that this is not exactly the case. As a matter of fact, the longer the observation period, the less variable is the correlation and it cannot be excluded that the sign of the correlation may differ as the observation period changes. Moreover, the issue that we are interesting to address is whether and to what extent, the implied correlation estimated by using equation 2 can help to predict the future values of the correlation between interest rate and risk.

4 Testing the Predictive Accuracy

We compute the 60-days correlation between the stock price of an Italian listed bank and the 3-month Euribor rate using the four measures of correlation depicted in the previous section (historical, EWMA with a decay factor set equal to 0.94, GARCH based correlation and implied correlation) from 01/01/2011 to 31/12/2011, for a total of 257 observations. All data are provided by Bloomberg TM database. The implied correlation is estimated with reference to a hypothetical constant maturity ATM 3 month call option, written on a security listed on the Italian stock market, calculated by means of the Black and Scholes (1973) formula. To this aim, we use the 3 month implied volatility provided by BloombergTM and, as risk free rate, the 3 month Euribor rate.

Figure 4 shows the comparison of each correlation forecast with the realized correlation. The evidence of positive correlation, for most of the observation period, between the 3 month Euribor and the stock price is consistent with the findings of Flannery and James (1984), given the inverse relation between bond prices and interest rates. Moreover, we notice that, the implied correlation is much more volatile with respect to the other measures, meaning that such correlation reacts faster to sudden changes of the quotations. However, if we look at the predictive accuracy, the most affordable measure seems to be the one estimated via multivariate GARCH(1,1). To test the predictive accuracy of the four types of correlation, we follow [19] and use three different methods: i) analysis of the forecast error; ii) partial optimality regression; iii) encompassing regression. The first method consists in analysing the correlation error ($\eta^j_{t,n}$), defined as the difference between the particular measure of n-days correlation forecast $\rho^j_{t,n}$, and the realized n-days correlation $rho^{REAL}_{t,n}$ that is $\eta^j_{t,n} = rho^j_{t,n} - rho^{REAL}_{t,n}$. After $\eta^j_{t,n}$ is calculated, we regress it on a constant. If the estimated constant is significantly different from zero, the correlation j is said to be a biased forecast of the realized correlation. The second method, partial optimal-

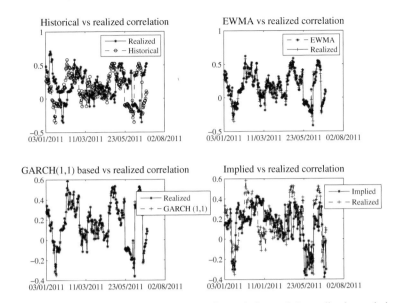

Fig. 4 Comparison between four measures of correlation and the realized correlation

ity regression consists in estimating, for each measure of correlation, the regression $rho_{t,n}^{REAL} = a_0 + a_1 \rho_{t,n}^j + e_t$. If the coefficients a_0 and a_1 are not significantly different from 0 and 1 respectively, the correlation measure j is said to be a partial optimal forecast of the realized correlation. Finally, the third method, encompassing regression, consists in estimating, for each measure of correlation, the equation $rho_{t,n}^{REAL} = b_0 + \sum_{j=1}^{m} b_j \rho_{t,n}^j + e_t$. Take $m = 2$ as an example. If b_0, b_1 and b_2 are not significantly different from 0, 1 and 0 respectively, than the correlation measure $j = 2$ encompasses the correlation measure $j = 1$, meaning that the information set included in the estimation of the former encompasses that included in the estimation of the latter. The table shows the regression coefficients, estimated by using the Newey and West (1987) standard errors, from the analysis of the forecast error (Method 1), the partial optimality regression (Method 2) and the encompass regression (Method 3). Standard deviation are in parentheses.

5 Results

The following able reports the regression coefficients for the tree type of tests. In performing the regressions, we use the Newey and West [22] standard errors to account for potential heteroskedasticity and autocorrelation.

The results seem to confirm the intuition behind the analysis of Fig. 4. If we look at the first column of table (Method 1), we notice that none of the correlation measures is said to be a biased forecast, even if only the EWMA and the GARCH correlation

Table 1 Regression Results

	Method1		Method2				Method3
Intercept	—	0.0087 (0.0126)	0.0041 (0.0048)	−0.0012 (0.0008)	−0.2163* (0.549)		0.0001 (0.0001)
IMPLIED	−0.0022 (0.0227)	0.5336** (0.0519)					−0.0006 (0.0006)
EWMA	0.0043 (0.0040)		0.9490*** (0.0162)				−0.1910 (0.0022)
GARCH	0.0007 (0.0006)			1.0033*** (0.0030)			1.1903** (0.0023)
HIST	0.0263 (0.0209)				0.2016** (0.0175)		0.0015 (0.0004)

*   the intercept is significantly different from zero at the 1% level.

** a_0 is significantly different from 1 at the 1% level.

*** the Wald test for the joint hypotheses that $b_0 = 0$ and $b_1 = 1$ cannot be rejected at the 1%.

are partially optimal. In fact, the coefficients a_0 and a_1 associated to both these measures of correlation are not significantly different from 0 and 1 respectively, while the coefficient a_1 associated to the implied and to the historical volatility measures is significantly different from 1 at 1% level. Moreover, the Wald test for the joint hypotheses that the coefficients a_0 and a_1 are equal to 0 and 1 respectively, can be rejected at the 1% level only for the GARCH and for the EWMA correlations. Finally, from the last column of table we notice that the coefficients associated to the correlation measures are not significantly different from 0, except for that associated to the GARCH correlation that is greater than 1. This evidence suggests that the GARCH measure of correlation encompasses all the other measures.

6 Conclusions

In this article we discuss an original numerical procedure that can efficiently be adopted to the aim of pricing hybrid securities. The procedure accounts for two sources of risk (the stock price and the spot interest rate) and, by means of an empirical evaluation we try to assess the relative contribution of the correlation component. To this aim, four correlation measures are concerned, namely, historical, exponentially weighted moving average correlation, a bivariate GARCH forecasted correlation and implied correlation obtained from the stock option prices. We show that the best pricing performances are associated to the implied correlation measure, even if also the adoption of the other correlation measures allows to obtain prices that are reasonably close to the observed market prices. However, the main drawback in computing the implied correlation from the bi-factorial procedure is related to its low predictive accuracy. Compared to other correlation measures (especially GARCH correlation), implied correlation cannot be considered an affordable forecast of the future realized correlation. However, the appreciable pricing performance and the lack evidence supporting the idea that implied correlation is a biased forecast of the realized correlation suggest that the numerical procedure we propose offers a reasonable compromise between computational burden and calibration efficiency. This is the case also because the necessary parameters are directly observable from the financial markets.

References

1. Amin, K.I., Jarrow, R.A.: Pricing options on risky assets in a stochastic interest rate economy. Mathematical Finance **2**(4), 217–237 (1992)
2. Barone-Adesi, G., Whaley, R.E.: Efficient Analytic Approximation of American Option Values. The Journal of Finance **42**(2), 301–320 (1987)
3. Black, F.: The pricing of commodity contracts. The Journal of Political Economy **3**(1–2), 167–179 (1976)
4. Black, F., Derman, E., Toy, W.: . A one-factor model of interest rates and its application to treasury bond options. Financial Analysts Journal **46**(1), 33–39 (1990)
5. Black, F., Scholes, M.: The pricing of options and corporate liabilities. The Journal of Political Economy **81**(3), 637–654 (1973)
6. Bollerslev, T., Engle, R.F., Wooldridge, J.M.: A Capital Asset Pricing Model with Time-Varying Covariances. Journal of Political Economy **96**(1), 116–131 (1988)
7. Cocozza, R., De Simone, A.: One numerical procedure for two risk factors modeling. Available at Munich Personal RePEc Archive: http://mpra.ub.uni-muenchen.de/30859/ (2011)
8. Cocozza, R., De Simone, A.: An Economical Pricing Model for Hybrid Products. Forthcoming in: Wehn, C., Hoppe, C., and Gregoriou, G.: Rethinking valuation and pricing models: Lessons learned from the crisis and future challenges. Elsevier Press (2012)
9. Cocozza, R., De Simone, A., Di Lorenzo, E., Sibillo, M. (2011). Participating policies: Risk and value drivers in a financial management perspectives.Proceedings of the 14th Conference of the ASMDA International Society, 7–10 June 2011
10. Cox, J.C., Ingersoll, J.E., Ross, S.A.: A theory of the term structure of interest rates. Econometrica **53**(2), 385–407 (1985)
11. Cox, J.C., Ross, S.A., Rubinstein, M.: Option pricing: A simplified approach. Journal of Financial Economics **7**(3), 229–263 (1979)
12. De Simone, A.: Pricing interest rate derivatives under different interest rate modeling: A critical and empirical analysis. Investment Management and Financial Innovations **7**(2), 40–49 (2010)
13. Fama, E.F.: Efficient Capital Markets: A Review of Theory and Empirical Work. The Journal of Finance **25**(2), 383–417 (1970). Papers and Proceedings of the Twenty-Eighth Annual Meeting of the American Finance Association New York, N.Y. December, 28–30, 1969 (May, 1970)
14. Flannery, M.J., James, C.M.: The Effect of Interest Rate Changes on the Common Stock Returns of Financial Institutions. The Journal of Finance **XXXIX**(4), 1141–1153 (1984)
15. Heat, D., Jarrow, R.A., Morton, A.: Bond Pricing and the Term Structure of Interest Rates: A New Methodology. Econometrica **60**(1), 77–105 (1992)
16. Ho, T.S.Y., Lee, S.-B.: Term structure movements and pricing interest rate contingent claims. The Journal of Finance **XLI**(5), 1011–1029 (1986)

17. Hung, M.-W., Wang, J.-Y.: Pricing convertible bonds subject to default risk. The journal of Derivatives **10**(2), 75–87 (2002)
18. Kim, Y.-J., Kunitomo, N.: Pricing options under stochastic interest rates: A new approach. Asia-Pacific Financial Markets **6**(1), 49–70 (1999)
19. Lopez, J.A., Walter, C.: Is implied correlation worth calculating? Evidence from foreign exchange options and historical data, Journal of Derivatives 7**a**, 65–82 (2000)
20. Merton, R.C.: Theory of rational option pricing. The Bell Journal of Economics and Management Science **4**(1), 141–183 (1973)
21. Miltersen, K.R., Schwartz, E.S.: Pricing of Options on Commodity Futures with Stochastic Term Structures of Convenience Yields and Interest Rates. Journal of Financial and Quantitative Analysis **33**(1), 33–59 (1998)
22. Newey, W.K., West K.D.: A Simple Positive Semidefinite, Heteroskedasticity and Autocorrelation Consistent Covariance Matrix. Econometrica **55**(3), 703-708 (1987)
23. Rabinovitch, R.: Pricing stock and bond options when the default-free rate is stochastic. Journal of Financial and Quantitative Analysis **24**(4), 447–457 (1989)
24. Scott, L.O.: Pricing stock options in a jump-diffusion model with stochastic volatility and interest rates: Applications of fourier inversion methods. Mathematical Finance **7**(4), 413–424 (1997)

Dynamic Strategies for Defined Benefit Pension Plans Risk Management

Ilaria Colivicchi, Gabriella Piscopo and Emanuele Vannucci

Abstract In the context of the decumulation phase of a defined benefit pension scheme, the aim of this paper is to describe the management of a pension provider which has to minimize a default probability and to maximize the expected surplus. Its management strategy is based on the possibility of change the risk level (i.e. the volatility of random returns) of the investment at an optimum time.

1 Introduction

This paper focuses on the risk assessment concerning the payment of a fixed rate life retirement annuity, coming from a pension scheme decumulation phase, in which the pension provider invests the residual amount after the payments of each annuity rate. The risk suffered by pension provider comes out from the uncertainty of both stochastic financial returns of the investment of the residual amount (financial risk) and of the pensioners' future lifetimes (longevity risk). There are many papers which have dealt with the financial and longevity risks in the decumulation phase of defined contribution pension schemes (see Albrecht and Mauer [1], Blake et al. [4], Gerrard et al. [8], Gerrard et al. [9], Milevsky and Young [12]), while with this paper we would underline the kind of risk suffered by those who have an obligation to pay

I. Colivicchi
Department of Mathematics for Decisions, University of Florence, Via delle Pandette 9, Florence, Italy
e-mail: ilaria.colivicchi@unifi.it

G. Piscopo (✉)
Department of Economics, University of Genoa, Via Vivaldi 5, 16126 Genoa, Italy
e-mail: piscopo@economia.unige.it

E. Vannucci
Department of Statistics and Applied Mathematics, University of Pisa, Via Ridolfi 10, Pisa, Italy
e-mail: e.vannucci@ec.unipi.it

M. Corazza, C. Pizzi (eds.), *Mathematical and Statistical Methods for Actuarial Sciences and Finance*, DOI 10.1007/978-3-319-02499-8_10, © Springer International Publishing Switzerland 2014

a fixed rate annuity. There are two opposing requirements in setting the constant rate of the life retirement annuity: on one hand the interest of the pensioner to have the highest possible rate given the initial invested amount; on the other hand the interest of the pension provider, which has to manage the risk of having to use its own capital in order to ensure the payment of the annuity. If the death of the pensioner occurs when the residual amount is still positive, there is a "gain" for the pension provider; conversely, if the residual amount hits zero-level while the pensioner is still alive, the pension provider has to release other own reserves to the payment of the pension benefits. The pension provider has to manage the investment of the amount paid by the pensioners to buy their pension plans, in order to minimize a "default probability" and to maximize a "discounted expected surplus". Its management strategy is based on the possibility of change the risk level (i.e. the volatility of random returns) of the investment at a certain time, that should be the optimum switching time. A previous work of the same authors (see for instance [6]) has already matched the problem of calculating the maximum constant rate of a retirement annuity, which allows the pension provider to achieve a given "default probability". This problem has been studied both for a single pensioner and for groups of same aged and different aged pensioners. From a practical point of view, we face this problem drawing inspiration from the actuarial literature on Variable Annuities (VA) (see Bauer at al. [3], Milevsky and Salisbury [11]). In particular, our original contribution lies in exploit valuation models developed for Guaranteed Lifelong Withdrawal Benefit Option embedded in VA's (Bacinello et al. [2], Piscopo [13], Haberman and Piscopo [10]) and to adapt them to pension schemes. We develop the model in a stochastic demographic framework and we refer to Black-Scholes financial scenario. We will present results inherent to the case of a single pensioner and for groups of same-aged pensioners. Numerical results will be proposed via Monte Carlo simulations. The paper is structured as follows. In Sect. 2 the demographic and financial scenario in which the model operates is described and the valuation problem is defined. In Sect. 3 we define the management strategy and an efficient way to compare different strategy at a certain optimum time. In Sect. 4, we propose some numerical examples to give a description of some sensitive analysis. The last section is devoted to some concluding remarks and to suggest possible future developments of the issue addressed.

2 The Model

Let $N_0 \equiv n_0$ be the deterministic number of x-aged pensioners at time 0 and N_i, for $i = 1, 2, \ldots$, the random number of pensioners living at time i. Assuming $H_0 := h_0$ be the sum that each pensioner pays to buy the life retirement annuity which provides a guaranteed rate g and an annual defined benefit equal to $G = gh_0$, withdrawing from the fund. Let $S_0 := s_0 = n_0 h_0$ the total amount initially invested by pension provider and S_i, for $i = 1, 2, \ldots$, be the random residual invested amount at time i. Let $R_i \equiv R$, for $i = 1, 2, \ldots$, the random return in the interval $(i-1, i]$. So, considering $G(i) = gN(i)$ the amount withdrawed by the fund at date i, the evolution of the residual invested amount is defined by the following iterative equation, for $i = 1, 2, \ldots$

$$S_i = S_{i-1}(1 + R) - G_i. \tag{1}$$

The mechanism is like that of Guaranteed Lifelong Withdrawal Benefit Option (GLWB) offered in Variable Annuity (VA) product. Our original contribution lies in exploit valuation models developed for GLWB's (Bacinello et al. [2], Piscopo [13], Haberman and Piscopo [10]) and adapt them to pension schemes.

Following the recent contributions on VA valuation, the random return R is described by a standard Brownian motion.

2.1 Default Probability and Surplus

Let observe that if the fund value resets to zero, the we have a sort of "default" and the pension provider has to intervene with own reserves. Formally the event of "default" occurs at time i if it holds

$$\min i : S_i \leq 0 \, i.e. \min i : S_{i-1}(1 + R) < gN_i. \tag{2}$$

This default event implies that the residual invested amount is not sufficient to pay the fixed life annuities rates to the living pensioners. We remark that in our previous work (2011), the problem matched is the maximization of the guaranteed rate g, given S_0 and a fixed maximum level α for the probabylity of default $P(d)$.

If the "default event" does not occur, then we have a positive discounted surplus, $U > 0$, for the pension provider (which could consider to share it with the others). If the "default event" occurs, then we have a negative discounted surplus, $U < 0$, by the pension provider (which is to be covered with some kind of own reserves).

3 Management Strategy

The management strategy of the pension provider is based on the opportunity to change the risk level of the investment, i.e. the volatility of random returns, at a certain time.

Let define R the random return (characterized by the parameters μ and σ) before the potential switching time, and with R_m (μ_m and σ_m) and R_l (μ_l and σ_l) respectively, the more and the less risky returns.

3.1 Management Strategy: Definition

We consider a switching strategy based on the theoretical values which may be obtained both by random returns and by the demographic aspect from the starting point to check date. In particular we consider to divide the results in 3 levels both for the random returns and for the demographic aspect. For random return we consider 2 levels for a generic date $i = 0, 1, 2, \ldots, R_{1,i}$ and $R_{2,i}$, such that for the random return from date 0 to date i, R_i, we have

$$1/3 = Prob.(R_i <= R_{1,i})$$
$$1/3 = Prob.(R_{1,i} < R_i <= R_{2,i})$$
$$1/3 = Prob.(R_i > R_{2,i}).$$

Let define

$R_i <= R_{1,i}$ *event* F_3
$R_{1,i} < R_i <= R_{2,i}$ *event* F_2
$R_i >= R_{2,i}$ *event* F_1.

For the demographic aspect we consider 2 levels for the number of survivors at a generic date $i = 0, 1, 2, \ldots, N_{1,i}$ and $N_{2,i}$, such that we have

$$1/3 = Prob.(N_i <= N_{1,i})$$
$$1/3 = Prob.(N_{1,i} < N_i <= N_{2,i})$$
$$1/3 = Prob.(N_i > N_{2,i}).$$

Let define

$N_i <= N_{1,i}$ *event* D_3
$N_{1,i} < N_i <= N_{2,i}$ *event* D_2
$N_i >= N_{2,i}$ *event* D_1.

In other words, in a first step we have analysed separately the financial and demographic components, dividing both financial return distribution and the distribution of the number of survivors in three bands, where in each band the $1/3$ of the probability mass is collected.

Let consider that in order to obtain a positive surplus, it is better to have less survivors, event D_3, and higher return, event F_1, and in the opposite case the worst case is given by events D_1 and F_3. In the light of this consideration, we define a naive management strategy based on the couple of events D_i, F_j for $i, j = 0, 1, 2, 3$ in the following way:

- switch towards R_m if $i + j <= 3$;
- keep R if $i + j = 4$;
- switch towards R_l if $i + j > 4$.

The idea is to check at a given checking date the financial and demographic structure of the portfolio and switch towards a more risky strategy if the whole management has produced good results, switch towards a more prudent strategy in the opposite case, or finally do not switch in the middle case.

Let t be the checking time, the evolution of the fund will be:

$$S_i = S_{i-1}(1+R) - G_i \ for \ i < t$$

and for $i \geq t$

$S_i = S_{i-1}(1+R_m) - G_i$ if $i+j <= 3$;
$S_i = S_{i-1}(1+R) - G_i$ if $i+j = 4$;
$S_i = S_{i-1}(1+R_l) - G_i$ otherwise.

3.2 Management Strategy: Aim

We assume that the aim of the management strategy, i.e. the choice of optimum check date, should be efficient in terms of a combination of default probability and expected surplus.

We define that a potential switching time t_i dominates another potential switching time t_j if it holds:

$E[U_{t_1}] \geq E[U_{t_2}]$
$P(d)_{t_1} \leq P(d)_{t_2}$

and at least one inequality holds in a strict sense.

4 The Simulation Algorithm

We consider a simulation algorithm in which the further lifetime of each pensioner is random and it is the starting point of the process.

To simulate the further lifetime we start considering death probabilities of each age after the retirement age e_1: let assume that such probabilities are indicated with

$$q_1, q_2, \ldots, q_\omega$$

where the generic q_j is obtained from the data of the survival at each age x, l_x, of a generic demographic table with last age ω, in the following way

$$q_j = \frac{l_{e_1+j-1} - l_{e_1+j}}{l_{e_1}}.$$

For j that goes from 1 to $\omega - e_1$ we have $\sum_{j=1}^{\omega-e_1} q_j = 1$.

It is sufficient to simulate a random number in the interval $[0, 1]$ to construct a simulation of further lifetime coherent with death probabilities: the further lifetime expressed in years is k if the random number $r \in [0, 1]$ belong to the interval

$$\sum_{j=1}^{k} q_j \leq r < \sum_{j=1}^{k+1} q_j.$$

Afterwards, assuming the independence between the demographic and the financial factors we go on with the simulation of the ongoing amount of the fund S_i. For each run of the simulation algorithm we record if the event of default has occurred and we estimate the probability of default with the proportion of the number of default on the total runs. We analyze the case of the management of a portfolio composed by more pensioners with the same risk characteristics. In this scenario we deal with a global fund for the payments of all pensions and we have a default only in the case that the global fund is null and there is at least one pension annuity which has to be paid, without considering any reversibility.

5 Numerical Examples

We assume the independence between the demographic and the financial factors, hence the simulation algorithm is divided in 2 parts:

- the vector N_i, from 1 to $\omega - x$, where ω is the last age of the demographic table for italian males called IPS 55, and x is the retirement age of the pensioners;
- the random return of the investment according to a B.S. financial scenario, with the values of the parameter μ, σ depending on the "riskiness" we have to consider at each time.

We consider a given scenario characterized by

$N_0 = 100$, $x = 64$, $h_0 = 100$;
$\mu = 0.05$, $\sigma = 0.15$, $r = 0.03$ (risk free discount factor);
$g = 3.14$ that is the maximum g under $P(d) < 0.003$, without management strategy [6].

For each scenario we have made 100,000 runs.

According to a CAPM approach, fixing a different value for μ, we have the corresponding value of σ. So the risk levels available for the pension provider management strategy can be given in terms of Δ_σ and we have $\sigma_m = \sigma + \Delta_\sigma$ and $\sigma_l = \sigma - \Delta_\sigma$.

We consider 3 different value of Δ_σ (0.02, 0.035, 0.05) and hence we have 7 different risk levels (A, B, C, D, E, F, G). In order to underline the effects of the risk level of the random return on the combination of default probability and expected surplus, we propose the results obtained without applying any management strategy to a portfolio of just one pensioner for sake of example, using the couples μ, σ in Table 1.

Table 1 Default Probability and Surplus under different scenarios

	μ	σ	$P(d)$	U
A	0.0388	0.1	0.0108	70.62
B	0.0417	0.115	0.0159	75.59
C	0.0452	0.13	0.0215	81.71
D	0.05	0.15	0.0302	92.05
E	0.0556	0.17	0.0399	105.61
F	0.0604	0.185	0.0453	118.50
G	0.0655	0.2	0.0528	134.40

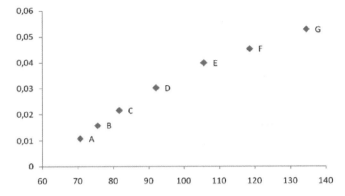

Fig. 1 Surplus and Probability of Default without management strategy

As it is clear from Fig. 1, without implementing a management strategy no one risk level dominates another.

At this time we start running the model with $N_0 = 100$ and $x = 64$ and we divide the results in 3 bands both for the random return and for the demographic risk.

Then we apply the management strategy described above and we obtain P(d) and U for each time t from $t = 0, \ldots, 10$. To synthetize the results, we propose just the case with $\Delta_\sigma = 0.035$ as follows.

Scenario with $\Delta_\sigma = 0.035$ and $N_0 = 100$.

Table 2 Probability of Default and time

t	$P(d)$	U
0	0.0101	14236.82
1	0.0072	16199.34
2	0.0060	16486.34
3	0.0061	16515.85
4	0.0060	16702.95
5	0.0064	16801.81
6	0.0065	16767.19
7	0.0071	16767.34
8	0.0074	16766.78

With this management strategy we find an efficient potential switching time for this scenario in $t = 5$. It also seems better for the pension provider to have the possibility of larger changes of the risk level: the results with $\Delta_\sigma = 0.035$ are generally better than the case with smaller Δ_σ.

6 Final Remarks

In this paper we have dealt with the problem of managing the investment strategy for a defined benefit pension plan. The choice of pensioner provider to switch towards more risky or prudent strategy has to take into account both the financial and the demographic factors and their interactive influence on the value of the accumulated surplus.

Further works are going to apper to improve the management strategy of the pension plan considering the introduction of the possibility of more than one switching time, the introduction of the life annuity reversibility aspect and the management of longevity risk.

References

1. Albrecht, P., Maurer, R.: Self-Annuitization, Consumption Shortfall in Retirement and Asset Allocation The Annuity Benchmark. Journal of Pension Economics and Finance **1**, 269–288 (2002)
2. Bacinello, A.R., Millossovich, P., Olivieri, A., Pitacco, E.: Variable Annuities Risk Identification and Risk Assessment. CAREFIN Research Paper **14** (2010)
3. Bauer, D., Kling, A.,Russ, J.:A Universal Pricing Framework for Guaranteed Minimum Benefits in Variable Annuities. ASTIN Bulletin **38**, 621–651 (2008)
4. Blake, D.,Timmermann, A.: International Asset Allocation with Time-Varying Investment Opportunities. The Journal of Business University of Chicago Press **78**, 71–98 (2005)
5. Colivicchi, I., Mulinacci, S., Vannucci, E.: A dynamic control strategy for pension plans in a stochastic framework. Giornale dell'Istituto Italiano degli Attuari **LXXII** (2009)
6. Colivicchi, I., Piscopo, G., Vannucci, E.: An equilibriumI model for defined benefit pension schemes in a stochastic scenario. 14th Applied Stochastic Models and Data Analysis Conference Edizioni ETS (2011)
7. Coppola, M., Di Lorenzo, E., Sibillo, M.: Risk sources in a life Annuities Portfolio Decomposition and Measurement Tools. Journal of Actuarial Practice **8**, 43–61 (2000)
8. Gerrard, R., Haberman, S., Vigna, E.: Optimal Investment Choices Post Retirement in a Defined Contribution Pension Scheme. Insurance Mathematics and Economics **35**, 321–342 (2004)
9. Gerrard, R., Haberman, S., Vigna, E.: The Management of De-cumulation Risks in a Defined Contribution Environment. North American Actuarial Journal **10**, 84–110 (2006)
10. Haberman, S., Piscopo, G.: The valuation of Guaranteed Lifelong Withdrawal Benefit Options in variable annuity contracts and the impact of mortality risk.*North American Actuarial Journal* **15**(1), 97–111 (2011)
11. Milevsky, M.A., Salisbury, T.S.: Financial valuation of guaranteed minimum withdrawal benefits. Insurance Mathematics and Economics **38**, 21–38 (2006)
12. Milevsky, M.A., Young, V.R.: Optimal Asset Allocation and The Real Option to Delay Annuitization It's Not Now-or-Never. The Schulich School of Business York University (2002)
13. Piscopo, G.: The fair price of Guaranteed Lifelong Withdrawal Benefit Option in Variable Annuity. Problems & Perspectives in Management **7**, 79–83 (2009)
14. Yang. S.S., Huang, H.: The Impact of Longevity Risk on the Optimal Contribution Rate and Asset Allocation for Defined Contribution Pension Plan. The Geneva Papers on Risk and Insurance Issues and Practice **34**, 660–681 (2009)

Particle Swarm Optimization for Preference Disaggregation in Multicriteria Credit Scoring Problems

Marco Corazza, Stefania Funari and Riccardo Gusso

Abstract In this paper we deal with the problem of preference disaggregation in credit scoring problems developed by using multicriteria analysis. In order to determine the values of the parameters that characterize the preference model of the decision maker, we adopt Particle Swarm Optimization, which is a biologically-inspired heuristics based on swarm intelligence. We test the ability of PSO to find the optimal values of the parameters on a real data set provided by an Italian bank.

1 Introduction

In this paper we deal with the problem of preference disaggregation in a credit scoring problem developed by using MURAME (MUlticriteria RAnking MEthod) [9], a particular multicriteria outranking method.

In the outranking methods, the preference model of the decision maker is usually characterized by several parameters (weights of criteria, preference, indifference and veto thresholds).

Such parameters could be obtained by using a direct procedure that requires that the decision maker explicitly determines the values of the parameters that express its preference structure. However, the explicit direct determination of these param-

M. Corazza (✉)
Department of Economics and Advanced School of Economics in Venice, Ca' Foscari University of Venice, Cannaregio 873, 30121 Venice, Italy
e-mail: corazza@unive.it

S. Funari
Department of Management, Ca' Foscari University of Venice, Cannaregio 873, 30121 Venice, Italy
e-mail: funari@unive.it

R. Gusso
Department of Economics, Ca' Foscari University of Venice, Cannaregio 873, 30121 Venice, Italy
e-mail: rgusso@unive.it

M. Corazza, C. Pizzi (eds.), *Mathematical and Statistical Methods for Actuarial Sciences and Finance*, DOI 10.1007/978-3-319-02499-8_11, © Springer International Publishing Switzerland 2014

eters cannot be considered realistic for several applications, so the use of preference disaggregation methods [10] is often more desirable.

In this latter case, the parameters of the model are determined from a given reference set of decisions. More precisely, the problem consists in finding the parameters that minimize the inconsistency between the model obtained with those parameters and the reference set of decisions revealed by the decision maker.

Because of the size and the complexity of the involved mathematical programming problem, some evolutionary algorithms (such as the variable neighborhood search or the differential evolution algorithm) have been used in the literature in order to handle the preference disaggregation problem in specific outranking methods (see for example [1, 7]).

The novelty of this contribution consists first in adopting an evolutionary algorithm based on swarm intelligence, that is the Particle Swarm Optimization (PSO) method [2], in order to deal with preference disaggregation problems in outranking models. Moreover, another distinguishing feature of this contribution is that it focuses on credit scoring problems developed by using MURAME, a particular outranking method [9]. We may note that the same approach could be applied also to credit scoring problems developed by using other multicriteria outranking techniques.

The paper is structured as follows. Section 2 presents the optimization problem involved in the preference disaggregation approach in a MURAME framework. Section 3 briefly summarizes the Particle Swarm Optimization (PSO) algorithm and illustrates its implementation in solving the preference disaggregation problem. Section 4 analyzes the ability of PSO to determine the optimal values of the parameters in a MURAME credit scoring problem. Section 5 presents some conclusions.

2 The Optimization Problem Involved in MURAME Preference Disaggregation

2.1 A Brief Description of MURAME

MURAME is a multicriteria method that allows to obtain a complete ranking of a set of alternatives $A = \{a_1, \ldots, a_i, \ldots, a_m\}$, on the basis of a set of criteria $\{crit_1, \ldots, crit_j, \ldots, crit_n\}$. In credit scoring problems, as the one considered in Sect. 4, the alternatives are the firms applicants for a loan and the criteria are the various aspects according to which the credit risk may be evaluated.

MURAME is based on a quite realistic preference structure of the decision maker, modeled by considering the following double threshold structure, which allows to consider the situation in which the decision maker (DM) prefers a given alternative, the case in which the DM is indifferent between two alternatives, but also the case of hesitation, in which the DM is not completely sure to prefer a given alternative (weak preference):

$$
\begin{array}{llll}
a_i \textbf{ P } a_k & \text{iff} & g_{ij} > g_{kj} + p_j \\
a_i \textbf{ I } a_k & \text{iff} & |g_{ij} - g_{kj}| \le q_j & \quad (1) \\
a_i \textbf{ Q } a_k & \text{iff} & g_{kj} + q_j \le g_{ij} \le g_{kj} + p_j
\end{array}
$$

where $a_i, a_k \in A$, g_{ij} denotes the score of the alternative a_i in relation to criterion $crit_j$, p_j denotes the preference threshold associated to $crit_j$ and q_j (with $0 \le q_j \le p_j$) the indifference threshold. $\textbf{P}$, $\textbf{Q}$ and $\textbf{I}$ denote the preference, the weak preference and the indifference relation, respectively.

The methodology can be implemented in two phases which take inspiration from two well known multicriteria methods, the ELECTRE III and PROMETHEE II (see [9] for a detailed explanation of MURAME).

Let us define the local concordance $C_j(a_i, a_k)$ and discordance $D_j(a_i, a_k)$ indexes:

$$
C_j(a_i, a_k) = \begin{cases}
1 & \text{if} \quad g_{kj} \le g_{ij} + q_j \\
0 & \text{if} \quad g_{kj} \ge g_{ij} + p_j \\
\frac{g_{ij} - g_{kj} + p_j}{p_j - q_j} & \text{otherwise}
\end{cases} \qquad (2)
$$

$$
D_j(a_i, a_k) = \begin{cases}
0 & \text{if} \quad g_{kj} \le g_{ij} + p_j \\
1 & \text{if} \quad g_{kj} \ge g_{ij} + v_j \\
\frac{g_{kj} - g_{ij} - p_j}{v_j - p_j} & \text{otherwise}
\end{cases} \qquad (3)
$$

where $v_j \ge p_j$ is the veto threshold (see [9]); and let $C(a_i, a_k)$ be the global concordance index, which delineates the dominance of a_i over a_k, obtained by aggregating the local concordance indexes

$$
C(a_i, a_k) = \sum_{j=1}^{n} w_j C_j(a_i, a_k) \qquad (4)
$$

where w_j represents the normalized weight associated to criterion $crit_j (j = 1, \ldots, n)$.

In the first phase the method aims at defining an outranking relation by building for each $a_i, a_k \in A$ an outranking (or credibility) index $O(a_i, a_k)$ defined as follows:

$$
O(a_i, a_k) = \begin{cases}
C(a_i, a_k) & \text{if} \quad D_j(a_i, a_k) \le C(a_i, a_k) \quad \forall j \\
C(a_i, a_k) \prod_{j \in T} \frac{1 - D_j(a_i, a_k)}{1 - C(a_i, a_k)} & \text{otherwise.}
\end{cases} \qquad (5)
$$

In formula (5) $T \subseteq \{1, \ldots, n\}$ denotes the subset of criteria for which $D_j(a_i, a_k) > C(a_i, a_k)$. We can see that the credibility index is equal to the global concordance $C(a_i, a_k)$, unless the performance of an alternative with respect to a single criterion is so bad that it poses a veto to the global outranking relation so that the outranking index decreases. If there is maximum discordance even only for a single criterion ($D_j(a_i, a_k) = 1$), the credibility index (5) will be zero.

In the second phase, the method computes for each alternative $a_i \in A$ the following final score (so-called net flow) $\varphi(a_i)$ that allows to produce a total preorder of the alternatives:

$$\varphi(a_i) = \sum_{k \neq i} O(a_i, a_k) - \sum_{k \neq i} O(a_k, a_i) \tag{6}$$

where $O(a_i, a_k)$ is the outranking index computed as in (5).

2.2 Preference Disaggregation and the Optimization Problem

As we have seen in Sect. 2.1, in order to apply the described MURAME-based model and to rank the alternatives, we have to determine the following parameters:

- the vector of the weights: $\mathbf{w} = (w_1, \ldots, w_n)$; $w_j \geq 0$, $j = 1, \ldots, n$ and $\sum_j w_j = 1$;
- the vector of indifference thresholds: $\mathbf{q} = (q_1, \ldots, q_n)$; $q_j \geq 0$, $j = 1, \ldots, n$;
- the vector of preference thresholds: $\mathbf{p} = (p_1, \ldots, p_n)$; $p_j \geq q_j$, $j = 1, \ldots, n$;
- the vector of veto thresholds: $\mathbf{v} = (v_1, \ldots, v_n)$; $v_j \geq p_j$, $j = 1, \ldots, n$.

Let us suppose to have a reference set of decisions provided by the decision maker, that is a set of past decisions regarding the alternatives, or regarding a subset $A' \subseteq A$ of the whole set of the alternatives.

Given the ordering of the alternatives in the reference set made by the decision maker, the preference disaggregation approach regards the problem of determining the set of parameters that minimizes a measure of inconsistency $f(\mathbf{w}, \mathbf{q}, \mathbf{p}, \mathbf{v})$ between the ordering produced by the model with that set of parameters and the given one. In order to infer the values of the parameters starting from a given reference set of decisions, the following mathematical programming problem has to be solved:

$$\begin{aligned}
\min_{\mathbf{w}, \mathbf{q}, \mathbf{p}, \mathbf{v}} \quad & f(\mathbf{w}, \mathbf{q}, \mathbf{p}, \mathbf{v}) \\
\text{s.t.} \quad & \mathbf{w} \geq \mathbf{0} \\
& \sum_{j=1}^{n} w_j = 1 \\
& \mathbf{q} \geq \mathbf{0} \\
& \mathbf{p} \geq \mathbf{q} \\
& \mathbf{v} \geq \mathbf{p} .
\end{aligned} \tag{7}$$

We can note that problem (7) can be reformulated in a simpler way by introducing the auxiliary variables $\mathbf{t} = \mathbf{p} - \mathbf{q}$ and $\mathbf{s} = \mathbf{v} - \mathbf{p}$, so that it becomes:

$$\begin{aligned}
\min_{\mathbf{w}, \mathbf{q}, \mathbf{t}, \mathbf{s}} \quad & f(\mathbf{w}, \mathbf{q}, \mathbf{t}, \mathbf{s}) \\
\text{s.t.} \quad & \mathbf{w}, \mathbf{q}, \mathbf{t}, \mathbf{s} \geq \mathbf{0} \\
& \sum_{j=1}^{n} w_j = 1 .
\end{aligned} \tag{8}$$

This apparently simple mathematical programming problems hides its complexity in the objective function $f(\mathbf{w}, \mathbf{q}, \mathbf{t}, \mathbf{s})$: indeed every choice of f requires that it produces an order of the alternatives and then that a measure of the consistency of the model is calculated. It is then hard to write an exact analytical expression for f in term of its variables $\mathbf{w}, \mathbf{q}, \mathbf{t}, \mathbf{s}$, so that the use of gradient methods for the optimization task is discouraged, and an evolutionary approach seems more appropriate.

In this study we consider two kinds of fitness function, in order to exploit all the information contained in the input data provided by the decision maker[1]: the first function allows to deal with the ordinal rank of the alternatives in the reference set, whereas the second one allows to handle the cardinal values.

The first fitness function is the S function defined as follows:

$$S(\mathbf{w}, \mathbf{q}, \mathbf{t}, \mathbf{s}) = \frac{6 \sum_{i=1}^{m'} (\bar{r}_i - r_i(\mathbf{w}, \mathbf{q}, \mathbf{t}, \mathbf{s}))^2}{m'^3 - m'} \tag{9}$$

where $\bar{r}_i$ is the rank of alternative a_i in the reference set assigned by the decision maker and $r_i(\mathbf{w}, \mathbf{q}, \mathbf{t}, \mathbf{s})$ the one determined by the MURAME-based model, with $m' \leq m$ being the total number of alternatives in the reference set. This fitness function is an application of the Spearman Rank Correlation Coefficient (see [12]) to measure the strength of correlation between the two orderings. Its values are in the interval $[0, 2]$, and $S(\mathbf{w}, \mathbf{q}, \mathbf{t}, \mathbf{s}) = 0$ clearly means that there is an exact correspondence between the ranking of the decision maker and that obtained by the model.

The second fitness function is represented by the following D function:

$$D(\mathbf{w}, \mathbf{q}, \mathbf{t}, \mathbf{s}) = \frac{\sum_{i=1}^{m'} (\lambda(\varphi(a_i; \mathbf{w}, \mathbf{q}, \mathbf{t}, \mathbf{s})) - s(a_i))^2}{m'} \tag{10}$$

where $s(a_i)$ is the (cardinal) score assigned by the decision maker to the alternative a_i and λ is a linear transformation that maps the net flow $\varphi(a_i; \mathbf{w}, \mathbf{q}, \mathbf{t}, \mathbf{s})$ computed as in Eq. (6) to the support of decision maker scores.

3 Particle Swarm Optimization for Preference Disaggregation

3.1 A Brief Description of PSO

Particle Swarm Optimization (PSO) is a bio-inspired iterative heuristics for the solution of nonlinear global optimization problems (see [2, 11]). The basic idea of PSO is to model the so called "swarm intelligence" [3] that drives groups of individuals belonging to the same species when they move all together looking for food. On this purpose every member of the swarm explores the search area keeping memory of its best position reached so far, and it exchanges this information with the neighbors in the swarm. Thus, the whole swarm is supposed to converge eventually to the best global position reached by the swarm members.

From a mathematical point of view every member of the swarm (namely a particle) represents a possible solution of an optimization problem, and it is initially positioned randomly in the feasible set of the problem. To every particle is also initially assigned a random velocity, which is used to determine its initial direction of movement.

---

[1] For more details about the nature of input data, we refer to Sect. 4.

Let us consider the following global optimization problem:

$$\min_{\mathbf{x}\in\mathbb{R}^d} f(\mathbf{x})$$

where $f : \mathbb{R}^d \mapsto \mathbb{R}$ is the objective function in the minimization problem. In applying PSO for its solution, we consider M particles; at the k-th step of the PSO algorithm, three vectors are associated to each particle $l \in \{1,\ldots,M\}$:

- $\mathbf{x}_l^k \in \mathbb{R}^d$, the position at step k of particle l;
- $\mathbf{v}_l^k \in \mathbb{R}^d$, the velocity at step k of particle l;
- $\mathbf{p}_l \in \mathbb{R}^d$, the best position visited so far by particle l;
- $\mathbf{p}_{g(l)} \in \mathbb{R}^d$ the best position in a neighborhood of the l-th particle.

Moreover, $pbest_l = f(\mathbf{p}_l)$ denotes the value of the objective function in the position $\mathbf{p}_l$ of the l-th particle.

The PSO algorithm, in the version with inertia weight [15], works as follows:

1. Set $k = 1$ and evaluate $f(\mathbf{x}_l^k)$ for $l = 1,\ldots,M$. Set $pbest_l = +\infty$ for $l = 1,\ldots,M$.
2. If $f(\mathbf{x}_l^k) < pbest_l$ then set $\mathbf{p}_l = \mathbf{x}_l^k$ and $pbest_l = f(\mathbf{x}_l^k)$.
3. Update position and velocity of the l-th particle, with $l = 1,\ldots,M$, according to the following equations:

$$\mathbf{v}_l^{k+1} = w^{k+1}\mathbf{v}_l^k + \mathbf{U}_{\phi_1} \otimes (\mathbf{p}_l - \mathbf{x}_l^k) + \mathbf{U}_{\phi_2} \otimes (\mathbf{p}_{g(l)} - \mathbf{x}_l^k) \tag{11}$$

$$\mathbf{x}_l^{k+1} = \mathbf{x}_l^k + \mathbf{v}_l^{k+1} \tag{12}$$

where $\mathbf{U}_{\phi_1}, \mathbf{U}_{\phi_2} \in \mathbb{R}^d$ and their components are uniformly randomly distributed in $[0,\phi_1]$ and $[0,\phi_2]$ respectively; parameters ϕ_1 and ϕ_2 are often called acceleration coefficients; the symbol $\otimes$ denotes component-wise product. The parameter w^k (the inertia weight) is generally linearly decreasing with the number of steps, i.e.:

$$w^k = w_{max} + \frac{w_{min} - w_{max}}{K}k$$

where K is the maximum number of steps allowed.

4. If a convergence test is not satisfied then set $k = k+1$ and go to 2.

For more details about PSO methodology, the specification of its parameters and of the neighborhood topology, we refer the reader to [2].

3.2 PSO Implementation

We propose to use PSO in order to solve the optimization problem which originates from the preference disaggregation process. In implementing the PSO algorithm described in Sect. 3.1, we have considered the so called *gbest* topology, that is $g(l) = g$ for every $l = 1,\ldots,M$, and g is the index of the best particle in the whole swarm, that is $g = \arg\min_{l=1,\ldots,M} f(\mathbf{p}_l)$. This choice implies that the whole swarm is used as the neighborhood of each particle.

Moreover, as stopping criterion, we decided that the algorithm terminated when the objective function did not have a decrease of at least 10^{-4} in a prefixed number of steps.

Since PSO was conceived for unconstrained problems, the algorithm above cannot prevent from generating infeasible particles' positions when constraints are considered. To avoid this problem, different strategies have been proposed in the literature, and most of them involve the repositioning of the particles [18] or the introduction of some external criteria to rearrange the components of the particles [6,16]. In this paper we follow the same approach adopted in [5], which consists in keeping PSO as in its original formulation and reformulating the optimization problem into an unconstrained one:

$$\min_{\mathbf{w},\mathbf{q},\mathbf{t},\mathbf{s}} P(\mathbf{w},\mathbf{q},\mathbf{t},\mathbf{s};\varepsilon) \tag{13}$$

where the objective function $P(\mathbf{w},\mathbf{q},\mathbf{t},\mathbf{s};\varepsilon)$ is defined as follows:

$$
\begin{aligned}
P(\mathbf{w},\mathbf{q},\mathbf{t},\mathbf{s};\varepsilon) = f(\mathbf{w},\mathbf{q},\mathbf{t},\mathbf{s}) + \frac{1}{\varepsilon}\Bigg[&\left| \sum_{j=1}^{n} w_j - 1 \right| + \sum_{j=1}^{n} \max\{0, -w_j\} \\
&+ \sum_{j=1}^{n} \max\{0, -q_j\} + \sum_{j=1}^{n} \max\{0, -t_j\} \\
&+ \sum_{j=1}^{n} \max\{0, -s_j\} \Bigg]
\end{aligned} \tag{14}
$$

with ε being the penalty parameter. For more details about this method and about the relationships between the solutions of the constrained problem (8) and those of the unconstrained problem (13), see [5, 8, 14, 17]. We only remark here that the penalty function $P(\mathbf{w},\mathbf{q},\mathbf{t},\mathbf{s};\varepsilon)$ is clearly nondifferentiable because of the ℓ_1-norm in (14). This also motivates the choice of using PSO for its minimization, since PSO evidently does not require the derivatives of $P(\mathbf{w},\mathbf{q},\mathbf{t},\mathbf{s};\varepsilon)$.

Since PSO is a heuristics, the minimization of the penalty function $P(\mathbf{w},\mathbf{q},\mathbf{t},\mathbf{s};\varepsilon)$ does not theoretically ensure that a global minimum of problem (8) is detected. Nevertheless, PSO often provides a suitable compromise between the performance of the approach (i.e. a satisfactory estimate of the global minimum solution for problem (8)) and its computational cost.

With regard to the initialization procedure, we can observe that, since we deal with an unconstrained problem, we could in principle generate initial values for the population in an arbitrary random way. However, like any metaheuristic, the performance of the algorithm could be improved by a careful choice of the initial population such that the feasible region is adequately explored by the particles at the initial step.

In this contribution, in order to obtain the initial weights, we generated $d_1 < \cdots < d_{n-1}$ random numbers uniformly distributed in $[0,1]$, and then we set:

$$w_1^0 = d_1, w_2^0 = d_2 - d_1, \ldots, w_n^0 = 1 - d_{n-1}.$$

To obtain the initial values of the variables q_j, t_j, s_j $(j = 1, \ldots, n)$, we generated three random numbers $u_j^1 < u_j^2 < u_j^3$ uniformly distributed in $[0, 2]$, and then we set:

$$q_j^0 = (\overline{g}_j - \underline{g}_j)\frac{u_j^1}{10}; \quad t_j^0 = (\overline{g}_j - \underline{g}_j)\frac{u_j^2}{10}; \quad s_j^0 = (\overline{g}_j - \underline{g}_j)\frac{u_j^3}{10}$$

where $\overline{g}_j = \max_{1 \leq i \leq m} g_{ij}$ and $\underline{g}_j = \min_{1 \leq i \leq m} g_{ij}$.

Moreover, for every particle $\mathbf{x}_l^0 = (\mathbf{w}_l^0, \mathbf{q}_l^0, \mathbf{t}_l^0, \mathbf{s}_l^0)$, the components of the initial velocity $\mathbf{v}_l^0$ are generated as random numbers uniformly distributed in $[-x_h^0, x_h^0]$ for every $h = 1, \ldots, 4n$.

4 Application to a Credit Scoring Problem

In this section we analyze the performance of the proposed approach by considering a credit scoring problem. We adopt the MURAME method in order to evaluate the creditworthiness of a set of firms, as proposed in [4], and use a real world data set provided by a major bank of north-eastern Italy, the Banca Popolare di Vicenza.

The database consists of around 12000 firms applicants for a loan, which represent the alternatives of the creditworthiness evaluation problem. The firms are divided into three groups, small, medium and large, with respect to their business turnover and the three groups are approximately of the same size. The evaluation criteria are represented by seven indicators $\{I_1, \ldots, I_7\}$ which have been computed by the bank starting from the balance sheets; their values are provided for both years 2008 and 2009.

In order to verify the ability of PSO to find the optimal values of the parameters of MURAME model, we employ a bootstrap analysis carried out on a subset of randomly selected 4000 firms.

Using data of year 2008, the firms have been first ordered using the MURAME methodology and the values of the parameters have been fixed according to the following specific rules, as proposed in [4]:

$$w_1 = \cdots = w_n = \frac{1}{n}, \quad q_j = (\overline{g}_j - \underline{g}_j)\frac{1}{6}, \quad p_j = 4q_j, \quad v_j = 5q_j \quad j = 1, \ldots, n.$$

Table 1 reports the values of the parameters so obtained.

Table 1 Actual parameters of MURAME

	I_1	I_2	I_3	I_4	I_5	I_6	I_7
w	0.1429	0.1429	0.1429	0.1429	0.1429	0.1429	0.1429
q	10.2283	0.0500	0.0700	0.2600	0.1383	0.0183	0.3532
p	40.9133	0.1999	0.2800	1.0400	0.5533	0.0730	1.4129
v	51.1417	0.2499	0.3500	1.3000	0.6917	0.0913	1.7662

Table 2 PSO parameters

ϕ_1	ϕ_2	K	ε	w_{min}	w_{max}
1.75	1.75	200	0.0001	0.4	0.9

A first set of bootstrap experiments has been conducted in order to determine the best values for the acceleration coefficients ϕ_1, ϕ_2, the maximum number of steps K and the value of the penalty parameter ε, while for the initial and final value of the inertia weight we used the ones most used in the literature. In these experiments we have considered only the training step (see beyond) and we have used as fitness function $\frac{S+D}{2}$. In order to determine the value of a given PSO parameter, first we have tested different values of that parameter while keeping fixed the remaining ones; for instance, in case of ε, we have considered $\varepsilon = 1, \varepsilon = 0.1, \varepsilon = 0.01$, etc. and for each of these values we have performed 10 runs of the algorithm; then we have chosen the value of the PSO parameter ($\varepsilon = 0.0001$ in the above mentioned case) which allowed to obtain on average the best results for the objective function. The same procedure has been adopted in determining the values of K, ϕ_1 and ϕ_2. Table 2 reports the values of the PSO parameters so obtained.

The bootstrap procedure that we have adopted is structured in the following steps.

(i)   A sample of $H = 100$ firms is randomly selected without replacement from the 4000 ones.

(ii) To the sample of firms selected as in step (i) we have applied the MURAME methodology with the actual parameters determined as in Table 1; an ordering of the H firms is obtained according to the net flow (6) associated to each firm.

(iii) To the sample of firms selected as in step (i) we have applied the PSO-based methodology, using the PSO parameters determined as in Table 2 and adopting both objective functions (9), (10). The optimal values of MURAME parameters are therefore determined through PSO, in order to minimize the discrepancy–as measured by the fitness function– between the score (ranking) obtained by the MURAME in step (ii) and that one computed by PSO (the so-called training step).

(iv) (Out-the-bootstrap). Another sample of firms of the same size $H = 100$ is randomly selected from the 4000 ones, excluding those belonging to the sample used for the training step. This sample is ordered with the MURAME method, by using both the actual MURAME parameters and those determined by PSO; the value of the fitness function is then computed in order to evaluate the distance between the two orderings.

The procedure is then repeated $N = 1000$ times in order to compute the necessary statistics of the values of the MURAME parameters and the performance measures. Table 3 reports the average (and the standard deviation) of the fitness function values obtained at each iteration of the procedure. The results are shown for both the fitness

Table 3 Results for in-the-bootstrap data

	M = 100		M = 200	
	S	D	S	D
Mean	0.0022	0.0025	0.0002	0.0003
Standard deviation	0.0011	0.0017	0.0001	0.0002

Table 4 Results for out-the-bootstrap data

	M = 100		M = 200	
	S	D	S	D
Mean	0.1234	0.1725	0.012	0.016
Standard deviation	0.0450	0.0612	0.032	0.035

functions S and D and for a population size of $M = 100$ and $M = 200$ number of particles. Table 4 presents the results obtained for the out-the-bootstrap phase.

It can be noted that while $M = 100$ particles are enough to obtain full consistency (an average value of the fitness function nearly zero), with both the performance measures considered, between the model produced by PSO and the actual ordering of the alternatives in the bootstrap data, the performance of the so obtained models on the out of the bootstrap data is not as good, and a higher number of particles ($M = 200$) is required to improve it.

Table 5 illustrates the average values of the parameters (and the standard deviations) determined by PSO in the bootstrap procedure.

It is interesting to remark that the discrepancies between the values of the parameters obtained with $M = 100$ and $M = 200$ are not very high and they concern mainly the values of the thresholds, while the weights are determined in a consistent way even with the smaller number of particles.

As for the relation between the actual parameters of Table 1 and the average values of parameters obtained in Table 5, it can be noticed that there are significant differences, similarly to what observed in [7] with regard to the use of the differential evolution algorithm. However, since the ordering performance of the model is high, this seems to suggest that there is a certain flexibility in the specification of the parameters of the model consistent with the ordering of the alternatives in the reference set.

5 Conclusions

The results obtained in this contribution showed a high consistency between the model obtained by MURAME with the actual parameters and the model produced by PSO.

Table 5 Average values (and standard deviation) of MURAME parameters, determined by bootstrap procedure

	I_1	I_2	I_3	I_4	I_5	I_6	I_7
M = 100							
w	0.1980	0.1284	0.0273	0.3405	0.2254	0.0035	0.0768
	0.0920	0.1230	0.1540	0.1420	0.0980	0.1640	0.1330
q	6.1076	0.0094	0.0149	0.2127	0.3249	0.1244	0.2521
	0.1170	0.0950	0.1620	0.0830	0.1390	0.1140	0.0930
p	274.3694	0.0410	0.0209	0.5882	0.5647	0.1997	0.6648
	0.1260	0.0840	0.1310	0.1340	0.1690	0.0920	0.1210
v	4324.4420	12.7064	3.2699	1.6909	5.3118	0.5246	6.0462
	0.1740	0.1420	0.1280	0.0980	0.1560	0.1650	0.1430
M = 200							
w	0.1394	0.1454	0.0457	0.3256	0.2465	0.0013	0.0961
	0.0540	0.0620	0.0430	0.0650	0.0390	0.0240	0.0310
q	2.6672	0.0135	0.1090	0.0198	0.0125	0.2745	0.0166
	0.0790	0.0420	0.0570	0.0670	0.0280	0.0340	0.0860
p	3.8324	0.0179	0.1755	0.2692	0.2274	0.3468	0.5485
	0.0620	0.0840	0.0530	0.0650	0.0310	0.0440	0.0220
v	109.6670	3.4567	0.2023	0.9412	4.5347	0.6543	1.4986
	0.0640	0.0720	0.0760	0.0830	0.0660	0.0750	0.0830

The differences encountered between the actual and the optimal values of parameters underline a certain flexibility degree in the specification of the parameters of the multicriteria credit scoring model consistent with the ordering of the alternatives in the reference set.

As future research we intend to analyze whether PSO is able to find, using a realistic computational time, the values of the parameters of the MURAME model that make them consistent with the actual classification of the firms provided by the bank, which acts as decision maker. In this case, additional information on the reference decisions is necessary, concerning the final scores that the bank has actually assigned to the firms in order to establish their ranking in terms of their credit quality.

Possible differences among the results obtained in the various groups of firms (small, medium and large) will be also investigated.

Acknowledgements The authors are grateful to Banca Popolare di Vicenza for its collaboration in the research project.

References

1. Belacel, N., Bhasker Raval, H., Punnen, A.P.: Learning multicriteria fuzzy classification method PROAFTN from data. Comput Oper Res **34**, 1885–1898 (2007)

2. Blackwell, T., Kennedy, J., Poli, R.: Particle swarm optimization - An overview. Swarm Intell **1**, 33–57 (2007)
3. Dorigo, M., Theraulaz, G.: From natural to artificial swarm intelligence. Oxford University Press, Oxford (1999)
4. Corazza, M., Funari, S., Siviero, F.: An MCDA-based approach for creditworthiness assessment. Working Papers. Department of Applied Mathematics, University of Venice (2008). http://ideas.repec.org/p/vnm/wpaper/177.html
5. Corazza, M., Fasano, G., Gusso, R.: Portfolio selection with an alternative measure of risk: Computational performances of particle swarm optimization and genetic algorithms. In: Perna C, Sibillo M (eds) Mathematical and Statistical Methods for Actuarial Sciences and Finance. Springer, Berlin (2012)
6. Cura, T.: Particle swarm optimization approach to portfolio optimization. Nonlinear Anal: R. World Appl. **10**, 2396–2406 (2009)
7. Doumpos, M., Marinakis, Y., Marinaki, M., et al: An evolutionary approach to construction of outranking models for multicriteria classification: The case of the ELECTRE TRI method. Eur J Oper Res **199**, 496–505 (2009)
8. Fletcher, R.: Practical methods of optimization. John Wiley & Sons, Glichester (1991)
9. Goletsis, Y., Askounis, D.T., Psarras, J.: Multicriteria judgments for project ranking: An integrated methodology. Econ Financ Model **8**, 127–148 (2001)
10. Jacquet-Lagrèze, E., Siskos,Y.: Preference disaggregation: 20 years of MCDA experience. Eur J Oper Res **130**, 233–245 (2001)
11. Kennedy, J., Eberhart, R.C.: Particle swarm optimization. Proceedings of the IEEE international conference on neural networks **IV**, 1942–1948 (1995)
12. Lehmann, E.L.: Nonparametrics: statistical methods based on ranks. Springer, Berlin (2006)
13. Mousseau, V., Slowinski, R., Zielniewicz, P.: ELECTRE TRI 2.0a: Methodological guide and user's documentation. Université de Paris-Dauphine (1999). http://http://l1.lamsade.dauphine.fr/mcda/biblio/PDF/mous3docl99.pdf
14. Di Pillo, G., Grippo, L.: Exact penalty functions in constrained optimization. SIAM J Control and Optim **27**, 1333–1360 (1989)
15. Shi, Y., Eberhart, R.: A modified particle swarm optimizer. Evolutionary Computation Proceedings. IEEE World Congress on Computational Intelligence, pp. 69–73 (1998)
16. Thomaidis, N., Angelidis, T., Vassiliadis, V., et al: Active portfolio management with cardinality constraints: An application of particle swarm optimization. New Math Nat Comput **5**, 535–555 (2009)
17. Zangwill, W.I.: Non-linear programming via penalty functions. Manag Sci **13**, 344–358 (1967)
18. Zhang, W.J., Xie, X.F., Bi, D.C.: Handling boundary constraints for numerical optimization by particle swarm flying in periodic search space. Proceedings of the 2004 Congress on Evolutionary Computation IEEE, 2307–2311 (2005)

Time Series Clustering on Lower Tail Dependence for Portfolio Selection

Giovanni De Luca and Paola Zuccolotto

Abstract In this paper we analyse a case study based on the procedure introduced by De Luca and Zuccolotto [8], whose aim is to cluster time series of financial returns in groups being homogeneous in the sense that their joint bivariate distributions exhibit high association in the lower tail. The dissimilarity measure used for such clustering is based on tail dependence coefficients estimated using copula functions. We carry out the clustering using an algorithm requiring a preliminary transformation of the dissimilarity index into a distance metric by means of a geometric representation of the time series, obtained with Multidimensional Scaling. We show that the results of the clustering can be used for a portfolio selection purpose, when the goal is to protect investments from the effects of a financial crisis.

1 Introduction

Several approaches to time series clustering are present in the literature. After the first studies, where dissimilarities between time series were merely derived by the comparison between observations or some simple statistics computed on the data (see for example [3]), more complex solutions have been proposed. An interesting review can be found in [29]. Piccolo [23] and Corduas and Piccolo [7] proposed a distance measure for time series generated by ARIMA processes, based on the comparison between the parameters of the corresponding Wold decomposition. Otranto [20] extended this approach to GARCH models, Galeano and Peña [12] considered the generalized distance between the autocorrelation functions of two time series while

G. De Luca (✉)
Department of Business and Quantitative Sciences, University of Naples Parthenope, Naples, Italy
e-mail: giovanni.deluca@uniparthenope.it

P. Zuccolotto
Department of Economics and Management, University of Brescia, Brescia, Italy
e-mail: paola.zuccolotto@sis-statistica.org

M. Corazza, C. Pizzi (eds.), *Mathematical and Statistical Methods for Actuarial Sciences and Finance*, DOI 10.1007/978-3-319-02499-8_12, © Springer International Publishing Switzerland 2014

Caiado [4] introduced a metric based on the normalized periodogram. Alternative methods, employing parametric and non-parametric density forecasts, was discussed in [1] and in [28], respectively. To give an idea of the great variety of approaches in this framework, it is finally worth recalling the frequency domain approach of Kakizawa [15] and Taniguchi and Kakizawa [27], the use of two-dimensional singular value decomposition by Weng and Shen [30], the procedure using a robust evolutionary algorithm of Pattarin et al. [21]. A comparison of sevaral parametric and non-parametric approaches can be found in [22]. But the list of citations could be even longer. In this paper we show a case study based on the use of the procedure proposed by De Luca and Zuccolotto [8] to cluster time series of returns of financial assets according to their association in the lower tail. Then, we show how this approach can be employed for portfolio selection, especially in a financial crisis perspective. With respect to the seminal paper of De Luca and Zuccolotto [8], here we propose, firstly, two indexes for evaluating the quality of the clusterization from the point of view of tail dependence, and, secondly, a more specific criterion for portfolio selection, based on the Omega index proposed by Keating and Shadwick [16]. The paper is organized as follows: in Sect. 2 the clustering procedure is briefly recalled, while the main results of the case study are summarized in Sect. 3. Concluding remarks follow in Sect. 4.

2 Tail Dependence-Based Clustering Procedure

The interest of researchers in modelling the occurring of extreme events has several empirical motivations, especially in contexts where it can be directly associated to risk measurement, such as, for example, financial markets. Recently, a great deal of attention has been devoted also to the study of association between extreme values of two or more variables. From a methodological point of view, the problem of quantifying this association has been addressed in different ways. One of the proposed approaches consists in analyzing the probability that one variable assumes an extreme value, given that an extreme value has occurred to the other variables (see [6]). This probability is known as lower or upper tail dependence and we will restrict its analysis to the bivariate case. Let Y_1 and Y_2 be two random variables and let $U_1 = F_1(Y_1)$ and $U_2 = F_2(Y_2)$ be their distribution functions. The lower and upper tail dependence coefficients are defined respectively as

$$\lambda_L = \lim_{v \to 0^+} P[U_1 \leq v | U_2 \leq v] \quad \text{and} \quad \lambda_U = \lim_{v \to 1^-} P[U_1 > v | U_2 > v].$$

In practice, the tail dependence coefficients are estimated from observed data after assuming a probabilistic framework. The copula functions are commonly used for financial data with the advantage that the tail dependence estimation is both simple and flexible. A two-dimensional copula function for two random variables Y_1 and Y_2 is defined as a function $C : [0,1]^2 \to [0,1]$ such that

$$F(y_1, y_2; \theta) = C(F_1(y_1; \vartheta_1), F_2(y_2; \vartheta_2); \tau),$$

for all y_1, y_2, where $F(y_1, y_2; \theta)$ is the joint distribution function of Y_1 and Y_2 (see [19]) and $\theta = (\vartheta_1, \vartheta_2, \tau)$. It is straightforward to show that the tail dependence coefficients can be expressed in terms of the copula function. In particular, the lower tail dependence coefficient, which will be hereafter the focus of the paper, is given by

$$\lambda_L = \lim_{v \to 0^+} \frac{C(v, v)}{v}.$$

In the analysis of the relationship between financial returns, the lower tail dependence coefficient gives an idea of the risk of investing on assets for which extremely negative returns could occur simultaneously. So, the lower tail dependence is strictly linked to the diversification of investments, especially in financial crisis periods. For this reason, De Luca and Zuccolotto [8] proposed to cluster time series of financial returns according to a dissimilarity measure defined as

$$\delta(\{y_{it}\}, \{y_{jt}\}) = -\log(\hat{\lambda}_L),$$

where $\{y_{it}\}_{t=1,\dots,T}$ and $\{y_{jt}\}_{t=1,\dots,T}$ denote the time series of returns of two assets i and j, and $\hat{\lambda}_L$ is their estimated tail dependence coefficient. In this way we obtain clusters of assets characterized by high tail dependence in the lower tail. From a portfolio selection perspective, it should then be avoided portfolios containing assets belonging to the same cluster. Given p assets, the clustering procedure proposed by De Luca and Zuccolotto [8] is composed by two steps. In the first step, starting from the dissimilarity matrix $\Delta = (\delta_{ij})_{i,j=1,\dots,p}$, an *optimal* representation of the p time series $\{y_{1t}\}, \dots, \{y_{pt}\}$ as p points $\mathbf{y}_1, \dots, \mathbf{y}_p$ in $\mathbb{R}^q$ is found by means of Multidimensional Scaling (MDS). The above mentioned term *optimal* means that, with MDS, the Euclidean distance matrix $D = (d_{ij})_{i,j=1,\dots,p}$, with $d_{ij} = \|\mathbf{y}_i - \mathbf{y}_j\|$, of the points $\mathbf{y}_1, \dots, \mathbf{y}_p$ in $\mathbb{R}^q$ has to fit as closely as possible the dissimilarity matrix Δ. The extent to which the interpoint distances d_{ij} "match" the dissimilarities δ_{ij} is measured by an index called *stress*, which should be as low as possible. MDS works for a given value of the dimension q, which has to be given in input. So, it is proposed to start with the dimension $q = 2$ and then to repeat the analysis by increasing q until the minimum stress of the corresponding optimal configuration is lower than a given threshold $\bar{s}$. In the second step, the k-means clustering algorithm is performed using the obtained geometric representation of the p time series. Among the several hierarchical and non-hierarchical clustering techniques which we may resort to, the k-means algorithm performed on the MDS geometrical representation has revealed, through simulation studies, a good performance in this context [8].

3 Case Study

In this case study we analyse the time series of the daily prices of the 24 stocks which have been included in FTSE MIB index during the whole period from January 3, 2006 to October 31, 2011.

3.1 Clustering

After transforming prices into log-returns, we preliminary removed autocorrelation and heteroskedasticity by means of univariate Student-t AR-GARCH models. For each couple of stocks we estimated a bivariate Joe-Clayton copula function [14],

$$C(u_1,u_2) = 1 - \{1 - [(1 - (1 - u_1)^\kappa)^{-\theta} + (1 - (1 - u_2)^\kappa)^{-\theta} - 1]^{-1/\theta}\}^{1/\kappa}$$

using the estimated distribution functions of the standardized residuals. After estimating the 276 lower tail dependence coefficients, which in the case of the Joe–Clayton copula are given by $\hat{\lambda}_L = 2^{-1/\hat{\theta}}$, we carried out MDS using the dissimilarity matrix $\Delta = (\delta_{ij})_{i,j=1,...,24}$. We set $\bar{s} = 0.005$. The minimum dimension allowing a final configuration with minimum stress lower than $\bar{s}$ resulted $q = 14$. In the second step, we performed a k-means clustering algorithm using the MDS point configuration in $\mathbb{R}^{14}$, $y_1, \ldots y_{24}$. The graph displayed in the left panel of Fig. 1 shows the pattern of the ratio of deviance between clusters over total deviance, as a function of the number of clusters k and its increments when considering the solution with k clusters, with respect to $k - 1$ clusters. This graph helps the researcher in deciding the optimal number of clusters. In this case we observe that moving from $k = 3$ to $k = 4$ allows an improvement of 32.7% in the quality of the clusterization, while from $k = 5$ onward, the increments are appreciably lower and more stable. For this reason, we feel that a good choice could be $k = 4$. In addition we have computed the following indexes proposed in the literature for determining the optimal number of clusters (see [9] for an exhaustive review): CH [5], H [13], RL [24], SS [25], M [17], BB [2], $TraceCovW$ [18], $TraceW$ [10,11], $TraceW^{-1}B$ and $|T| = |W|$ [11]. All the indexes except CH suggest to choose the solution with $k = 4$ clusters. So, we judge this choice the most reliably founded as it combines subjective remarks and

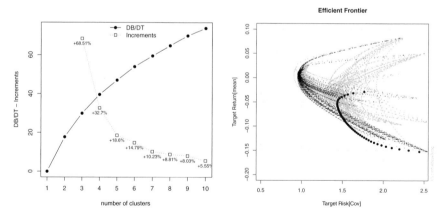

Fig. 1 *Left*: Deviance between clusters over total deviance. *Right*: Efficient frontiers of the 640 selections based on clustering (black, thin), efficient frontiers of 1000 selections built with 4 randomly selected stocks (gray) and efficient frontier of the stocks belonging to cluster 4 (black, bold)

Table 1 Cluster composition by stocks

Cluster 1	Cluster 2		Cluster 3	Cluster 4	
Atlantia	Autogrill	Fiat	SNAM	Banca MPS	Fondiaria
ENEL	Finmeccanica	Lottomatica	Terna	Generali	Intesa SP
ENI	Luxottica	Pirelli		Mediobanca	Mediolanum
Saipem	Stmicroelectronics	Telecom		Mediaset	Banca PM
				UBI	Unicredit

Table 2 Cluster composition by sectors (with cluster labels)

Cluster 1 *Power, energy and mobility*	*Cluster 2* *Living goods*
Oil and natural gas (2)	Travels and free time (2)
Industrial services and products (1)	Cars and components (2)
Public services (1)	Tecnology and Communications (2)
	House, personal utilities, fashion (1)
	Industrial services and products (1)

Cluster 3 *Public services*	*Cluster 4* *Banks and insurance*
Public services (2)	Banks (6)
	Insurance (3)
	Media (1)

objective criteria. The cluster composition by stocks and by economic activity sectors are displayed in Table 1 and Table 2, respectively. We observe that there are some affinities in the sectors of stocks belonging to the same cluster, so we propose to label the clusters as shown in Table 2.

In order to measure the extent to which the clusterization has been able to group stocks with high tail dependence, separating them from the others, we propose to compute two indexes, which we call average tail dependence coefficient within and between clusters, respectively. On the one hand, the average tail dependence coefficient within cluster c is given by

$$\bar{\lambda}_c^W = \frac{2}{n_c(n_c - 1)} \sum_{i_c=1}^{n_c} \sum_{j_c=i_c+1}^{n_c} \hat{\lambda}_L^{i_c j_c},$$

where n_c is the number of stocks belonging to cluster c, i_c and j_c are the ith and the jth stocks of cluster c and $\hat{\lambda}_L^{i_c j_c}$ is their estimated tail dependence coefficient. The index $\bar{\lambda}_c^W$ measures the extent to which the clusters are internally homogeneous from the point of view of tail dependence and should be as high as possible. On the other hand, the average tail dependence coefficient between cluster c and the others

Table 3 Average tail dependence within and between clusters

Cluster (c)	1	2	3	4
$\bar{\lambda}_c^W$	0.2770	0.2719	0.2975	0.4249
$\bar{\lambda}_c^B$	0.2224	0.2425	0.1211	0.2475

is given by

$$\bar{\lambda}_c^B = \frac{1}{n_c n_{\bar{c}}} \sum_{i_c=1}^{n_c} \sum_{i_{\bar{c}}=1}^{n_{\bar{c}}} \hat{\lambda}_L^{i_c i_{\bar{c}}},$$

where $n_{\bar{c}}$ is the number of stocks not belonging to cluster c, $i_{\bar{c}}$ is the ith stock outside cluster c and $\hat{\lambda}_L^{i_c i_{\bar{c}}}$ is the estimated tail dependence coefficient if stocks i_c and $i_{\bar{c}}$. This index measures the extent to which the clusters are externally separated from the point of view of tail dependence and should be as low as possible. In general, a good clusterization should have $\bar{\lambda}_c^W > \bar{\lambda}_c^B$ for all $c = 1, \cdots, k$. Results for this case study are displayed in Table 3.

3.2 Portfolio Selection

We used the obtained clustering to construct portfolios composed by as many stocks as the number k of clusters. The stocks are selected by imposing the restriction that each stock belongs to a different cluster [8]; with the $k = 4$ above mentioned clusters, 640 different selections can be made according to this criterion. This strategy should protect the investments from parallel extreme losses during crisis periods, because the clustering solution is characterized by a moderate lower tail dependence between clusters. Using the popular Markowitz portfolio selection procedure, in the right panel of Fig. 1, we plotted the efficient frontiers of all the possible 640 selections (black, thin), compared to those of 1000 portfolios (gray) built with 4 randomly selected stocks and to the efficient frontier of the portfolio built using all the stocks belonging to cluster 4 (black, bold), the cluster with the highest average tail dependence within cluster $\bar{\lambda}_c^W$. The efficient frontiers of the 640 selections dominate the main part of the others. After selecting the minimum variance portfolio of each frontier, we evaluated the performance of the 640 resulting portfolios in the following 36 days, thus with an out-of-sample perspective. Fig. 2 displays the returns of the 640 portfolios (black) in the period November 1–December 20, with respect to the price of October, 31. These returns are compared to (*i*) the returns of the minimum variance portfolios built using 4 randomly selected stocks (left panel, gray), (*ii*) the returns of the minimum variance portfolio built using all the stocks (right panel, bold gray, dashed line) and to (*iii*) the returns of the minimum variance portfolio built using all the stocks belonging to cluster 4 (right panel, bold gray). We observe a good performance of the 640 portfolios with respect to the selected competitors. At this step, a criterion should be given in order to select one of these 640 portfolios. To this purpose, we propose to resort to a proper index, such as for example the Sharpe

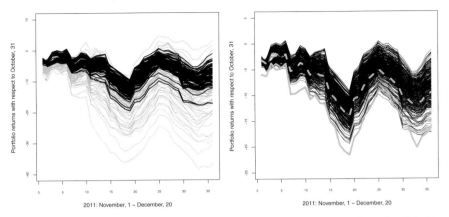

Fig. 2 *Left*: Returns of the 640 minimum variance portfolios based on clustering (black), returns of the minimum variance portfolios built using 4 randomly selected stocks (gray). *Right*: Returns of the 640 minimum variance portfolios based on clustering (black), returns of the minimum variance portfolios built using all the stocks (gray, dashed line) and using the stocks of cluster 4 (bold gray)

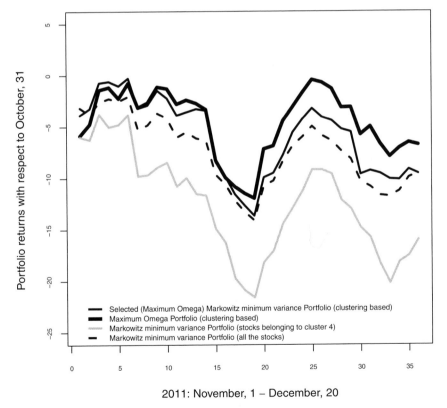

Fig. 3 Returns of the selected portfolios

Ratio [26] or the Omega index [16]. In order to take into account the whole returns distribution, we decide to use the Omega index, given by

$$\Omega(L) = \frac{\int_L^b (1 - F(r))dr}{\int_a^L F(r)dr},$$

where $F(r)$ denotes the cumulative probability distributions of the portfolio returns, $(a,b) \in \mathbb{R}$ their domain, L a reference return, often set equal to 0. As a first approach, we select, among the 640 minimum variance portfolios, the one with the highest value of Ω. Alternatively, instead of restricting the choice to the minimum variance portfolios, we may explore all the portfolios lying on the 640 efficient frontiers plotted in the right panel of Fig. 1.

Figure 3 displays the returns of the two portfolios corresponding to these two approaches, compared to the competitors described above. The composition of the portfolios in Fig. 3 is given in Table 4. Both the portfolios tend to outperform the others, the one exploring the whole efficient frontier being the best one in the long period. The results obtained with the Sharpe Ratio are similar.

Table 4 Composition of the portfolios in Fig. 3

Selected (Maximum Ω) Markowitz minimum variance Portfolio (clustering based)

Saipem	Fiat	Terna	Intesa
0.0390	0.0290	0.9062	0.0258

Maximum Ω Portfolio (clustering based)

Saipem	Fiat	Terna	UBI
0.5384	0	0	0.4616

Markowitz minimum variance Portfolio (stocks belonging to cluster 4)

Banca Mps	Fondiaria	Generali	Intesa	Mediobanca
0	0	0.3222	0	0.2663
Mediolanum	Mediaset	Banca PM	UBI	Unicredit
0	0.3246	0	0.0869	0

Markowitz minimum variance Portfolio (all the stocks)

Atlantia	Autogrill	Banca Mps	ENEL	ENI	Fiat
0.0438	0.0399	0	0	0	0
Finmeccanica	Fondiaria	Generali	Intesa	Lottomatica	Luxottica
0.0400	0	0	0	0.0374	0.0113
Mediobanca	Mediolanum	Mediaset	Pirelli	Banca PM	Saipem
0.0765	0	0	0	0	0
SNAM	Stm	Telecom	Terna	UBI	Unicredit
0.4456	0	0 0.3055	0	0	

4 Concluding Remarks

In this chapter, we have clustered 24 stocks included in FTSE MIB index according to the association among extremely low returns. For each couple of stocks we have estimated a bivariate copula function and computed the lower tail dependence coefficient. Then, following a two-step clustering procedure integrating the use of Multi Dimensional Scaling and the k-means clustering algorithm, we have formed four groups of stocks. The obtained clustering has been used to build portfolios according to the criterion of selecting one stock from each cluster. We have shown that in an out-of-sample period characterized by financial crisis, the returns of some of these portfolios are less unfavourable with respect to the return of the minimum variance portfolio built using all the stocks or to the returns of portfolios built with randomly selected stocks. Moreover, portfolios composed of stocks belonging to the same cluster can exhibit a very bad performance. Finally, we have proposed a criterion for selecting one portfolio out of the hundreds of possible choices deriving from the rule of taking one stock from each cluster. The procedure has revealed effective on the data of the described case study. Future research could be developed in several directions. For example, the clustering procedure could be adapted to a time varying perspective, with the result of a dynamic clustering and portfolio selection. In addition, some effort should be devoted to refine the portfolio selection procedure in order to more specifically focus on the tails of the returns distribution, coherently with the proposed tail dependence clustering.

Acknowledgements This research was funded by a grant from the Italian Ministry od Education, University and Research to the PRIN Project entitled "Multivariate statistical models for risks evaluation" (2010RHAHPL_005).

References

1. Alonso, A.M., Berrendero, J.R., Hernández, A., Justel, A.: Time series clustering based on forecast densities. Computational Statistics and Data Analysis **51**, 762–776 (2006)
2. Ball, G.H., Hall, D.J.: ISODATA, A novel method of data analysis and pattern classification. Tech. Rep. NTIS No.AD 699616, Stanford Research Institute (1965)
3. Bohte, Z., Cepar, D., Kosmelj, K.: Clustering of time series. In: Barritt, M.M., Wishart, D. (eds.) COMPSTAT 1980, Proceedings in Computational statistics, pp. 587–593. Physica-Verlag, Wien (1980)
4. Caiado, J., Crato, N., Peña, D.: A periodogram-based metric for time series classification. Computational Statistics and Data Analysis **50**, 2668–2684 (2006)
5. Calinski, R.B., Harabasz, J.: A dendrite method for cluster analysis. Communications in Statistics **3**, 1–27 (1974)
6. Cherubini, U., Luciano, E., Vecchiato, W.: Copula methods in finance. Wiley, New York (2004)
7. Corduas, M., Piccolo, D.: Time series clustering and classification by the autoregressive metrics. Computational Statistics and Data Analysis **52**, 1860–1872 (2008)
8. De Luca, G., Zuccolotto, P.: A tail dependence-based dissimilarity measure for financial time series clustering. Advances in Classification and Data Analysis **5**, 323–340 (2011)
9. Dimitriadou, E., Dolničar, S., Weingessel, A.: An examination of indexes for determining the number of clusters in binary data sets. Psychometrika **67**, 137–159 (2002)

10. Edwards, A.W.F., Cavalli-Sforza, L.: A method for cluster analysis. Biometrics **21**, 362–375 (1965)

11. Friedman, H.P., Rubin, J.: On some invariant criteria for grouping data. Journal of the American Statistical Association **62**, 1159–1178 (1967)

12. Galeano, P., Peña, D.: Multivariate analysis in vector time series. Resenhas **4**, 383–404 (2000)

13. Hartigan, J.A.: Clustering algorithms. Willey, New York (1975)

14. Joe, H.: Multivariate models and dependence concepts. Chapman & Hall/CRC, New York (1997)

15. Kakizawa, Y., Shumway, R.H., Taniguchi, M.: Discrimination and clustering for multivariate time series. Journal of the American Statistical Association **93**, 328–340 (1998)

16. Keating, C., Shadwick, W.F.: A universal performance measure. Journal of performance measurement **6**, 59–84 (2002)

17. Marriot, F.H.C.: Practical problems in a method of cluster analysis. Biometrics **27**, 501–514 (1971)

18. Milligan, G.W., Cooper, M.C.: An examination of procedures for determining the number of clusters in a data set. Psychometrika **50**, 159–179 (1985)

19. Nelsen, R.: An introduction to copulas. Springer, New York (2006)

20. Otranto, E.: Clustering heteroskedastic time series by model-based procedures. Computational Statistics and Data Analysis **52**, 4685–4698 (2008)

21. Pattarin, F., Paterlini, S., Minerva, T.: Clustering financial time series: an application to mutual funds style analysis. Computational Statistics and Data Analysis **47**, 353–372 (2004)

22. Pertega Diaz, S., Vilar, J.A.: Comparing Several Parametric and Nonparametric Approaches to time series clustering: a simulation study. Journal of Classification **27**, 333–362 (2010)

23. Piccolo, D.: A distance measure for classifying ARMA models. Journal of Time Series analysis **11**, 153–164

24. Ratkowsky, D.A., Lance, G.N.: A criterion for determining the number of groups in a classification. Australian Computer Journal **10**, 115–117 (1978)

25. Scott, A.J., Symons, M.J.: Clustering methods based on likelihood ratio criteria. Biometrics **27**, 387–397 (1971)

26. Sharpe, W. F.: Mutual Fund Performance. Journal of Business **39**, 119–138 (1966)

27. Taniguchi, M., Kakizawa, Y.: Asymptotic theory of statistical inference for time series. Springer, New York (2000)

28. Vilar, J.A., Alonso A.M., Vilar, J.M.: Non-linear time series clustering based on nonparametric forecast densities. Computational Statistics and Data Analysis **54**, 2850–2865 (2010)

29. Warren Liao, T.: Clustering of time series – a survey. Pattern recognition **38**,1857–1874 (2005)

30. Weng, X., Shen, J.: Classification of multivariate time series using twodimensional singular value decomposition. Knowledge-Based Systems **21**, 535–539 (2008)

Solvency Analysis of Defined Benefit Pension Schemes

Pierre Devolder and Gabriella Piscopo

Abstract Defined Benefit Pension Schemes (DB) are affected by a lot of different risks able to put in danger the viability of the system. A solvency analysis seems therefore to be essential as in insurance but it must take into account the specificities of pension liabilities. In particular, pension funds are characterized by a long term aspect and a limited need of liquidity. In this perspective, the purpose of this paper is to combine the three major risks affecting a DB plan (market, inflation and longevity risks) and to look at their effect on the solvency of the pension fund.

1 Introduction

In Solvency II, solvency requirements for insurance companies are based on the idea that risk can be handled if a sort of buffer capital is available to deaden the impact of financial and demographic instability [6, 13]. The purpose of this paper is to propose a similar but adapted risk based approach for a defined benefit pension scheme (DB plan). Following the IAS norms, we use as funding technique the so called projected unit credit cost method (see for instance [2]) in order to compute contributions and actuarial liabilities. Two main risks are then considered in a stochastic environment: investment risk and inflation risk. In a first part of the paper, longevity risk is not considered. Afterwards, mortality before retirement is introduced in a deterministic way and a complete risk model is proposed. We do not consider in this paper longevity risk after retirement because we only valuate pension schemes pay-

P. Devolder
Institute of Statistics, Biostatistics and Actuarial Sciences (ISBA), Université Catholique de Louvain, Voie du Roman Pays 40, B-1348 Louvain-la-Neuve, Belgium
e-mail: Pierre.Devolder@uclouvain.be

G. Piscopo (✉)
Department of Economics, University of Genoa, Via Vivaldi 5, 16126 Genoa, Italy
e-mail: piscopo@economia.unige.it

M. Corazza, C. Pizzi (eds.), *Mathematical and Statistical Methods for Actuarial Sciences and Finance*, DOI 10.1007/978-3-319-02499-8_13, © Springer International Publishing Switzerland 2014

ing lump sum at retirement age. We are aware that this lump sum analysis is but a first step and we will to extend the results presented here, later for annuities. Let us remark nevertheless that many papers on pension only deals with the accumulations phase because often the decumulation phase is managed independently from the accumulation step, which justifies a specific solvency analysis for this first step as discussed here. Moreover, some countries allow paying to retires lump sums (for instance quite common and general in Belgium) and for this kind of scheme the analysis presented here seems relevant. The long term aspect of pension liability is taken systematically into account by analyzing the risks not only on a one year horizon, as in Solvency 2, but until maturity (retirement age). An ALM approach for the assets and liabilities of the scheme is proposed and various classical risk measures are applied to the surplus of the pension fund (probability of default, value at risk). The effect of the duration of the liability is clearly illustrated and is surely one of the key factors if we want to consider solvency measures for pension plans. This result is line with classical considerations about the time effect on risky investments (see for instance [3–5, 8, 11, 12].

The paper is organized as follows. In Sect. 2 we describe the general framework of the model in terms of asset and liability structure of the DB scheme. Section 3 is based on a static risk measurement and analyzes the influence of the time horizon on the probability of default at maturity. In Sect. 4, we compute a solvency level using a classical value at risk approach. In Sect. 5, we introduce the longevity risk using a deterministic approach based on the difference between the real mortality and the a priori mortality. Numerical examples are presented in Sect. 6. Finally, some conclusions are traced.

2 Model for a DB Pension Scheme

We consider a Defined Benefit pension scheme based on final salary. At time $t = 0$ an affiliate aged x is entering the scheme with an initial salary $S(0)$.

At retirement age, at time $t = T$, a lump sum will be paid, expressed as a multiple of its last wage, for instance:

$$B = NbS \tag{1}$$

where:
N is the years of service credited by the scheme;
S is the final salary;
B is benefit to pay (lump sum);
b is the coefficient (for instance 1%).

In order to compute the contributions to the pension fund, we will use the IAS norms and in particular the projected unit credit cost method as funding technique.

We need then the following assumptions:

1. a fixed discount rate: the risk free rate r;

2. a salary scale: the salary at time t denoted by $S(t)$ will follow a stochastic evolution given by:

$$dS(t) = \mu S(t)dt + \eta S(t)dz(t), \tag{2}$$

where
μ is the avarage salary increase;
η is the volatility on salary evolution;
z is standard Brownian motion.

Using a best estimate approach, the contribution for the first year of service (or the normal cost) is then given by:

$$NC_0 = bS(0)e^{(\mu-r)T} \tag{3}$$

(present value at the risk free rate of the average projected benefit at retirement age).
At time t $(t = 1, 2, \ldots, T-1)$, the normal cost will have the same form:

$$NC_t = bS(t)e^{(\mu-r)(T-t)}. \tag{4}$$

We can also introduce a loading factor β on the contribution; the normal cost becomes then:

$$NC_t = bS(t)(1+\beta)e^{(\mu-r)(T-t)}. \tag{5}$$

The actuarial liability, corresponding to the present value at time $t(t = 0, 1, , T-1)$ of the future liabilities derived from past services, is given by:

$$AL_t = (t+1)bS(t)(1+\beta)e^{(\mu-r)(T-t)}. \tag{6}$$

This value is equal to the capitalized sum of the past normal costs, when reality follows the actuarial assumptions. On the asset side we assume each contribution is invested in a Geometric Brownian motion (see for instance [9]) whose evolution is solution of:

$$dA(s) = \delta A(s)ds + \sigma A(s)dw(s), \tag{7}$$

where
δ is mean return of the investment fund;
σ is volatility of the return;
w is standard Brownian motion.

The two sources of risk (inflation and market risks) are off course correlated:

$$corr(w(t), z(t)) = \rho t.$$

By actuarial fairness, the first actuarial liability must correspond to the first normal cost (liability linked to the first year of service):

$$A(0) = NC_0.$$

Then the corresponding final asset is given by (projection between $t = 0$ and $t = T$):

$$A_0(T) = NC_0 e^{\delta - \frac{\sigma^2}{2} T + \sigma w(t)} = bS(0)(1 + \beta) e^{\mu + \delta - r - \frac{\sigma^2}{2} T + \sigma w(t)}. \tag{8}$$

More generally we could consider the risk between time t and time T by computing the future evolution till maturity of the investment of the actuarial liability AL existing at time t in the reference asset A (investment risk between time t and time T):

$$A_t(T) = AL_t e^{\delta - \frac{\sigma^2}{2}(T-t) + \sigma w(T) - w(t)} = bS(t)(1 + \beta) e^{\mu + \delta - r - \frac{\sigma^2}{2}(T-t) + \sigma w(T) - w(t)} \tag{9}$$

with initial condition $A_t(t) = AL_t$. In an ALM approach, these asset values must be compared to their respective liability counterparts.
We obtain successively:

- for the final liability corresponding to the first year contribution:

$$L_0(T) = bS(0) e^{\mu - \frac{\eta^2}{2} T + \eta z(T)}; \tag{10}$$

- for the final liability corresponding to the actuarial liability until time t:

$$L_t(T) = bS(t) e^{\mu - \frac{\eta^2}{2}(T-t) + \eta(z(T) - z(t))}. \tag{11}$$

In the next sections, we will compare the final assets given by Eq. (9) with the final liabilities given by Eq. (11).

3 Probability of Default

A first interesting question is to look at the probability of default at maturity without any extra resource (i.e. the risk to have not enough assets at maturity to pay the required pension benefit). In particular we can consider this probability as a function of the residual time $T - t$. This probability computed at time t ($t = 0, 1, .., T - 1$) is given by:

$$\varphi(t, T) = P(A_t(T) < L_t(T)) = P(Y(t, T) < M) \tag{12}$$

where

$$Y(t, T) = \sigma(w(T) - w(t)) - \eta(z(T) - z(t)) = N(0, \bar{\sigma}^2(T - t))$$
$$\bar{\sigma}^2 = \sigma^2 + \eta^2 - 2\rho\sigma\eta \tag{13}$$
$$M = (r - \delta + \frac{\sigma^2}{2} - \frac{\eta^2}{2})(T - t) - ln(1 + \beta).$$

So finally the probability of default at maturity depends on the residual time and is given by:

$$\varphi(t,T) = \Phi(a(T-t))$$

$$a(s) = \frac{(r - \delta + \frac{\sigma^2}{2} - \frac{\eta^2}{2})\sqrt{s} - \frac{ln(1-\beta)}{\sqrt{s}}}{\bar{\sigma}} \tag{14}$$

with $\Phi = $ *distribution function* $N(0,1)$.

Let us remark that other rational choices could be used for this multi period required safety level.

4 Value at Risk Approach

In order to control this probability of default, we could as in Solvency 2 introduce a solvency level based on a value at risk approach [1,7,10].

We will use the following notations:

SC is the solvency capital using a value at risk methodology;
VaR is the Value a Risk;
$\alpha(N)$ is a chosen safety level for a horizon of N years (for instance 99.5% on one year in Solvency 2).

For this safety level we can choice the following value based on yearly independent default probabilities (probabilities of default of $(1-\alpha)$ independently each year):

$$\alpha_N = \alpha^N. \tag{15}$$

We will assume that the solvency capital is invested in the reference investment fund; so we can define this solvency capital $SC(t,T)$ at time t $(t = 0,1,..,T-1)$ for the investment and inflation risks between time t and time T as solution of:

$$P\{A_t(T) + SC(t,T)\frac{A_t(T)}{A_t(t)} < L_t(T)\} = 1 - \alpha_{T-t}. \tag{16}$$

Using (9) and (11), this condition becomes:

$$P\{tbS(t)(1+\beta) + SC(t,T)e^{\delta - \frac{\sigma^2}{2}(T-t) + \sigma w(T) - w(t)} < tbS(t)e^{\mu - \frac{\eta^2}{2}(T-t) + \eta(z(T) - z(t))}\}$$
$$= 1 - \alpha_{T-t}.$$

After direct computation, we obtain the following value for the solvency capital:

$$SC(t,T) = tbS(t)\{e^{(\mu - \delta)(T-t) + z_{\alpha(T-t)}\sigma\sqrt{T-t} + \frac{(\sigma^2 - \eta^2)(T-t)}{2}} - e^{(\mu - r)(T-t)}(1+\beta)\} \tag{17}$$

where $z_\beta = \beta$ is the quantile of the normal distribution on such that $\Phi(z_\beta) = \beta$. We can express the solvency capital as a percentage of the actuarial liability AL given by Eq. (6) (solvency level in percent):

$$SC^{\%}(t,T) = \frac{SC(t,T)}{AL_t} = \frac{1}{1+\beta}e^{-(\delta-r)(T-t)+z_{\alpha(T-t)}\bar{\sigma}\sqrt{T-t}+\frac{(\sigma^2-\eta^2)(T-t)}{2}} - 1. \quad (18)$$

We can observe that this relative level does not depend on the average salary increase.

In particular if we look at a one year risk (as in Solvency 2), we get:

$$SC^{\%}(0,1) = \frac{SC(0,1)}{AL_0} = \frac{1}{1+\beta}e^{-(\delta-r)+z_{\alpha}\bar{\sigma}+\frac{(\sigma^2-\eta^2)}{2}} - 1.$$

5 Introduction of the Longevity Risk

Until now, two risk factors have been considered: investment and inflation. However, another risk source has to be introduced to outline a more complete risk model for pension plans: longevity. In this paper, since we are dealing with the case of payment of a lump sum at retirement and not a pension annuity, the pension provider has to evaluate just the probability that the affiliate dies before retirement, ignoring his remaining lifetime afterwards. In order to take into account this case, the formulae developed in the previous sections have to be modified as follows. First of all, normal cost and actuarial liabilities becomes:

$$NC_t = bS(t)(1+\beta)e^{(\mu-r)(T-t)}{}_{T-t}p_{x+t} \quad (19)$$

$$AL_t = (t+1)bS(t)(1+\beta)e^{(\mu-r)(T-t)}{}_{T-t}p_{x+t} \quad (20)$$

where ${}_{T-t}p_{x+t}$ is the probability at age x to survive until T, calculated according to an a priori valuation mortality table, used by pension provider to estimate the actuarial liabilities. Longevity risk arises when the real ex-post survival probabilities differs from the a priori ones.

Let ${}_{T-t}\tilde{p}_{x+t}$ be the real survival probability; the liability at maturity seen from time t becomes:

$$L_t(T) = tbS(t)e^{\mu-\frac{\eta^2}{2}(T-t)+\eta(z(T)-z(t))}{}_{T-t}\tilde{p}_{x+t}. \quad (21)$$

Therefore, the probability of default computed at time t ($t = 0,1,..,T-1$) is given by:

$$\varphi(t,T) = P(A_t(T) < L_t(T)) = P(Y(t,T) < M)$$
$$M = (r-\delta+\frac{\sigma^2}{2}-\frac{\eta^2}{2})(T-t) - ln(1+\beta) + ln(\frac{{}_{T-t}\tilde{p}_{x+t}}{{}_{T-t}p_{x+t}}) \quad (22)$$

and $Y/t,T)$ is computed according to Eq. (13).

So finally as in the previous section the probability of default at maturity depends mainly on the residual time and is given by:

$$\varphi(t,T) = \Phi(\frac{(r-\delta+\frac{\sigma^2}{2}-\frac{\eta^2}{2})(T-t) - ln(1+\beta) + ln(\frac{{}_{T-t}\tilde{p}_{x+t}}{{}_{T-t}p_{x+t}})}{\bar{\sigma}\sqrt{T-t}}). \quad (23)$$

After direct computation, we obtain the following value for the solvency capital:

$$SC^\%(0,1) = \frac{SC(0,1)}{AL_0} = \frac{1}{1+\beta} e^{-(\delta-r)+z_\alpha\bar\sigma + \frac{(\sigma^2-\eta^2)}{2}} \frac{T-t\tilde{p}_{x+t}}{T-tp_{x+t}} - 1. \qquad (24)$$

The general framework traced can be specified through the introduction of a given mortality model, as in the case of the well known Gompertz life table, which is based on the assumption of exponential mortality intensity. For sake of example, valuation and real life table can be derived as two different Gompertz tables:

$$T-tp_{x+t} = exp\left(-\int_t^T \mu_{x+s}ds\right) = exp\left(-\mu_{x+s}\frac{e^{\gamma(T-t)}-1}{\gamma}\right)$$

$$T-t\tilde{p}_{x+t} = exp\left(-\int_t^T \tilde{\mu}_{x+s}ds\right) = exp\left(-\tilde{\mu}_{x+s}\frac{e^{\kappa(T-t)}-1}{\kappa}\right).$$

Under this assumption, equations (23) and (24) are transformed into (25) and (26):

$$\varphi(t,T) = \Phi\left(\frac{(r-\delta+\frac{\sigma^2}{2}-\frac{\eta^2}{2})(T-t)-ln(1+\beta)+\mu_{x+s}\frac{e^{\gamma(T-t)}-1}{\gamma}-\tilde{\mu}_{x+s}\frac{e^{\kappa(T-t)}-1}{\kappa}}{\bar\sigma\sqrt{T-t}}\right) \qquad (25)$$

$$SC^\%(0,1) = \frac{SC(0,1)}{AL_0} = \qquad (26)$$
$$\frac{1}{1+\beta} e^{-(\delta-r)(T-t)+z_\alpha\bar\sigma\sqrt{T-t}+\frac{(\sigma^2-\eta^2)(T-t)}{2}} e^{\mu_{x+s}\frac{e^{\gamma(T-t)}-1}{\gamma}-\tilde{\mu}_{x+s}\frac{e^{\kappa(T-t)}-1}{\kappa}} - 1.$$

6 Numerical Example

In this section, we carry out an applicative analysis, in which we compute the described quantities under given scenarios, in order to highlight how the solvency position of a DB changes during the time. To this aim, we have set the following financial parameters:

risk free rate	$r = 2\%$
mean return of the fund	$\delta = 6\%$
volatility of the fund	$\sigma = 10\%$
average increase of salary	$\mu = 5\%$
volatility of the salary	$\eta = 5\%$
correlation	$\rho = 50\%$
no safety loading is considered	$\beta = 0$

With respect to the demographic assumptions, we have assumed that the affiliate subscribes the pension plan at 35 years and the retirement age is 65. The valuation table used is the Italian male mortality table of the year 2006 downloaded from the Human Mortality Database, from which the values are derived. We have implemented the model without considering mortality and then with the introduction of mortality risk through two different ex post mortality tables. In the first case (a), we have considered that the force of mortality in the real table is smaller than the a priori mortality to take into account the effects of survival improvements. In the second case (b) we have considered the opposite situation. Under the scenario a) and b) the mortality tables are derived modifying the a priori mortality intensity according to the following assumptions:

a) $\tilde{\mu}'_{x+t} = \mu_{x+t}(1 - \Delta t)$;
b) $\tilde{\mu}''_{x+t} = \mu_{x+t}(1 + \Delta t)$ with $\Delta = 2\%$.

Both the valuation table and the real tables are fitted to the Gompertz law, producing the following parameters:

$\gamma = 0.004768$ for the valuation table;
$\kappa' = 0.003038$ for the real table under the assumption a);
$\kappa'' = 0.005512$ for the real table under the assumption b).

Figure 1 shows the probability of default as a function of the residual time $T - t$ under three different hypothesis: no mortality (Eqs. (12)–(14)), survival improvements (a) and survival decreasing (b) (Eq. (23)):

    We can see clearly a time effect: for short residual time to maturity this probability is quite high but it decreases rapidly for long residual time. Moreover, the longevity risk overdraws the time effect: in the case of survival improvements the probability of default is higher than that calculated ignoring the mortality and decreases more rapidly during the time. On the contrary, in the case of survival decreasing the probability of default is lower than that calculated ignoring the mortality and decreases less rapidly during the time.

    In a second step of our application, we have computed the solvency capital under the same assumptions drawn so far. In addition, we have set:

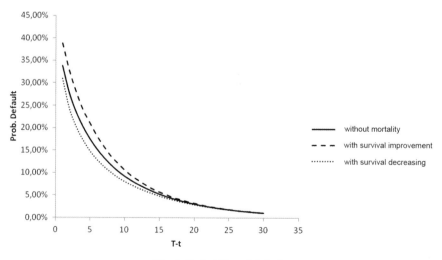

Fig. 1 Probability of default

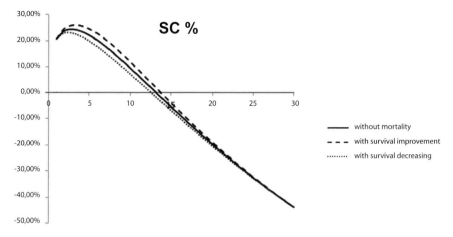

Fig. 2 Solvency capital

safety level on one year: $\alpha = 99.5\%$;
safety level on N years: $\alpha_N = \alpha^N$.

Figure 2 shows then the evolution of the solvency level in percent as a function of the residual time $T - t$. Negative values for the SCR correspond to cases where no additional solvency is needed.

We can also observe as in Fig. 1 a time effect. As expected, in face of an increase of the probability of default in the case of survival improvements the solvency capital increases too; on the contrary, in the case of survival decreasing the actuarial liabilities decreases and a lower solvency capital is needed.

7 Final Remarks

The solvency analysis is an important issue in the risk valuation of a pension plan.. A similar approach as in Solvency 2 turns out to be necessary also for pension funds, but must take into account a long term aspect and a limited need for liquidity.

In this paper we have highlighted the importance to extend solvency evaluation to pension funds through an integrated analysis of the risks that have influence on pension assets and liabilities. In this context, the determination of the solvency capital has been influenced by the way to measure the underlying risks and to integrate time in the process; this time aspect is particularly important for long term liabilities.

Further works are planned to extend the framework outlined in this paper to more general cases, as the generalization to stochastic longevity models, the introduction of pension annuity rather than a lump sum paid at retirement and the extension of longevity risk to the decumulation phase.

References

1. Artzner, P., Delbaen, F., Eber, J.M., Heath, D.: Coherent Measures of Risk. Mathematical Finance **9**, 203–228 (1999)
2. Berin, B.N.: The fundamentals of pension mathematics. Society of Actuaries. Schaumburg, Illinois (1989)
3. Bodie, Z.: On the Risks of Stocks in the Long Run. Financial Analysts Journal **51**(3), 68–76 (1995)
4. Bodie, Z., Merton, R.C., Samuelson, W.F.: Labor Supply Flexibility and Portfolio Choice in a Lifecycle Model. Journal of Economic Dynamics and Control **16**(3), 427–449 (1992)
5. Campbell, J., Viceira, L.: Strategic Asset Allocation Portfolio Choice for Long term Investors. Oxford University Press (2002)
6. Grosen, A., Jorgensen, P.: Life Insurance Liabilities at Market Value: An Analysis of Insolvency Risk, Bonus Policy and Regulatory Intervention Rules in a Barrier Option framework. Journal of Risk and Insurance **69**(1), 63–91 (2002)
7. Hardy, M., Wirch, J.L.: The Iterated CTE: a Dynamic Risk Measure. North American Actuarial Journal **8**, 62–75 (2004)
8. Lee, W.: Diversification and Time, Do Investment Horizons matter? Journal of Portfolio Management **4**, 60–69 (1990)
9. Merton, R.: Continuous Time Finance. Wiley (1992)
10. Pflug, G., Romisch, W.: Modeling, Measuring and Managing Risk. World Scientific (2007)
11. Samuelson, P.: The Long Term Case for Equities. Journal of Portfolio Management **21**(1), 15–24 (1994)
12. Thorley, S.: The Time Diversification Controversy. Financial Analyst Journal **51**(3), 18–22 (1995)
13. Wirch, J.L., Hardy, M.R.: A synthesis of risk measures for capital adequacy. Insurance: Mathematics and Economics **25**, 337–347 (1999)

Stochastic Actuarial Valuations in Double-Indexed Pension Annuity Assessment

Emilia Di Lorenzo, Albina Orlando and Marilena Sibillo

Abstract The paper deals with the performance analysis of a portfolio of partici-pating survival-indexed annuities within a riskiness context, set out by the adverse deviations of the demographic and financial bases. The Authors deepen the inter-actions between the risk due the random fluctuations of the dynamic of the capital returns and the risk due to the systematic random fluctuations of the lifetime evolu-tionary trend.

1 Introduction

Solvency and financial health are two basic aims an insurance company should pur-sue within Solvency II regulatory framework. The guidelines continuously proposed by the involved Institutions in refined and clarifier versions stabilize the relevance of the capital amount to be allocated for achieving safe financial positions. Prod-uct choices, in particular, together with investment decisions, can be considered a good flywheel to harness, in a context in which several stochastic variables impact on the financial values. Our interest in this paper concerns the life annuity section in the particular case of participating life annuity contracts, i.e. contracts with interest

E. Di Lorenzo
Department of Economic and Statistical Sciences, via Cinthia Monte S. Angelo, University of Napoli "Federico II", 80126 Naples, Italy
e-mail: diloremi@unina.it

A. Orlando
CNR, Istituto per le Applicazioni del Calcolo Mauro Picone, via P. Castellino 111, 80126 Naples, Italy
e-mail: a.orlando@iac.cnr.it

M. Sibillo (✉)
Department of Economics and Statistics, Campus Universitario, University of Salerno, 84084 Fisciano (SA), Italy
e-mail: msibillo@unisa.it

M. Corazza, C. Pizzi (eds.), *Mathematical and Statistical Methods for Actuarial Sciences and Finance*, DOI 10.1007/978-3-319-02499-8_14, © Springer International Publishing Switzer-land 2014

rate guarantee with a contextual right of participation to the firm profit. Nowadays insurance/pension sector is called to face significant strategic challenges, within competitive scenarios of increasing complexity by virtue of the current socio-economic context. Regulatory structures, criticalities of welfare systems, growing demand of products aimed to safeguard for those who ask for investment instruments and possibilities fitting an elderly age, complexity of the new macroeconomic and demographic scenarios, constitute the most relevant risk drivers, but, at the same time, a prospective driving force for the economic growth. With regard to annuity products, we will focus on the impact of financial variables (return on investment of premiums/contributions and reserves) and the systematic component of demographic risk arising from overall betterment of the survival trend. Aiming to a reasonable balancing between insurer's profitability goals and marketing viewpoint, we focus on products that, on the one hand, allow the insurer to transfer risk and, on the other hand, provide attractive features for policyholders. Several authors payd attention on annuity products variously "survival indexed", this allowing longevity risk sharing (see [6,14,15]); in particular we focus on the proposal put forward in [6] and later extended in [7]), where the contract involves a profit sharing system for annuitants. With reference to the afore mentioned policies, we will consider a life annuity contract in which benefits are linked to the financial and demographic dynamics. The paper investigates, in particular, the portfolio performance by analyzing the trend of balance sheet indexes, which capture the interactions between the financial risk drivers – resulting from the random fluctuations of the return on the investments made by the insurer – and the demographic risk drivers- originated, from a "micro" point of view, with mortality accidental fluctuations and, from a "macro" point of view, with the systematic impact of longevity risk.

2 An Actuarial Approach in a Variable Annuity Profit Analysis

In the life annuity field, the high volatility of the financial markets, together with the bettering in the human expected lifetime, produces the need of offering new flexible products able to match the "pure" financial ones, contextually guaranteeing a correct covering action to not perverting the very basic characteristic of an insurance product. The new contractual architectures have to avoid the loss of competitive factors, safeguarding the company solvability and saving its profit aims. For solving this elaborate mosaic and inspecting the product foreseeable evolution in time, a good tool can be the income profitability analysis. Our paper is aimed to the inspection inside the actuarial income trend for addressing management choices, in particular in the specific architecture of complex products as the one we are deepening here. This study needs a careful reflection on the risk drivers' assumptions in light of their stochastic nature. A life annuity indexed on both the main risk drivers, fundamental in the forward perspective, proposes some peculiar parameter assumptions strongly affecting the income trend. Among the others, we recall the participation quota, the contract cap and floor for the indexes, the threshold for the period financial result depending on the administrative expenses assigned to each contract, under which the insurer will distribute the bonus. The stochastic hypotheses are particularly crucial in this assessment, considering moreover the very long contract durations characterizing these kinds of life contracts. Nowadays the significance of participating life annuities is growing, often being inserted in a voluntary life annuity demand as supplementary pension or pension fund: the demand and the supply are increasing especially in economic scenarios like in many western countries, affected by the financial crisis and for this reason engaged in social security system restrictions. Among the varied different profitability measures, in the paper we will consider the Actuarial Return on Equity (AROE). As the ROE, it is referred to one year time interval, often coinciding with the balance sheet dates. The AROE has the form of a ratio of profit to equity as in the following formula (3), this last being referred not to the Company surplus but to that specific part of it supporting the product or the line of products under consideration. Using AROE in its meaning of risk measure, basic is how the equity at the denominator is valued (cf. [8]). As well as profits at the numerator, the choice of the quantity at the denominator implies different interpretations and meanings of the ratio. We will put at the denominator of the AROE ratio the sum built up by means of the premiums collected in the considered business line, in this specific case the capital arising from the premiums paid by the insureds and invested by the insurer, less the benefit paid in the period; the AROE value at time t will have at the denominator this sum valued at the beginning of the t-th year in a forward perspective, in the sense that the insurer information flow is available at time 0. The profits at the numerator and the actuarial "equity" at the denominator are both results of stochastic valuations. Inside them we can recognize prospective and retrospective financial operations, stochastic for the financial assumptions on them. These valuations concern amounts that are stochastic also in the number of payments, being random the insured's survival at the payment times.

3 A Class of Variable Annuities

Our interest concerns the life annuity section in the particular case of participating life annuity contracts, i.e. contracts with interest rate guarantee with a contextual right of participation to the firm profit. The usual architecture of these contracts is established on the basis of the financial risk control, being the financial context in very long time periods the pre-eminent risk driver the insurer wants to focus. Nevertheless we have to notice that the long term risks connected to this kind of obligations are not completely drained with the financial risk control, being very working the systematic demographic betterment in mortality. Now we explore the construction of life annuity with profit participation appropriately survival-indexed, obtaining a product double indexed on both the financial and demographic risk driver.

The survival-linking procedure is fulfilled (cf. [6]) multiplying each installment b_t by the scale factor

$$SF_{x,t} = \frac{{}_t p_x^{proj}}{{}_t p_x^{obs}}, \qquad (1)$$

where ${}_t p_x^{proj}$ is the survival probability for an insured aged x to reach the age $x+t$, inferable from an opportune projected survival table and used as the technical base, whilst ${}_t p_x^{obs}$ is the analogous survival probability the insurer observes year by year.

Always following ([6]), minimum and maximum thresholds for the $SF_{x,t}$'s can be set in order to control too much marked projection degrees, hence:

$$\tilde{b}_t = b_t max\{min\{SF_{x,t}, SF_{max}\}, SF_{min}\} \qquad (2)$$

where b_t is the basic installment, SF_{max} and SF_{min} are properly determined evaluating cap and floor within a suitable marketing viewpoint. On the other side the profit sharing process, involving the differences between income, capital gains and losses (cf. [3, 7]), is achieved by means of an embedded option (cf. [4]) such that, if the period financial result R_{t+1} at the end of the interval $[t, t+1]$, net of the administrative expenses (γ) is positive, an additional bonus equal to a percentage α of $(R_{t+1} - \gamma)$ is immediately paid to the annuitants or added to the future installments.

In the following we develop the performance analysis of a portfolio of participating survival-indexed annuities within a riskiness context set out by the adverse deviations of the demographic and financial bases, deepening the interactions between the risk due the random fluctuations of the dynamic of the capital returns and the risk due to the systematic random fluctuations of the lifetime evolutionary trend. In particular we focus on deferred life annuities (say T the deferment period), premium payment until the time τ, annual installments at the beginning of each year, prospective additional bonus immediately paid.

We consider the ratio (say Actuarial Return on Equity, AROE therein after) of profit to the surplus the business provides, that is:

$$AROE = \frac{R_{t+1} - \alpha(R_{t+1} - \gamma)^+}{\sum_{j=0}^{\infty} X_j v^{sign(j-t)}(t, j)} \qquad (3)$$

with

$$X_j = (P_{j+1} \mathbf{1}_{(j \leq K_x < T)} - \tilde{b}_j) sign(t - j) \qquad (4)$$

and

$$R_{t+1} = (V_t + P_{t+1} \mathbf{1}_{(t+1 < \tau)}) v^{-1}(t, t+1) - (\tilde{b}_{t+1} + V_{t+1}) \mathbf{1}_{(K(x) > t+1)} \qquad (5)$$

in which V_t and P_t are respectively the mathematical reserve and the premium at time t, $\mathbf{1}_h$ is the indicator function assuming value 1 if the event h happens, 0 otherwise, $v(t,s)$ is the present value in t of 1 due in s. Within a strategic decision context, knowing the evolution of the AROE stochastic process suggests, for instance, how to address investment policies, how to choose the participating level, how to set boundaries to the indexing system. In the next section we show the explanatory potential of the AROE process by means of a quantile analysis within a managerial perspective.

4 Numerical Illustrations

In this section we will present our assumptions for implementing the AROE stochastic simulation procedure applied to the double indexed life annuity we present in the paper. From the financial point of view, as we pointed out previously, in the valuation process two different directions, prospective and retrospective, are kept on and this circumstance brings towards precise choices of two different financial processes. The distinct description of the two interest rate dynamics comes from the different perception of the randomness if thinking in a prospective or in a retrospective point of view. It is coherent to assume the uncertainty markedly linked to the market volatility in the retrospective operations, partially less controllable by the insurer than in the accumulation process involved in the prospective operations, in which the prudential character of the insurer's investments is much more intensive. For these reasons we will describe the evolution in time of the interest rate structure in the discounting valuations by means of the Hull and White process (HW from here on), arbitrage free and made consistently with the current financial market structure within a fair valuation perspective.

It is governed by the following stochastic differential equation:

$$dr = (\theta(t) - \alpha r)dt + \sigma dz, \tag{6}$$

where $\theta(t)$ is chosen to match the current term structure of interest rates and z is a Wiener process (for more details cf. [5]). We have implemented the HW model using the efficient methods proposed in [10] approximating the term structure by a tree-building simulation procedure [11, 12], more suitable of the closed forms, even if available, in cases of complex payoffs (cf. [5]). On the other hand the accumulation process can well be described by the Vasicek mean reversion model. As known, the basic assumption is that the instantaneous spot rate follows the process with constant coefficients under the statistical measure used for historical estimation described by the SDE:

$$dr = \alpha(\mu - r)dt + \sigma dW_t \tag{7}$$

with α, μ and σ positive constants and W_t a standard Wiener process.

Under the risk neutral measure used for valuation and pricing we assume the Vasicek process parametrized using one more parameter in the drift modeling, the market price of risk (cf. [2]). To calibrate the Vasicek and the HW models we refer to the monthly yield over the period January 2002–January 2012, on a basket of Treasury Italian bonds listed on the electronic bond and government market and having a residual greater than one year.

The demographic survival system is designed inside the contractual structure by means of two different assumptions, one for the projected probabilities $_t p_x^{proj}$ at the numerator of the index $SF_{x,t}$ in (1), and the other for the observed survival probabilities $_t p_x^{obs}$ at the denominator. In the following numerical application, the projected survival probabilities are got as in [5] in the case of a male population while the observed ones are those in the survival table SIM2006.

Being the AROE a stochastic variable, we resort to a simulation procedure obtaining the empirical distribution of AROE values for each t. The simulation procedure was implemented obtaining 10000 AROE values for each t: then the expected values and the quantiles $q(0.99)$ and $q(0.95)$ were obtained. Figure 1 shows, in the first subplot, the maximum and minimum values obtained for each t by the simulation procedure and the expected value calculated as a mean of the 10000 simulated values. In the second subplot of Fig. 1, the quantile values for each t and for each confidence level considered are shown together with the expected values. The results refere to a participation quota (see formula (3)) $\alpha = 20\%$.

We observe a general decreasing behavior of the AROE index with time, more marked in the first half of the policy duration, by which the contract performance can be measured. Moreover a considerable lowering of the AROE values happens when the participating quota moves from 20% to 80% (see Tables 1 and 2).

In particular, within a managerial perspective, (cf. also [4]) the quantile analysis allow to assess a sort of break-even scenario (depending on the investment policy, the choice of the participating level and in general all the strategic variables) which presents the "worst" AROE acceptable in compliance with insurer's opinion expressed by means of a certain confidence level.

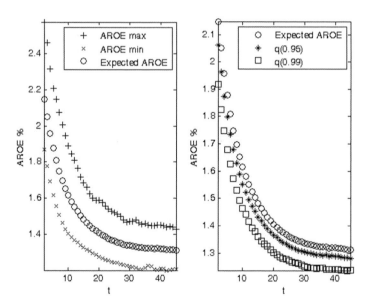

Fig. 1 Simulation procedure results: $\alpha = 20\%$

Table 1 Expected AROE and quantiles $\alpha = 0.2$

t	$E[AROE]\%$	q(.95)	q(0.99)
1	2.1526	1.9265	1.8029
5	1.8141	1.6448	1.5336
10	1.5852	1.4530	1.3737
20	1.3984	1.3008	1.2466
30	1.3362	1.2564	1.2082
40	1.3206	1.2389	1.1957
45	1.3138	1.2332	1.1931

Table 2 Expected AROE and quantiles $\alpha = 0.8$

t	$E[AROE]$	q(.95)	q(0.99)
1	0.6242	0.5979	0.5119
5	0.5066	0.4853	0.4246
10	0.4277	0.4112	0.3672
20	0.3623	0.3515	0.3188
30	0.3394	0.3312	0.3039
40	0.3324	0.3246	0.2991
45	0.3299	0.3224	0.2973

References

1. AGE Platform Europe in partnership with the Committee of the Regions and the European Commission: How to promote active ageing in Europe. EU support to local and regional actors (2011)
2. Brigo, D., Mercurio, F.: Interest Rate Models – Theory and Practice, 2nd ed. Series: Springer Finance. Berlin, Heidelberg, New York (2006)
3. Cocozza, R., De Simone, A., Di Lorenzo, E., Sibillo, M.: Participating policies: risk and value drivers in a financial management perspective. Proceedings of the 14th Applied Stochastic Models and Data Analysis Conference, pp. 41–48. Rome (June 7–10, 2011)
4. D'Amato, V., Di Lorenzo, E., Orlando, A., Russolillo, M., Sibillo, M.: Profit Participation Annuities: A Business Profitability Analysis within a Demographic Risk Sensitive Approach. Seventh International Longevity Risk and Capital Markets Solutions Conference. Goethe University Frankfurt, Germany (September 8–10, 2011). http://www.longevity-risk.org/programme.html
5. D'Amato, V., Di Lorenzo, E., Haberman, S., Russolillo, M., Sibillo, M.: The Poisson log-bilinear Lee Carter model: Applications of efficient bootstrap methods to annuity analyses. North American Actuarial Journal **15**(2), 315–333 (2011)
6. Denuit, M., Haberman, S., Renshaw, A.: Longevity-indexed life annuities. North American Actuarial Journal **4**(1), 97–111 (2011)
7. Di Lorenzo, E., Orlando, A., Sibillo, M.: Survival betterment as competitive leverage in insurance sector: profitability analysis for a class of participating variable annuities. Working paper (2012)

8. Easton, A.E., Harris, J.F.: Actuarial Aspects of Individual and Annuity Contracts, second edition. Actex Publication Inc (2007)
9. Green Paper Confronting demographic change: a new solidarity between the generations. http://share-dev.mpisoc.mpg.de/data-access-documentation.html
10. Kijima, M., Nagayama, I.: Efficient numerical procedures for the Hull-White extended Vasicek model. Journal of Financial Engineering **3**(4), 275–292 (September/December 1994)
11. Hull, J., White, A.: Numerical procedures for implementing term structure models I: Single-Factor Models. Journal of Derivatives **2**(1), 7–16 (Fall 1994)
12. Hull, J., White, A.: Numerical procedures for implementing term structure models II: Two-Factor Models. Degli-Esposti/chapter Journal of Derivatives **2**(2), 37–48 (Winter 1994)
13. Morgan, E.: The role and structure of profit participation products (2010). http://it.milliman.com/Pubblicazioni/articoli/.../role-structure-profit-participation.pdf
14. Piggott, J., Valdez, E., Detzel, B.: The Simple Analytics of a Pooled Annuity Fund Journal of Risk and Insurance **72**, 497–520 (2005)
15. Richter, A., Weber, F.: Mortality-indexed annuities. Managing longevity risk via product design North American Actuarial Journal **4**(1), 212–236 (2011)
16. Rosti, L.: Il Costo Demografico. In: Bombelli, C., Finzi, E. (eds.) "Over 45". Guerini e Associati, Milano (2006)
17. New Swiss Re sigma study "World insurance in 2010" reveals growth in global premium volume and capital (July 2011)
18. http://europa.eu/legislation_summaries/about/index_it.htm

Testing for Normality When the Sampled Distribution Is Extended Skew-Normal

Cinzia Franceschini and Nicola Loperfido

Abstract The extended skew-normal (ESN) distribution includes the normal one as a special case. Unfortunately, its information matrix is singular under normality, thus preventing application of standard likelihood-based methods for testing the null hypothesis of normality. The paper shows that for univariate ESN distributions the sample skewness provides the locally most powerful test for normality among the location and scale invariant ones. The generalization to multivariate ESN distributions considers projections of the data onto the direction corresponding to maximal skewness. Related computational problems simplifies for the multivariate ESN distribution, where the direction maximizing skewness is shown to have a simple parametric interpretation.

1 Introduction

The multivariate extended skew-normal distribution (ESN, hereafter) was introduced by [12] and later independently rediscovered by [4, 7, 10]. Its pdf is

$$f(z;\xi,\Omega,\eta,\tau) = \frac{\phi_p(z;\xi,\Omega)}{\Phi(\tau)} \cdot \Phi\left\{\tau\sqrt{1+\eta^T\Omega^{-1}\eta} + \eta^T(z-\xi)\right\}, \quad (1)$$

where $z,\xi,\eta \in \Re^d, \tau \in \Re$ $\Omega \in \Re^d \times \Re^d$, $\Omega > O$, $\Phi(\cdot)$ is the cdf of a standard normal distribution and $\phi_p(z;\xi,\Omega)$ is the pdf of a d-dimensional normal distribution with mean ξ and variance Ω. The distribution of z is said to be ESN with location

C. Franceschini
Facoltà di Giurisprudenza, Università degli Studi di Urbino "Carlo Bo", Via Matteotti 1, 61029 Urbino (PU), Italy
e-mail: cinziafranceschini@msn.com

N. Loperfido (✉)
Dipartimento di Economia, Società e Politica, Università degli Studi di Urbino "Carlo Bo", Via Saffi 42, 61029 Urbino (PU), Italy
e-mail: nicola.loperfido@uniurb.it

M. Corazza, C. Pizzi (eds.), *Mathematical and Statistical Methods for Actuarial Sciences and Finance*, DOI 10.1007/978-3-319-02499-8_15, © Springer International Publishing Switzerland 2014

parameter ξ, scale parameter Ω, nonnormality parameter η and truncation parameter τ, that is $z \sim ESN(\xi, \Omega, \eta, \tau)$. This parameterization, with minor modifications, was used in [13] for inferential purposes and this paper gives further motivations for its use. Other parameterizations appear in [1, 2, 7, 10, 19]. The multivariate normal and multivariate skew-normal [11] distributions are ESN with $\tau = 0$ and $\eta = 0_d$, respectively. [8, 9] review theoretical properties of the ESN.

The ESN appears in several areas of statistical theory: Bayesian statistics [27], regression analysis [14] and graphical models [13]. It also appears in several areas of applied statistics: environmetrics, medical statistics, econometrics and finance. Environmental applications of the ESN include modelling data from monitoring stations aimed at finding large values of pollutants [19] and uncertainty analysis related to the economics of climate change control [30]. In medical statistics, the ESN has been used as a predictive distribution for cardiopulmonar functionality [15] and for visual acuity [20]. Financial applications mainly deal with portfolio selection [2, 4] and the market model, which relates asset returns to the return on the market portfolio [1]. In econometrics, the ESN is known for its connection with bias modelling in the Heckman's model [16] and with stochastic frontier analysis [22].

Unfortunately, the information matrix of the ESN is singular when $\eta = 0_d$, i.e. when it is a normal distribution. This prevents straightforward application of standard likelihood-based methods to test the null hypothesis of normality. The problem is well-known for the skew-normal case and has been successfully dealt with via the centred parametrization, which could be useful in the ESN case, too, but satisfactory theoretical results for this distribution appear to be difficult to obtain [6]. Problems with the information matrix of the ESN are made worse by the truncation parameter τ, which indexes the distribution only when it is not normal. As a direct consequence of the above arguments the rank of the information matrix is at least two less than the full, thus preventing application of results in [28].

An alternative approach could be based on the sample skewness, since it provides the locally most powerful test for normality among the location and scale invariant ones, when the underlying distribution is assumed to be univariate skew-normal [29]. [23] proposed a multivariate generalization based on data projections onto directions maximizing skewness. The test has been criticized for involving prohibitive computational work [5] and because calculation of the corresponding population values seemed impossible [25]. Both problems vanish in the SN case [21]. In the first place, the direction maximizing skewness has a simple parametric interpretation, and hence can be estimated either by maximum likelihood or method of moments. In the second place, the population values of skewness indices proposed by [23, 24] coincide and have a simple analytical form.

The paper generalizes results in [21, 29] to univariate and multivariate ESN distributions, respectively. It is structured as follows. Section 2 shows that sample skewness provides the locally most powerful test for normality among the location and scale invariant ones, under the assumption of extended skew-normality. Section 3 shows that the nonnormality parameter η identifies the linear functions of ESN random vector with maximal skewness, and discusses the related inferential implications. Sections 4 and 5 contain a numerical example and some concluding remarks, respectively.

2 The Univariate Case

Let $X_1,\ldots,X_n$ be a random sample from an univariate extended skew-normal distribution with nonnegative skewness:

$$f(z;\xi,\omega,\eta,\tau) = \frac{1}{\omega}\phi\left(\frac{z-\xi}{\omega}\right)\cdot\frac{\Phi\left\{\tau\sqrt{1+\eta^2/\omega^2}+\eta(z-\xi)\right\}}{\Phi(\tau)}, \qquad (2)$$

where $z,\xi,\tau \in \mathfrak{R}$, $\omega > 0$, $\eta \geq 0$ and ϕ denotes the pdf of a standard normal distribution. Moreover, let $\overline{X}$, S^2 and G_1 be the sample mean, the sample variance and the sample skewness, respectively:

$$\overline{X} = \frac{1}{n}\sum_{i=1}^{N}X_i, \; S^2 = \frac{1}{n-1}\sum_{i=1}^{N}(X_i-\overline{X})^2, \; G_1 = \frac{1}{n}\sum_{i=1}^{N}\left(\frac{X_i-\overline{X}}{S}\right)^3. \qquad (3)$$

Interest lies in the most powerful test of given size for $H_0 : \eta = 0$ against $H_0 : 0 < \eta < \varepsilon$, where ε is a small enough positive constant, based on statistics which do not depend on location and scale changes, that is functions of

$$\frac{X_1-\overline{X}}{S},\ldots,\frac{X_n-\overline{X}}{S}. \qquad (4)$$

The following theorem shows that such tests are characterized by rejection regions of the form $R = \{(X_1,\ldots,X_n) : G_1 > c\}$, where c is a suitably chosen constant.

Theorem 1 *Let $X_1,\ldots,X_n$ be a random sample from an univariate extended skew-normal distribution with nonnegative skewness. Then the locally most powerful location and scale invariant test for normality rejects the null hypothesis when the sample skewness exceeds a given threshold value.*

Proof Without loss of generality it can be assumed that the location and scale parameters of the sampled distribution are zero and one, respectively. Let $\xi_i(x)$ and $m(x)$ denote the i-th derivative of $\log \Phi(x)$ and the inverted Mill's ratio, respectively:

$$\xi_i(x) = \frac{\partial^i \log \Phi(x)}{\partial^i x} \; ; \; m(x) = \frac{\phi(x)}{\Phi(x)}. \qquad (5)$$

Let also denote by U standard normal truncated from below at $-\tau$: $U = Y|Y > -\tau$, where Y is standard normal. Straightforward calculus techniques lead to the following equations:

$$\mu = E(U) = \xi_1(\tau), \; E\left\{(U-\mu)^2\right\} = 1 - \xi_2(\tau), \; E\left\{(U-\mu)^3\right\} = \xi_3(\tau). \qquad (6)$$

Let first prove that $\xi_3(x)$ is a strictly positive function. The function $\xi_1(x) = m(x)$ is strictly decreasing, since it is well-known that its first derivative $\xi_2(x) = -m(x)\{x+m(x)\}$ is strictly negative. Equivalently, $-m(x)$ is strictly increasing. The function $x + m(x)$ is strictly increasing, too, since its first derivative $1 - m(x)\{x+m(x)\}$ is the variance of U. The product of two increasing functions is also increasing, so that the function $\xi_2(x) = -m(x)\{x+m(x)\}$ is strictly increasing. Equivalently, its first derivative $\xi_3(x)$ is a strictly positive function, and this completes the first part of the proof.

In order to complete the proof, recall the following representation theorem:

$$X = \frac{Z}{\sqrt{1+\eta^2}} + \frac{\eta U}{\sqrt{1+\eta^2}} \sim ESN_1(0,1,\eta,\tau), \tag{7}$$

where Z is standard normal and independent of independent of U [17] or, equivalently [27],

$$X|U = u \sim N\left(\frac{\eta u}{\sqrt{1+\eta^2}}, \frac{1}{1+\eta^2}\right). \tag{8}$$

Hence $X \sim ESN_1(0,1,\eta,\tau)$ can be represented as a location mixture of normal distributions, with a truncated normal as mixing distribution. The third cumulant of a standard normal distribution truncated from below at $-\tau$ is always positive, being equal to $\xi_3(\tau)$. [31] show that these are sufficient conditions for the sample skewness to give the locally most powerful location and scale invariant test for normality against one-sided alternatives.

The above result generalizes the one in [29], who proved local optimality of the test only for $\tau = 0$, i.e. for the skew-normal distribution. Surprisingly enough, the optimality property of the test statistic is unaffected by the parameter τ, even if its sampling distribution under the alternative hypothesis does.

3 The Multivariate Case

We shall now consider the problem of testing multivariate normality when the sampled distribution is assumed to be ESN. One possible way for doing it is evaluating skewness of all linear combinations of the variables, and reject the normality hypothesis if at least one of them is too high in absolute value. This argument, based on the union-intersection approach, inspired [23] to introduce the test statistic

$$\max_{c \in \mathfrak{R}_0^d} \left\{ \frac{1}{n}\sum_{i=1}^n \left(\frac{c^T x_i - c^T \bar{x}}{\sqrt{c^T S c}}\right)^3 \right\}^2 \tag{9}$$

where $\mathfrak{R}_0^d$ is the set of all real, nonnull, d-dimensional vectors, while $\bar{x}$, S and X denote the sample mean, the sample variance and the data matrix X whose rows are the vectors $x_1^T,\ldots,x_n^T$. Closure properties of the ESN distribution under linear transformations and optimality properties of sample skewness for the univariate ESN encourage the use of the above statistic for testing multivariate normality within the ESN class.

Its analogue for a random vector z with expectation μ, nonsingular variance Σ, and finite third-order moments is:

$$\max_{c \in \Re_0^d} \frac{E^2\left[\left\{c^T(x-\mu)\right\}^3\right]}{\left(c^T \Sigma c\right)^3}. \tag{10}$$

[25] argued that difficulties in evaluating the above measure for well-known parametric families of multivariate distributions posed a severe limitation to its use. To the best of the author's knowledge, the skew-normal distribution, i.e. ESN with null truncation parameter, is the only known example of statistical model for which Malkovich and Afifi's skewness has a straightforward parametric interpretation: the direction maximizing skewness is proportional to the nonnormality parameter [21]. Theorem 2 in this section generalizes the result to all multivariate ESN distributions, regardless of the truncation parameter's value.

Another criticism to Malkovich and Afifi's skewness came from [5], who pointed out the involved computational difficulties. Indeed, the method proposed by [23] for computing their statistic appear to be based more heuristics rather than on a formal theory. Theorem 2 also suggests an approach based on standard maximum likelihood estimation rather than maximization of a $d-$variate cubic form subject to quadratic constraints. The maximum likelihood estimate for the shape parameter converges to the shape parameter itself, by well-known asymptotic arguments. Moreover, the direction maximizing sample skewness converges to the direction maximizing population's skewness, when it is unique [21], as it happens in the ESN case. Hence the direction maximizing sample skewness and the direction of the maximum likelihood estimate for the shape parameter will converge to each other. From the practical point of view, when the sample size is large enough, the former direction can be satisfactorily approximated by the latter one. Technical aspects of maximum likelihood estimation for the multivariate ESN are dealt with in [13].

Theorem 2 *The vector $c \in \Re^d$, $d > 1$, maximizing the skewness of $c^T x$, where $x \sim ESN(\xi, \Omega, \eta, \tau)$ and $\eta \neq 0_d$, is proportional to the nonnormality parameter η.*

Proof First recall some results regarding cumulants of an extended skew-normal random vector $x \sim ESN(\xi, \Omega, \eta, \tau)$ [2]:

$$\mu = \xi + \delta \zeta_1(\tau), \ \Sigma = \Omega + \zeta_2(\tau)\delta\delta^T, \ \kappa_3(x) = \zeta_3(\tau)\delta \otimes \delta \otimes \delta^T, \tag{11}$$

where $\zeta_i(\tau) = \partial^i \log \Phi(\tau)/\partial^i \tau$, $\delta = \Omega\eta/\sqrt{1+\eta^T\Omega\eta}$ and $\mu, \Sigma, \kappa_3(x)$ denote the mean, the variance and the third cumulant of x, respectively. The extended skew-normal class is closed under affine transformations, so that the distribution of the standardized random vector $z = \Sigma^{-1/2}(x-\mu)$ is extended skew-normal, too: $z \sim ESN(\xi_z, \Omega_z, \eta_z, \tau)$, with $\delta_z = \Omega_z\eta_z/\sqrt{1+\eta_z^T\Omega_z\eta_z} = \Sigma^{-1/2}\delta$.

Let λ and λ_z be unit length vectors maximizing the skewness of a linear combination of components of x and z, respectively:

$$\lambda = \arg\max_{c \in \mathscr{C}} \beta_1 \left(c^T x\right); \qquad \lambda_z = \arg\max_{c \in \mathscr{C}} \beta_1 \left(c^T z\right), \qquad (12)$$

where $\mathscr{C}$ is the set of d-dimensional random vectors of unit length and $\beta_1(Z)$ is the squared skewness of the random variable Z. It follows that $\lambda \propto \Sigma^{-1/2} \lambda_z$. Definitions of $\mathscr{C}$ and z imply that

$$\beta_1 \left(c^T z\right) = E^2 \left\{ \left(c^T z\right)^3 \right\} \qquad c \in \mathscr{C}. \qquad (13)$$

Apply now linear properties of cumulants [26, p. 32] to obtain

$$E^2 \left\{ \left(c^T z\right)^3 \right\} = \left\{ (c \otimes c)^T \kappa_3(z) c \right\}^2, \qquad (14)$$

where $\kappa_3(z) = \zeta_3(\tau) \delta \otimes \delta \otimes \delta^T$ is the third cumulant of z. Then

$$E^2 \left\{ \left(c^T z\right)^3 \right\} = \zeta_3^2(\tau) \left(c^T \delta\right)^6 \qquad (15)$$

by ordinary properties of the Kronecker product. Hence λ_z is proportional to δ_z or, equivalently, λ is proportional to $\Sigma^{-1} \delta$. Basic formulae for matrix inversion lead to

$$\lambda \propto \Sigma^{-1} \delta = \left\{ \Omega^{-1} - \frac{\Omega^{-1} \delta \delta^T \Omega^{-1}}{\zeta_3^{-1}(\tau) + \delta^T \Omega^{-1} \delta} \right\} \text{ and to} \qquad (16)$$

$$\delta = \Omega^{-1} \delta \left\{ 1 - \frac{\delta^T \Omega^{-1} \delta}{\zeta_3^{-1}(\tau) + \delta^T \Omega^{-1} \delta} \right\}. \qquad (17)$$

Since $\eta = \Omega^{-1} \delta / \sqrt{1 - \delta^T \Omega \delta}$, the vector maximizing the skewness of x is proportional to the parameter η, and this completes the proof.

4 A Numerical Example

In this section we shall use projections which maximize skewness to highlight interesting data features. The approach is exploratory in nature, and differs from the inferential apprach of the previous sections. Skewness maximization provides a valid criterion for projection pursuit [18], which has never been applied to financial data, to the best of our knowledge.

Each observation is the closing price of an European financial market, as recorded by MSCI Inc., a leading provider of investment decision support tools. The included countries are Austria, Belgium, Denmark, Finland, France, Germany, Greece, England, Ireland, Italy, Norway, Holland, Portugal, Spain, Sweden, Switzerland. The first and last closing prices were recorderd during June 24, 2004 and June 23, 2008, respectively. Data are arranged in a matrix where each row corresponds to a day and each column to a country. Hence the size of the data matrix is 1305×16, which is quite large.

Table 1 reports the skewnesses and the kurtoses for each country. All prices are mildly skewed: their third standardized moments are never greater than 0.688 and exceed 0.4 in Finland, Germany and Holland only. Also, all prices are platykurtic: their fourth standardized moments are never greater than 2.33 and exceed 2.00 in Finland and England only. Both features are illustrated in the box plots of Fig. 1: despite obvious differences in location and spread, all box plots suggest mild skew-

Table 1 Skewness, kurtosis and weight in the linear combination of each country

Country	Skewness	Kurtosis	Weight
Austria	−0.1905	1.7816	0.0500
Belgium	−0.1336	1.9531	−0.1210
Denmark	0.3671	1.7606	0.0536
Finland	0.6879	2.2944	−0.3675
France	0.1224	1.7403	0.0161
Germany	0.4500	1.8817	0.1291
Greece	0.0992	1.8210	−0.4474
England	0.2091	2.3252	−0.3091
Ireland	−0.1451	1.9206	0.5079
Italy	0.3539	1.7341	−0.0107
Norway	0.2455	1.8896	0.0622
Holland	0.4072	1.9636	0.2464
Portugal	0.3465	1.8087	0.3298
Spain	0.1910	1.9257	−0.0367
Sweden	0.0106	1.5700	0.0321
Switzerland	0.0458	1.9069	−0.3137

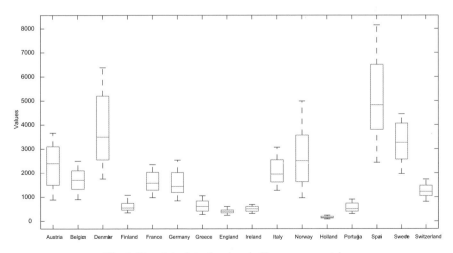

Fig. 1 Boxplot of stock prices in European countries

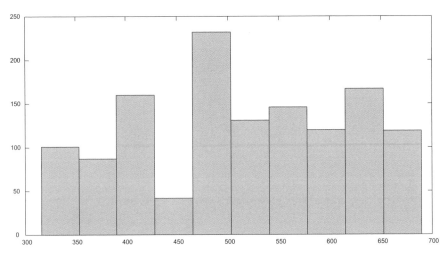

Fig. 2 Histogram of Italian Prices

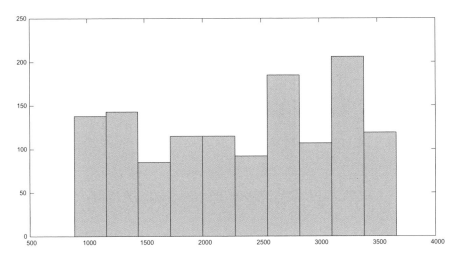

Fig. 3 Histogram of Austrian Prices

ness and absence of ouliers. Histograms of closing prices for different countries are multimodal and light tailed. We were unable to report all histograms due to space constraints. However, they are well exemplified by the histograms of Austria (Fig. 3) and Italy (Fig. 2). Austrian and Italian stock prices are linearly related, as shown in the scatter plot of Fig. 4. Again, the graph does not suggest the presence of outliers. Data exhibit a completely different structure when projected onto the direction which maximizes their skewness. The histogram of the projected data (Fig. 5) is definitely unimodal, markedly skewed and very heavy tailed. From the economic viewpoint, it is interesting to notice that the fifty greatest projected value correspond to the latest

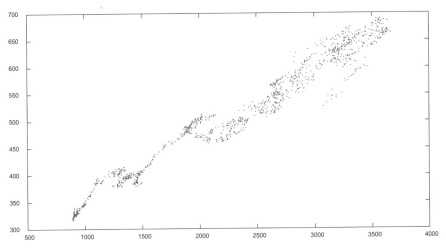

Fig. 4 Scatterplot of Italian Prices versus Austrian Prices

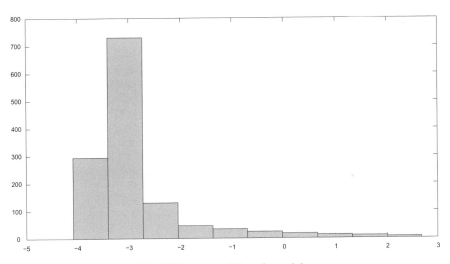

Fig. 5 Histogram of Transformed data

fifty days of the time period under consideration, when Europe began to suffer from the financial crisis.

5 Concluding Remarks

We have proposed two approaches for testing normality when the sampled distribution is assumed to be ESN. The first approach, recommended in the univariate case, is motivated by local optimality. The second approach, recommended in the multivariate case, is motivated by union-intersection arguments. Both approaches use some measure of skewness to overcome problems posed by likelihood-based methods.

Alternatively, multivariate normality can be tested via Mardia's measure of skewness [24], which presents some advantages over Malkovich and Afifi's one in the general case [5, 25]. However, these advantages vanish in the ESN case, as shown in Sect. 3. Moreover, Malkovich and Afifi's index has a straightforward application in terms of projection pursuit, since univariate skewness is a well-known projection pursuit index [18].

A natural question to ask is whether results in the paper hold for more general classes of distributions, as the ones discussed in [3]. Simulation studies (not shown here) suggest that the answer is in the positive, but the problem deserves further investigation and constitutes a direction for future research.

Acknowledgements The author would like to thank Marc Hallin, Christophe Ley and Davy Pandevene for their insightful comments on a previous version of the paper.

References

1. Adcock, C.J.: Capital asset pricing for UK stocks under the multivariate skew-normal distribution. In: Genton, M. G. (ed.) Skew-Elliptical Distributions and Their Applications: A Journey Beyond Normality, pp. 191–204. Chapman and Hall / CRC, Boca Raton (2004)
2. Adcock, C.J.: Extensions of Stein's Lemma for the Skew-Normal Distribution. Communications in Statistics – Theory and Methods **36**, 1661–1671 (2007)
3. Adcock, C.J.: Asset pricing and portfolio selection based on the multivariate extended skew-student-I distribution. Annals of operations research **176**, 221–234 (2010)
4. Adcock, C.J., Shutes, K.: Portfolio selection based on the multivariate skew normal distribution. In: Skulimowski, A. (ed.) Financial modelling (2001)
5. Andrews, D.F., Gnanadesikan, R., Warner, J.L.: Methods for assessing multivariate normality. In: Krishnaiah, P.R. (ed.) Proc. International Symposisum Multivariate Analysis **3**, pp. 95–116. Academic Press, New York (1973)
6. Arellano-Valle, R.B., Azzalini, A.: The centred parametrization for the multivariate skew-normal distribution. Journal of Multivariate Analysis **99**, 1362–1382 (2008)
7. Arnold, B.C., Beaver, R.J.: Hidden truncation models. Sankhya, series A**62**, 22–35 (2000)
8. Arnold, B.C., Beaver, R.J.: Skewed multivariate models related to hidden truncation and/or selective reporting (with discussion). Test **11**, 7–54 (2002)
9. Arnold, B.C., Beaver, R.J.: Elliptical models subject to hidden truncation or selective sampling. In: Genton, M.G. (ed.) Skew-Elliptical Distributions and Their Applications: A Journey Beyond Normality, pp. 101–112. Chapman and Hall / CRC, Boca Raton, FL (2004)
10. Azzalini, A., Capitanio, A.: Statistical applications of the multivariate skew-normal distributions. Journal of the Royal Statistical Society B **61**, 579–602 (1999)
11. Azzalini, A., Dalla Valle, A.: The multivariate skew-normal distribution. Biometrika **83**, 715–726 (1996)
12. Birnbaum, Z.W.: Effect of linear truncation on a multinormal population. Annals of Mathematical Statistics **21**, 272–279 (1950)

13. Capitanio, A., Azzalini, A., Stanghellini, E.: Graphical models for skew-normal variates. Scandinavian Journal of Statistics **30**, 129–144 (2003)
14. Copas, J.B., Li, H.G.: Inference for non-random samples (with discussion). Journal of the Royal Statistical Society B **59**, 55–95 (1997)
15. Crocetta, C., Loperfido, N.: Maximum Likelihood Estimation of Correlation between Maximal Oxygen Consumption and the Six-Minute Walk Test in Patients with Chronic Heart Failure. Journal of Applied Statistics **36**, 1101–1108 (2009)
16. Heckman, J.J.: Sample selection bias as a specification error. Econometrica **47**, 153–161 (1979)
17. Henze, N.: A Probabilistic Representation of the "Skew-Normal" Distribution. Scandinavian Journal of Statistics **13**, 271–275 (1986)
18. Huber, P.J.: Projection pursuit (with discussion). Annals of Statistics **13**, 435–475 (1985)
19. Loperfido, N., Guttorp, P.: Network bias in air quality monitoring design. Environmetrics **19**, 661–671 (2008)
20. Loperfido, N.: Modeling Maxima of Longitudinal Contralateral Observations. TEST **17**, 370–380 (2008)
21. Loperfido, N.: Canonical Transformations of Skew-Normal Variates. TEST **19**, 146–165 (2010)
22. Kumbhakar, S.C., Ghosh, S., McGuckin, J.T.: A generalized production frontier approach for estimating determinants of inefficiency in US dairy farms. Journal of Business and Economice Statistics **9**, 279–286 (1991)
23. Malkovich, J.F., Afifi, A.A.: On tests for multivariate normality. Journal of the American Statistical Association **68**, 176–179 (1973)
24. Mardia, K.V.: Measures of multivariate skewness and kurtosis with applications. Biometrika **57**, 519–530 (1970)
25. Mardia, K.V.: Assessment of Multinormality and the Robustness of Hotelling's T^2 Test. Applied Statistics **24**, 163–171 (1975)
26. McCullagh, P.: Tensor Methods in Statistics. Chapman and Hall, London UK (1987)
27. O'Hagan, A., Leonard, T.: Bayes estimation subject to uncertainty about parameter constraints. Biometrika **63**, 201–202 (1976)
28. Rotnitzky, A., Cox, D.R., Bottai, M., Robins, J.: Likelihood-based inference with singular information matrix. Bernoulli **6**, 243–284 (2000)
29. Salvan, A.: Test localmente più potenti tra gli invarianti per la verifica dell'ipotesi di normalità. In: Atti della XXXIII Riunione Scientifica della Società Italiana di Statistica, volume II, pp. 173–179, Bari. Società Italiana di Statistica, Cacucci (1986)
30. Sharples, J.J., Pezzey, J.C.V.: Expectations of linear functions with respect to truncated multinormal distributions-With applications for uncertainty analysis in environmental modelling. Environmental Modelling and Software **22**, 915–923 (2007)
31. Takemura, A., Matsui, M., Kuriki, S.: Skewness and kurtosis as locally best invariant tests of normality. Technical Report METR 06-47 (2006). Available at http://www.stat.t.u-tokyo.ac.jp/~takemura/papers/metr200647.pdf

On the RODEO Method for Variable Selection

Francesco Giordano and Maria Lucia Parrella

Abstract In this work, we work around an iterative estimation procedure which has been proposed recently by Lafferty and Wasserman. The procedure is called RODEO and can be used to select the relevant covariates of a sparse regression model. A drawback of the RODEO is that it fails to isolate some relevant covariates, in particular those which have linear effects on the model, and for such reason it is suggested to use the RODEO on the residuals of a LASSO. Here we propose a test which can be integrated to the RODEO procedure in order to fill this gap and complete the final step of the variable selection procedure. A two-stage procedure is therefore proposed. The results of a simulation study show a good performance of the new procedure.

1 Introduction

Estimating a high-dimensional regression function is notoriously difficult, due to the *curse of dimensionality*. The local polynomial estimator is particularly affected by this problem, since it becomes unfeasible when the number of covariates is high. However, for some applications a sparse condition can be formulated, which assumes that the true regression function only depends on a small number of the total covariates. In such cases, an estimation procedure which is capable of isolating the relevant variables can reach rates of convergence which are satisfactory. The aim of

F. Giordano (✉)
Department of Economics and Statistics, University of Salerno, Via Giovanni Paolo II, 84084, Fisciano (SA), Italy
e-mail: giordano@unisa.it

M.L. Parrella
Department of Economics and Statistics, University of Salerno, Via Giovanni Paolo II, 84084, Fisciano (SA), Italy
e-mail: mparrella@unisa.it

M. Corazza, C. Pizzi (eds.), *Mathematical and Statistical Methods for Actuarial Sciences and Finance*, DOI 10.1007/978-3-319-02499-8_16, © Springer International Publishing Switzerland 2014

this work is to propose a two-stage procedure, based on local polynomials, which can be used to select the relevant covariates of a sparse regression model, avoiding the *curse of dimensionality* problem. The first stage of the procedure uses the RODEO method of [1]. This is a greedy regression algorithm which is based on the innovative idea of using bandwidth selection to make variable selection. A drawback of the RODEO procedure is that it fails to isolate some relevant covariates, in particular those which have a linear effect on the dependent variable. For such reason, Lafferty and Wasserman suggest to use the RODEO on the residuals of a LASSO. Anyway, the use of the LASSO can introduce some spurious correlation in the residuals which can compromise the results of the final variable selection step. Moreover, the LASSO is crucially dependent on the correct selection of the regularization parameters, which are generally difficult to set, and this is in contrast with the simplicity of the RODEO method. This is the reason why we append a second stage to the RODEO procedure, with a relevance test for the identification of the linearities, which represents the contribution of this work. We present a simulation study which gives evidence of the empirical performance of our proposal.

This paper is organized as follows. In the following section we introduce the framework and the basic concepts of the multivariate local polynomial regression. In the third section, we describe our proposal: the relevance test based on the asymptotic theory of the local linear estimators. In the last section, we present the results of a simulation study, showing the oracle properties of the procedure.

2 Description of the Nonparametric Framework

Let $(X_1, Y_1), \ldots, (X_n, Y_n)$ be a set of $\mathbb{R}^{d+1}$-valued random vectors, where the Y_i represent the dependent variables and the X_i are the $\mathbb{R}^d$-valued covariates of the following nonparametric regression model

$$Y_i = m(X_i) + \varepsilon_i, \qquad i = 1, 2, \ldots \tag{1}$$

The function $m(X_i) = E(Y|X_i) : \mathbb{R}^d \to \mathbb{R}$ is the multivariate conditional mean function. The errors ε_i are supposed to be normally distributed, with mean equal to zero and variance equal to σ^2. Moreover, they are mutually independent, and independent from X_i.

Our goal is to estimate the function $m(x) = E(Y|X = x)$. We suppose that the point of estimation x is not a boundary point. We also assume that the number of covariates d is high but that only some covariates are relevant. The analysis of this setup raises the problem of the *curse of dimensionality*, which usually concerns nonparametric estimators, but also the problem of *variable selection*, which is necessary to pursue dimension reduction. The RODEO method has been recently proposed by [1] to deal with such kind of setup.

The Local Linear Estimator (LLE) is an appealing nonparametric tool whose properties have been studied deeply. It corresponds to perform a locally weighted least squares fit of a linear function, equal to

$$\hat{m}(x;H) \equiv \hat{\alpha}(x) = \arg\min_{\alpha,\beta} \sum_{i=1}^{n} \left\{ Y_i - \alpha(x) - \beta^T(x)(X_i - x) \right\}^2 K_H(X_i - x), \quad (2)$$

where the function $K_H(u) = |H|^{-1} K(H^{-1}u)$ gives the local weights and $K(u)$ is the Kernel function, a d-variate probability density function. The $d \times d$ matrix H represents the smoothing parameter, called the *bandwidth matrix*. It controls the variance of the Kernel function and regulates the amount of local averaging on each dimension, that is the local smoothness of the regression function. Here and in the following, we denote with e_j a unit vector, with all zeroes and a one in the j-th position. Using the matrix notation, the solution of the minimization problem in the (2) can be written in closed form:

$$\hat{m}(x;H) = e_1^T (\mathbf{X}^T \mathbf{W} \mathbf{X})^{-1} \mathbf{X}^T \mathbf{W} \mathbf{Y}, \quad (3)$$

where

$$\mathbf{X} = \begin{pmatrix} 1 & (X_1 - x)^T \\ \vdots & \vdots \\ 1 & (X_n - x)^T \end{pmatrix}, \quad \mathbf{W} = \begin{pmatrix} K_H(X_1 - x) & \cdots & 0 \\ \vdots & \ddots & \vdots \\ 0 & \cdots & K_H(X_n - x) \end{pmatrix}.$$

The practical implementation of the LLE is not trivial in the multivariate case, since its good properties are subject to the correct identification of the bandwidth matrix H. An asymptotically optimal bandwidth usually exists and can be obtained taking account of a bias-variance trade-off. In order to simplify the analysis, often H is taken to be of simpler form, such as $H = hI_d$ or $H = diag(h_1, \ldots, h_d)$, where I_d is the identity matrix, but even in such cases the estimation of the optimal H is difficult, because it is computationally cumbersome and because it involves the estimation of some unknown functionals of the process.

In general, throughout this paper we use the same notation and consider the same assumptions as in [1]. In particular, we assume that:

- the kernel K_H is a product kernel, based on a univariate non-negative bounded kernel function K, with compact support and zero odd moments; moreover

$$\int uu^T K(u)du = \mu_2 I_d,$$

where $\mu_2 \neq 0$ and I_d is the identity matrix of dimension d;
- the sampling density of the covariate X_i is uniform on the unit cube;
- the partial derivatives of the function $m(x)$ satisfy the assumptions (A2) and (A3) of [1].

3 The Two-Stage Variable Selection Procedure

The variable selection procedure is based on two stages. In the first stage, the RODEO method of [1] is used in order to estimate the multivariate bandwidth matrix H^*. The estimated bandwidth matrix has the properties of isolating the irrelevant variables in a very peculiar way: the bandwidths associated to the irrelevant variables are set to a high value, so that all the data observed along those directions are used for the regression and a local constant fit is made. [1] show that the estimated bandwidth H^* is of the optimal order, and that the the final nonparametric regression estimator is rate $n^{-4/(4+r)+\varepsilon}$, which is closed to the optimal one, where $r << d$ is the number of relevant variables. Unfortunately, also the linearities are classified among the irrelevant covariates, and for this reason we propose here to add a second stage to the RODEO procedure, which points to identify such relevant covariates (we denote these as *linear covariates*, to distinguish them from the *nonlinear covariates*, which are those covariates which have a nonlinear effect on the dependent variable).

3.1 First Stage – Multivariate Bandwidth Estimation

The first stage of the procedure is based on the hard threshold version of the RODEO method, proposed by [1]. It runs as follows.

1. Select a constant $0 < \beta < 1$ and the initial bandwidth (the constant c_0 determines the width of the grid)
$$h_0 = \frac{c_0}{\log\log n}.$$

2. Initialize the bandwidth matrix, and activate all covariates:
   - $\hat{H} = diag(\hat{h}_1, \ldots, \hat{h}_d)$, where $\hat{h}_j = h_0$, $j = 1, 2, \ldots, d$;
   - $\mathscr{A} = \{1, 2, \ldots, d\}$.

3. While $\mathscr{A}$ is not empty, do for each $j \in \mathscr{A}$:
   - Compute the statistic Z_j (estimated derivative expectation) and its estimated variance s_j^2

$$Z_j = \frac{\partial \hat{m}(x; \hat{H})}{\partial h_j} = \sum_{i=1}^{n} G_j(X_i, x, \hat{H}) Y_i, \qquad s_j^2 = \hat{\sigma}^2 \sum_{i=1}^{n} G_j^2(X_i, x, \hat{H}),$$

   where $\hat{\sigma}^2$ is any consistent estimator of the variance σ^2 (see [1] for the definition of $G_j(X_i, x, \hat{H})$).
   - Compute the threshold $\lambda_j = s_j \sqrt{2 \log n}$.
   - If $|Z_j| > \lambda_j$, then set $\hat{h}_j \leftarrow \beta \hat{h}_j$; otherwise remove j from $\mathscr{A}$.

4. Output the multivariate bandwidth matrix $H^* = diag(\hat{h}_1, \ldots, \hat{h}_d)$.

3.2 Second Stage – Relevance Test Through LP Asymptotics

Let $\mathscr{C}$ be the set of variables for which the RODEO method gives a final estimated bandwidth $\hat{h}_j \equiv h_0, \forall j \in \mathscr{C}$. It can be shown that this set $\mathscr{C}$ includes, at the beginning, the *linear covariates* and the irrelevant variables. So the set $\mathscr{B} = \overline{\mathscr{C}}$ includes the *nonlinear covariates*. Denote with $m'_j(\mathbf{x})$ the first partial derivative of the function $m(x)$ with respect to the variable j, that is $m'_j(\mathbf{x}) = \partial m(\mathbf{x})/\partial x_j$. We use the asymptotic normality of the local linear partial derivative estimator, shown in [2], in order to test the hypotheses

$$\mathbb{H}_0 : m'_j(\mathbf{x}) = 0 \quad \text{vs} \quad \mathbb{H}_1 : m'_j(\mathbf{x}) \neq 0$$

for $j \in \mathscr{C}$. The second stage of the variable selection procedure runs as follows:

1. Using the estimated bandwidth matrix H^*, compute

$$\mathbf{B} = (\mathbf{X}^T \mathbf{W} \mathbf{X})^{-1} \mathbf{X}^T \mathbf{W}, \qquad \mathbf{V} = \hat{\sigma}^2 \mathbf{B} \mathbf{B}^{-1}.$$

2. Set $\mathscr{B} = \overline{\mathscr{C}}$. Do for each $j \in \mathscr{C}$:

   - Compute the statistic N_j and its estimated variance v_j^2

$$N_j = \mathbf{e}_{j+1}^T \hat{m}(x; \hat{H}), \qquad v_j^2 = \mathbf{e}_{j+1}^T \mathbf{V} \mathbf{e}_{j+1}.$$

   - If $|N_j| < v_j z_{1-\alpha/2}$ then remove j from $\mathscr{C}$, where z_α is the α-quantile of the standard normal density.

3. Output the sets $\mathscr{B} \cup \mathscr{C}$, including the relevant covariates (*nonlinear* and *linear*).

4 Simulation Study

We applied the procedure described in Sect. 3 to several models observing similar results. Here we present the results concerning the following two models:

Model A: $Y_i = 5X_i^2(8)X_i^2(9) + 2X_i(10)X_i(1) + X_i(2)X_i(3)X_i(4) + \varepsilon_i$
Model B: $Y_i = 5X_i^2(8)X_i^2(9) + 2X_i^2(10)X_i(1) + \varepsilon_i$

All the models have been derived as variations of the models used in example 1 of [1]. As a consequence, the choice of the parameters is generally oriented on the choice made in their paper. Here, specifically, we want to highlight the capacity of our two-stage procedure to identify all the covariates, including those which are overlooked by the RODEO method, so we insert some linearities in the simulated models (i.e., covariates 1,2,3,4 and 10 of model A and covariate 1 of model B).

To be consistent with the RODEO setup, for all the models we generate the covariates from the uniform density. We consider normal errors with variance equal to $\sigma^2 = 0.5$; 200 Monte Carlo replications are generated for each model. We consider different lengths $n = (250, 500, 750, 1000, 1250)$ and dimension $d = (10, 15, 20, 25)$ of the datasets. The point of estimation is $x = (1/2, \ldots, 1/2)$. Figures 1 and 2 show graphically the results for models A and B respectively, using $n = 750$ and $d = 10$,

Table 1 Mean square error ($\times 100$) of the estimator $\hat{m}(x; H^*)$, for different values of n and d

	Model A				Model B			
	$d = 10$	$d = 15$	$d = 20$	$d = 25$	$d = 10$	$d = 15$	$d = 20$	$d = 25$
$n = 250$	2.31	2.52	2.60	2.73	4.92	4.95	5.02	5.30
$n = 500$	1.33	1.44	1.36	1.39	3.08	3.40	3.44	3.30
$n = 750$	0.86	0.97	0.86	1.05	2.19	2.46	2.15	2.59
$n = 1000$	0.68	0.75	0.64	0.75	1.84	1.92	1.75	2.04
$n = 1250$	0.58	0.63	0.61	0.64	1.53	1.71	1.58	1.80

while Table 1 summarizes numerically the performance of the method for increasing values of n and d.

Model A considers two *nonlinear covariates* and 5 *linear covariates*, for a total of 7 relevant variables among the d regressors. The positions of the relevant covariates have been chosen randomly. Here there are two kinds of linear relationships: multiplicative (i.e. covariate $X(1)$ is multiplied to $X(10)$) and additive (i.e. the term $2X_i(10)X_i(1)$ is added to the others). The two plots on the top of Fig. 1 present the results of the two stages of the procedure. On the left, the box-plots summarizes the bandwidths estimated by the RODEO method for each covariate. Note the peculiar behavior of such box-plots, which present almost no variability for $n = 750$. The RODEO method correctly identifies the two nonlinear covariates (i.e. $X(8)$ and $X(9)$), for which the optimal bandwidth assume a small value, but it does not make any distinction among the other variables. This is to be expected, since the optimal bandwidths for such covariates, under the assumptions considered here, is infinitely large. On the other hand, the plot on the right shows the percentages of rejection of the null hypothesis, $\mathbb{H}_0 : X(j)$ *is irrelevant*. Note that such percentages represent the size of the test ($= 0.05$) for the irrelevant variables (i.e., from $X(5)$ to $X(7)$) and the power of the test for the relevant variables.

The two plots on the bottom of Fig. 1 summarize the results of the final nonparametric estimations of the function (i.e., the first box-plot on the left in each plot) and its gradient (i.e., the others 10 box-plots in each plot). Note that the true values of the partial derivatives are highlighted through dotted horizontal lines, while the true value of the function $m(x)$ is shown through a solid horizontal line. The only difference between the plot on the left and the plot on the right is that in the first the LLE is made using all the regressors, while in the second it is made after removing all the non-relevant variables. As we can note, there are non substantial difference among the two situations. This confirm the oracle property of the function estimator shown by [1].

Figure 2 shows the results for model B. The performance of the method is similar to that shown in Fig. 1 for model A.

Three main comments can be formulated, valid in general for all the models. The first comment is that the power of the relevance tests shown in the plot on the top-right of each figure are influenced, as expected, by the distance between the true value of the partial derivatives and the value hypothesized under $\mathbb{H}_0$. In fact, note

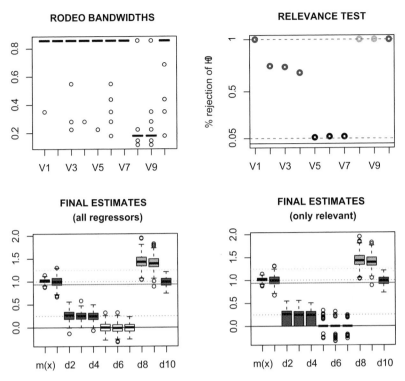

Fig. 1 Results of the variable selection procedure applied to model A, for $n = 750$ and $d = 10$, over 200 Monte Carlo replications. *Top-left*: final bandwidths estimated with RODEO (first stage). *Top-right*: percentage of rejection of the relevance test (second stage). *Bottom*: box-plots of the final estimates of the function $m(x)$ and its partial derivatives, obtained using all the regressors (*left*), or using only the relevant variables (*right*)

from the box-plots on the bottom of each figure that, the more the distance of such coefficients from zero ($\equiv \mathbb{H}_0$), the higher the power of the test.

Secondly, if one is interested in the function estimation itself instead of the variable selection, we can conclude that it is not necessary to remove the irrelevant variables from the final nonparametric regression (note the equivalence of the two plots on the bottom of each figure), because the high bandwidth associated to such non-relevant variables automatically neutralize their effects on the nonparametric regression, as confirmed by the theoretical results shown by [1]; as a result, we can also neglect the problem of multiple testing for the determination of the effective size of the test, since the relevance tests can be interpreted as individual tests.

Finally, the third general comment is that we observe in all the cases a remarkable bias in the estimations of $m(x)$, compared to its variance. This is an unacceptable situation, if one has the objective of testing the value of $m(x)$. The reason for this situation is that the bandwidth estimated by the RODEO method is of the correct

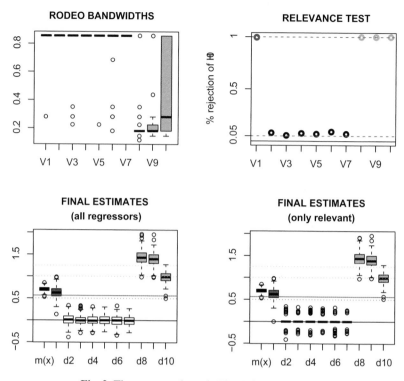

Fig. 2 The same results as in Fig. 1, but for model *B*

order (i.e., asymptotically consistent), but it is not the optimal one. Therefore, this suggests a further development of the method to be analyzed in the next future.

References

1. Lafferty, J., Wasserman, L.: RODEO: sparse, greedy nonparametric regression. The Annals of Statistics **36**, 28–63 (2008)
2. Lu, Z.-Q.: Multivariate locally weighted polynomial fitting and partial derivative estimation. Journal of Multivariate Analysis **59**, 187–205 (1996)

Portfolio Allocation Using Omega Function: An Empirical Analysis

Asmerilda Hitaj, Francesco Martinelli and Giovanni Zambruno

Abstract It is widely recognized that expected returns and covariances are not sufficient to characterize the statistical properties of securities in the context of portfolio
selection. Therefore different models have been proposed. On one side the Markowitz model has been extended to higher moments and on the other side, starting from
Sharpe ratio, a great attention has been addressed to the correct choice of the risk (or
joint risk-performance) indicator. One such indicator has been proposed recently in
the financial literature: the so-called Omega Function, that considers all the moments
of the return distribution and whose properties are being investigated thoroughly.
The main purpose of this paper is to investigate empirically, in an out-of-sample perspective, the portfolios obtained using higher moments and the Omega ratio. Moreover we analyze the impact of the target threshold (when the Omega Ratio is used)
and the impact of different preferences for moments and comoments (when a higher-
moments approach is used) on portfolio allocation. Our empirical analysis is based
on a portfolio composed of 12 Hedge fund indexes.

1 Introduction

One of the key problems in modern Quantitative Finance is the determination of
portfolios which prove optimal with respect to some risk measure, possibly jointly

A. Hitaj
Department of Statistics and Quantitative Methods, University of Milano Bicocca, Milan, Italy
e-mail: asmerilda.hitaj1@unimib.it

F. Martinelli
UBI Banca, Bergamo, Italy
e-mail: francesco.martinelli@ubibanca.it

G. Zambruno (✉)
Department of Statistics and Quantitative Methods, University of Milano Bicocca, Milan, Italy
e-mail: giovanni.zambruno@unimib.it

M. Corazza, C. Pizzi (eds.), *Mathematical and Statistical Methods for Actuarial Sciences and
Finance*, DOI 10.1007/978-3-319-02499-8_17, © Springer International Publishing Switzerland 2014

with some other criteria. This paper addresses this problem, in particular in our empirical analysis we use the Omega function as a risk-adjusted performance measure and implement a procedure that incorporates a joint optimization based on the first four moments of the portfolio return distribution.

Reference to higher-order moments can be traced back in Kendall and Hill [13]. In recent years many authors have proposed different methods to incorporate higher moments in portfolio allocation, see e.g. [1,7,9,14] and the references therein. Such models require knowledge of the investor's utility function which is needed for the truncated Taylor expansion.

It is well known that each investor has his own utility function and determining it is not an easy job. For this reason Davies et al. in [3] consider a different way to introduce higher moments into portfolio allocation: actually they implement a multiple-criteria decision approach. We will present this approach in more details in Sect. 2. The advantage of this approach is that it does not assume a particular utility function for the investor.

Another way to introduce higher moments in portfolio allocation is to consider a risk-adjusted performance measure that takes into account the preferences of the investors for skewness and kurtosis. In this work we will consider the Omega ratio proposed by Shadwick and Keating in [12]. This measure is an evolution of the Sharpe ratio [16]. One way to define it is the ratio of the expected returns conditional on exceeding a given threshold, over the same conditional on not exceeding it. The numerator is thus meant to be a measure of "good" returns, while the denominator represents the "bad" ones. Varying the threshold one obtains the Omega function. The main advantage of the Omega measure is that it involves all moments of the return distribution, including skewness and kurtosis and even higher moments as well. Recently the Omega ratio has been generalized by Farinelli and Tibiletti in [18] and in [5] they show the superiority of this measure with respect to the Sharpe ratio in forecasting performances. In literature different risk adjusted performance measures have been proposed in the attempt to overcome the drawbacks of the Sharpe ratio, see e. g. [2,4,8,11,15,17,19] etc.

In the empirical part we combine the model proposed in [3] and the Omega ratio proposed in [12] in order to account for investor preferences to higher moments. The aim of this work is to analyse, in an out-of-sample perspective, which model performs better. According to our knowledge this is the first empirical analysis that compares these two procedures for portfolio allocation. Then we analyse the sensitivity of the optimal portfolio with respect to the threshold and the different preferences for higher moments. In the empirical analysis we use a rolling window strategy of 48 months in-sample and 3 months out-of-sample[1] for the portfolio selection and calculate the in-sample optimal weights using in turn the Omega ratio and the multi objective approach. We perform an analysis of the optimal weights obtained in-sample with each method; subsequently we discuss the results of the out-of-sample

---

[1] We chose the rolling window strategy 48-3 as this is commonly used in real hedge fund world where the data are scarce.

portfolio returns. Our analysis has been carried on using a portfolio of 12 hedge fund indexes.

The paper is organized as follows: in Sect. 2 we briefly explain how the portfolio selection with four moments is operated. In Sect. 3 we report the definition of the Omega function. Section 4 describes the empirical analysis, where we analyze the characteristics of our portfolios and discuss the results obtained; Sect. 5 draws some conclusions. In the Appendix we report selected numerical results.

2 Higher Moments Portfolio Allocation Using a Multi-Objective Approach

Different methodologies exist that incorporate higher moments in the portfolio decision process. In this section we will review briefly the Polynomial[2] Goal Programming (PGP henceforth) approach, applied in Finance by Davies et al. in [3]. We consider an investor who selects his portfolio from n risky assets, under no transaction costs and no short sales allowed. Using the PGP approach the investor should maximize the odd moments (mean and skewness) and minimize the even moments (variance and kurtosis).

Denote with $R = (R_1, R_2, \ldots, R_n)$ the return vector, where R_i is the return of asset i, and by $w = (w_1, w_2, \ldots, w_n)$ the vector of weights, where w_i is the fraction of the initial endowment invested in the i-th asset. Under the assumption that the first four moments $(\overline{R}_i, \sigma_i^2, S_i$ and $K_i)$ exist for all risky assets we can calculate the first four moments of the portfolio, that are:

$$E[R_P] = w' \, \overline{R}; \quad \sigma_P^2 = w' \, M_2 \, w; \quad S_P = w' \, M_3 \, (w \otimes w); \quad K_P = w' \, M_4 \, (w \otimes w \otimes w)$$

where M_2, M_3 and M_4 are the tensor matrices of co-variance, co-skewness and co-kurtosis resp., and $\otimes$ denotes the Kronecker product.

Define the set of feasible portfolios as $\mathscr{S} = \{w | \sum_{i=1}^{n} w_i = 1, \, 0 \leq w_i\}$. The portfolio selection over the first four moments can be formulated as follows:

$$\begin{cases} \max \, E[R_P] = w' \, \overline{R} & \min \, \sigma_P^2 = w' \, M_2 \, w \\ \max \, S_P = w' \, M_3 \, (w \otimes w) & \min \, K_P = w' \, M_4 \, (w \otimes w \otimes w). \\ \text{s.t.} \quad w \in \mathscr{S} \end{cases} \tag{1}$$

A general way to solve the multiobjective problem is a two-step procedure:

In the first step we solve separately each single-optimization problem, that is maximize the mean and skewness and minimize the variance and kurtosis:

$$\mu_P^* = \max \left\{ w' \, \overline{R} | w \in \mathscr{S} \right\}; \qquad \left(\sigma_P^2 \right)^* = \min \left\{ w' M_2 \, w | w \in \mathscr{S} \right\};$$
$$S_P^* = \max \left\{ w' \, M_3 \, (w \otimes w) | w \in \mathscr{S} \right\}; \qquad K_P^* = \min \left\{ w' \, M_4 \, (w \otimes w \otimes w) | w \in \mathscr{S} \right\}.$$

---

[2] Actually, the term 'polynomial' refers to the formulation whereby the aspiration levels are determined, not to the objective function of the main program.

Solving separately these four problems we determine the aspiration levels of the investor for the portfolio mean, variance, skewness and kurtosis $(\mu_P^*, (\sigma_P^2)^*, S_P^*, K_P^*)$.

In the second step we build the PGP, where the portfolio allocation decision is given by the solution of the PGP which minimizes the generalized Minkowski distance from the aspiration levels, namely:

$$
\begin{cases}
\min_{w_i, d_j} Z = \left| \dfrac{d_1}{\mu_P^*} \right|^{\gamma_1} + \left| \dfrac{d_2}{(\sigma_P^2)^*} \right|^{\gamma_2} + \left| \dfrac{d_3}{S_P^*} \right|^{\gamma_3} + \left| \dfrac{d_4}{K_P^*} \right|^{\gamma_4} \\
w \in \mathcal{S}
\end{cases}
\tag{2}
$$

where $d_1 = \mu_P^* - w'\, \overline{R}$, $d_2 = w'\, M_2\, w - (\sigma_P^2)^*$, $d_3 = S_P^* - w'\, M_3(w \otimes w)$, $d_4 = w'\, M_4(w \otimes w \otimes w) - K_P^*$ and γ_1, γ_2, γ_3 and γ_4 represent the investor's subjective parameters: the greater the γ_i, the more important the corresponding moment appears to the investor. Note that in this case we do not assume a particular utility function for the investor, apart from the preferences for all moments discussed before.

3 Omega Performance Measure

Omega is a performance measure recently introduced by Keating and Shadwick in [12] which accounts for all the distributional characteristics of a returns series while requiring no parametric assumption on the distribution. Precisely, it considers the returns below and above a specific threshold and consists of the ratio of total probability-weighted gains to losses in a way that fully describes the risk-reward properties of the distribution. It can therefore be interpreted as a performance measure. The mathematical definition is as follows:

$$
\Omega_{F_X}(\tau) = \frac{\int_\tau^b (1 - F_X(x))\, dx}{\int_a^\tau F_X(x)\, dx}
$$

where $F_X(x) = \Pr(X \leq x)$ is the cumulative distribution function of the portfolio returns defined on the interval $[a, b]$ and τ is the loss threshold selected by the investor. For any investor, returns below his specific loss threshold are considered as losses and returns above it as gains. A higher value of Omega is always preferred to a lower value.

In the empirical analysis, in order to find the portfolio allocation optimal according to Omega as a performance measure, we solve the following problem:

$$
\begin{cases}
\max_{w_i} \Omega_{F_X}(\tau) \\
\text{s.t. } \sum_{i=1}^N w_i = 1,\ 0 \leq w_i.
\end{cases}
\tag{3}
$$

It is well known that the objective function in Omega-optimal portfolio is a non-convex problem and different algorithms have been used to solve this problem (see e.g. [6,10], etc). In the empirical analysis we use the MATLAB Global Optimization Toolbox to solve this problem.

4 Empirical Analysis and Results

In this empirical analysis we have considered a portfolio composed of 12 Hedge fund Indexes[3]. The observations are monthly and span the period January 1994 to December 2011. In total for each Hedge fund Index we got 217 monthly observations. In Table 1 we have reported general statistics and the Jarque-Bera test for each hedge fund index. As we can observe almost all the Hedge funds under consideration display negative skewness and kurtosis higher than 3, while looking at the results obtained for the Jarque-Bera test we can say that the hypothesis of normality is rejected for each time series of returns except 'Managed Future Hedge Fund Index' on the whole period under consideration. Therefore we can conclude that the Hedge Funds on which our portfolios are built display returns not normally distributed. For this reason, and for purposes of comparison, we will start our portfolio allocation decision with a model that considers only the first two moments and than extend it to the third and finally to the fourth moment. The model used in this analysis is the PGP explained before, where we consider different weights for moments and comoments: for $i = 1, \ldots, 4$ we set in turn $\gamma_i = k$ where $k = 1, \ldots, 5$. Also we have computed the portfolio allocation resulting from the maximization of the Omega function. As explained before, this performance measure has the advantage of taking into account the whole distribution of assets returns and has the drawback of being dependent on the reference point. For this reason, in the portfolio allocation decision we have used the Omega performance measure using different reference returns, moving from 1% to 15%, on annual basis, with a step of 0.5.

For the portfolio allocation we have used a rolling-window strategy of 48 months as the in-sample period and 3 months as the out-of-sample period. Using the different approaches for portfolio allocation, we find the optimal weights in each in-sample period and keep these constant until a new rebalance takes place.

We have performed an in-sample and out-of-sample analysis, determining the impact of assigning different weights to moments (comoments), and considering different reference points for the Omega ratio.

4.1 In-Sample Analysis

In total we have 55 in-sample-periods: for each we have computed: Optimal weights using the PGP with 2, 3 and 4 moments, labeled respectively *Mean Var allocation*, *Mean Var Skew allocation* and *Mean Var Skew Kurt allocation*, in Sect. 5. For these portfolios we have considered different cases, where $\gamma_i = 1, \ldots, 5$ for $i = 1, \ldots, 4$. Finally we have computed the optimal weights using the Omega ratio where different values for the annual target threshold have been considered: $\tau = 1\%$ to $\tau = 15\%$ with a step of 0.5.

Table 4 reports selected[4] results obtained in the in-sample analysis. We report the results: for the mean-variance PGP in case $[(\gamma_1 = 1, \gamma_2 = 1), (\gamma_1 = 1, \gamma_2 = 3)$

---

[3] The data have been collected through the Dow Jones Credit Suisse Hedge Fund Index.

[4] The complete results are available upon request.

and $(\gamma_1 = 3, \gamma_2 = 1)$]; for the mean-variance-skewness in case $[(\gamma_1 = 1, \gamma_2 = 1, \gamma_3 = 1)$, $(\gamma_1 = 1, \gamma_2 = 3, \gamma_3 = 1)$ and $(\gamma_1 = 3, \gamma_2 = 1, \gamma_3 = 3)$]; for the mean-variance-skewness-kurtosis in case $[(\gamma_1 = 1, \gamma_2 = 1, \gamma_3 = 1, \gamma_4 = 1)$, $(\gamma_1 = 1, \gamma_2 = 3, \gamma_3 = 1,$ $\gamma_4 = 3)$ and $(\gamma_1 = 3, \gamma_2 = 1, \gamma_3 = 3, \gamma_4 = 1)$]; for the Omega-optimal portfolio in case $\tau = 1\%, 3\%, 7\%, 8\%, 10\%$ and 15% on annual basis.

Once the in-sample optimal weights are obtained we calculate general statistics (mean, standard deviation, quartiles (Q_1, Q_2, Q_3), minimum and maximum value) for each portfolio component across time. Looking at these results it clearly appears that in the case the Omega ratio is used for portfolio allocation, the higher the reference value the less diversified is the portfolio (looking for example at Q_3, the higher the reference point the lower the number of assets whose Q_3 is significantly different from zero). Further, looking at the results obtained with reference to a portfolio allocation using PGP and varying the exponents γ_i, or when we move from a portfolio allocation with 2 moments to the ones with 3 or 4 moments no substantial difference occurs between these statistics.

Concluding the in-sample analysis, we can observe that a fair degree of diversification occurs only when operating according to the PGP or through the Omega ratio with low threshold values.

4.2 Out-of-Sample Analysis

Once we have calculated the optimal weights in the in-sample period, we keep them constant in the next out-of-sample period and for each portfolio we compute the out-of-sample returns, of which we present general statistics. In Table 2 we report these results for the selected cases considered in Sect. 4.1. As is well known, any investor will prefer a portfolio with higher mean and skewness and lower variance and kurtosis. As we can observe in this table, it is not clear, according to the mean-variance-skewness-kurtosis criterion, which portfolio performs best: actually there is no portfolio with higher mean and skewness and lower variance and kurtosis at the same time.

From the out-of-sample returns we computed the out-of-sample performances for each portfolio. In Fig. 1 we have plotted the out-of-sample performances obtained using the PGP with two, three and four moments and using the Omega ratio with different reference values. From these figures it appears that the higher the weights of the even moments in the PGP, the greater the difference in the out-of-sample performances. In addition, looking at the out-of-sample performances using the Omega ratio, we can say the higher the reference point the less stable the out-of-sample performance. In Fig. 2 we compare the out-of-sample performances obtained with Omega ratio and PGP with 2, 3 and 4 moments resp.[5]. It appears that the out-of-sample performances obtained with Omega and PGP are closer to each other when we consider a low reference value for the Omega ratio and the PGP with more than

---

[5] For the Omega ratio we have reported only the out-of-sample performances for $\tau = 3\%, 7\%, 8\%$ on annual basis.

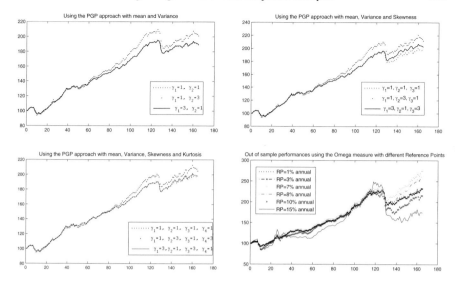

Fig. 1 Portfolio out-of-sample performances obtained using, respectively, the PGP with 2, 3, 4 moments and Omega ratio with different reference point. We observe a decrease in the out-of-sample performance in observation 128, which belong to July 2008 (period of the Hedge fund crisis starting with the default of Madoff Hedge Fund)

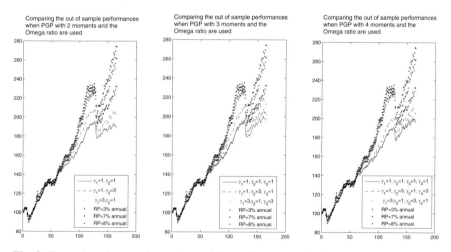

Fig. 2 Comparing the out-of-sample portfolio performances obtained with Omega and the PGP with 2, 3 and 4 moments

two moments. Looking at these graphs we cannot argue which method is the best for portfolio allocation.

In order to give an advice on which procedure to use for portfolio allocation we rank, in Table 3, selected out-of-sample portfolios using Sharpe[6], Sortino[7] and Omega considering as reference values $1\%, 3\%, 7\%$ and 8%. From Table 3 we observe that independently from the performance measure used for the ranking of the portfolios, when the reference point is very low (in this case 1%), no approach clearly dominates the others. This fact seems consistent with what we have seen in Fig. 2 where the performance of the out-of-sample portfolios obtained with Omega was very close to that obtained using the PGP approach.

When the reference point is high, independently from the performance measure used for the ranking, we observe that the rank of the Omega portfolios is higher than the one of the PGP.

Concluding we can say that if one has to choose between the PGP and the Omega portfolio, the latter is better as it displays higher ranks most of the time. Comparing the PGP approaches, we can observe that almost always the PGP with 3 or 4 moments have a higher rank than the PGP with two moments. If only the PGP with 3 and the PGP with 4 moments are compared the rank is not univocal and may appear counterintuitive since one would expect that using a wider set of information (including kurtosis) results in a better specification. This is possibly due to the fact that in the estimation of moments and comoments we are using the sample approach, which is characterized by high estimation error. To overcome this drawback one should use robust estimators for moments and comoments (see [14] and the references therein). In terms of computation costs we can say that when higher moments are used for portfolio allocation the number of parameters to estimate is high (see [14]), which requires a much larger dataset and an increase of computer time; in case of Omega-portfolio the only thing needed is a good optimization algorithm.

5 Conclusions

In this paper we have performed an empirical analysis based on a Hedge fund portfolio: its aim was to analyse the impact, in-sample and out-of-sample, of choosing an approach based on moments and comoments (we consider PGP approach) as compared to the one based on the Omega ratio. We considered the PGP with 2, 3 and 4 moments (each case featuring different shapes of weights in order to assess their relative importance) and the Omega ratio (using different reference values: $1\%, 1.5\%, 2\%, \ldots, 15\%$ on annual basis).

In the in-sample analysis we reached the conclusion that choosing a high reference value, when Omega ratio is used for portfolio allocation, does not make sense since in almost all periods the portfolio will be concentrated in one asset. Instead, when using a Omega ratio with a low reference value, or the PGP approach with 2, 3 or 4 moments the optimal portfolio is well diversified.

---

[6] $Sharpe = \frac{\mu_P - \tau}{\sigma_P}$.

[7] $Sortino = \frac{\mu_P - \tau}{\sqrt{LPM_2(\tau)}}$.

In out-of-sample perspective, in terms of mean-variance-skewness-kurtosis, there is no portfolio offering at the same time the highest mean and skewness and the lowest variance and kurtosis. While comparing the out-of-sample performances (see Fig. 2) we conclude that the lower the reference value (for the Omega ratio) or the higher the number of moments considered (in the PGP approach), the closer the out-of-sample performances of the portfolios obtained under the Omega ratio and the PGP approach.

We have then ranked the different portfolios, obtained in the out-of-sample period, using different performance measures (Sharpe ratio, Sortino ratio and Omega ratio) (see Table 3). We conclude that for a low reference value the portfolio ranking changes with the performance measure (see the case when the reference value is 1% annual). As long as the reference value increases the portfolio ranking becomes more stable and it appears that in this case the choice of the performance measure used for optimization does not matter that much. In this case the best ranked portfolios are the Omega-portfolios followed by those obtained with the PGP 3 or 4 moments and finally the PGP with 2 moments. In the case of the PGP approach the results are counterintuitive, because we would have expected a higher rank for the PGP with 4 moments, then the PGP with 3 moments and the lowest rank for the PGP with 2 moments. This unexpected result is probably due to the estimation error of moments and comoments. The superiority of the Omega-portfolio with respect to the PGP approach is also supported by lower computational costs since there are no parameters to estimate.

These conclusions are valid for the considered dataset, but: how do these results modify if we change the dataset, or the length of the in-sample or out-of-sample period? How do these results change if we consider robust estimators for moments and comoments? How can we choose the best weights for moments and comoments when the PGP approach is used? Future research will address these questions.

Appendix

Table 1 General statistics for each Hedge Fund index

General statistics for each time series of returns					
Dow Jones Credit Suisse Hedge Fund Index	Annual mean	Annual STD	Skewness	Kurtosis	JB test
'Hedge Fund Index'	0.087	0.076	−0.327	5.437	57.295
'Convertible Arbitrage'	0.075	0.072	−2.997	21.048	3254.979
'Dedicated Short Bias'	-0.034	0.168	0.449	3.745	12.252
'Emerging Markets'	0.073	0.152	−1.201	9.714	457.581
'Event Driven'	0.092	0.065	−2.451	15.221	1560.520
'Event Driven Distressed'	0.102	0.068	−2.388	15.579	1629.257
'Event Driven Multi-Strategy'	0.087	0.070	−1.944	11.603	802.213
'Event Driven Risk Arbitrage '	0.067	0.042	−1.097	8.028	270.850
'Fixed Income Arbitrage'	0.052	0.060	−4.700	36.613	10963.463
'Global Macro'	0.121	0.097	−0.246	6.888	138.245
'Long/Short Equity'	0.092	0.099	−0.218	6.205	94.174
'Managed Futures'	0.060	0.117	−0.079	2.979	0.227

Table 2 General statistics for out-of-sample portfolios returns.

The first column indicates the weights of moments and comoments in case the PGP approach is used and the reference point considered when the allocation is determined using Omega ratio. The second column shows the methodology used for portfolio allocation, for e.g 'Mean Var Skew Kurt' means that we have used the PGP with 4 moments in finding the portfolio allocation; 'Omega' means that we have used the Omega ratio in finding the portfolio allocation

		General statistics on the out of sample Portfolios returns			
		Annual Mean	Annual STD	Skewness	Kurtosis
$\gamma_1 = 1. \gamma_2 = 1$	Mean Var	0.0479	0.0355	−1.5656	7.3539
$\gamma_1 = 1. \gamma_2 = 3$	Mean Var	0.0533	0.0409	−1.6780	7.6363
$\gamma_1 = 3. \gamma_2 = 1$	Mean Var	0.0480	0.0351	−1.5809	7.5566
$\gamma_1 = 1. \gamma_2 = 1. \gamma_3 = 1$	Mean Var Skew	0.0518	0.0347	−1.5601	7.4558
$\gamma_1 = 1. \gamma_2 = 3. \gamma_3 = 1$	Mean Var Skew	0.0585	0.0390	−1.5140	7.0248
$\gamma_1 = 3. \gamma_2 = 1. \gamma_3 = 3$	Mean Var Skew	0.0525	0.0346	−1.5232	7.1759
$\gamma_1 = 1. \gamma_2 = 1. \gamma_3 = 1. \gamma_4 = 1$	Mean Var Skew Kurt	0.0502	0.0348	−1.4295	6.9050
$\gamma_1 = 1. \gamma_2 = 3. \gamma_3 = 1. \gamma_4 = 3$	Mean Var Skew Kurt	0.0551	0.0378	−1.3738	6.4693
$\gamma_1 = 3. \gamma_2 = 1. \gamma_3 = 3. \gamma_4 = 1$	Mean Var Skew Kurt	0.0505	0.0348	−1.4247	6.8774
'Using as reference point 1% annual'	Omega 1	0.0636	0.0449	−1.4952	7.1230
'Using as reference point 3% annual'	Omega 3	0.0643	0.0483	−1.6499	7.7270
'Using as reference point 7% annual'	Omega 7	0.0780	0.0538	−1.4242	7.3894
'Using as reference point 8% annual'	Omega 8	0.0757	0.0584	−1.8443	9.3481
'Using as reference point 10% annual'	Omega 10	0.0626	0.0744	−1.3939	7.3592
'Using as reference point 15% annual'	Omega 15	0.0572	0.1144	−0.8470	7.6858

Table 3 Ranking of portfolios using different performance measures and different reference points

Ranking of the portfolios using different performance measures

Using Reference Point of 1% annual

Sharpe Ratio		Sortino Ratio		Omega Ratio	
Omega 7	0.3521	Omega 7	0.3796	Omega 7	2.5313
MVS 131	0.3489	MVS 131	0.3614	MVS 131	2.4278
MVS 313	0.3460	MVS 111	0.3585	MVS 313	2.3953
MVS 111	0.3393	MVSK 1111	0.3547	Omega 1	2.3777
MVSK 1313	0.3377	MVS 313	0.3404	MVS 111	2.3698
Omega 1	0.3348	Omega 1	0.3353	MVSK 1313	2.3413
MVSK 3131	0.3322	MVSK 1313	0.3351	Omega 8	2.3391
Omega 8	0.3303	Omega 8	0.3318	MVSK 3131	2.3208
MVSK 1111	0.3137	MVSK 3131	0.3255	MVSK 1111	2.3119
MV 31	0.3055	MV 11	0.3190	MV 31	2.1800
MV 11	0.3017	MV 13	0.2991	MV 11	2.1556
MV 13	0.2977	MV 31	0.2733	MV 13	2.1536
Omega 10	0.1982	Omega 10	0.2167	Omega 10	1.7353
Omega 15	0.1160	Omega 15	0.1415	Omega 15	1.3936

Using Reference Point of 3% annual

Sharpe Ratio		Sortino Ratio		Omega Ratio	
Omega 7	0.2463	Omega 7	0.2680	Omega 7	1.9415
Omega 8	0.2163	Omega 8	0.2304	Omega 8	1.8215
Omega 1	0.2081	Omega 1	0.2142	Omega 1	1.7399
MVS 131	0.2032	MVS 131	0.2124	MVS 131	1.6954
MVS 111	0.1851	MVS 111	0.1877	MVSK 1313	1.6174
MVSK 1313	0.1816	MVSK 1313	0.1856	MVS 313	1.6069
MVS 313	0.1754	MVS 313	0.1822	MVS 111	1.5850
MVSK 3131	0.1647	MVSK 1111	0.1771	MVSK 3131	1.5411
MVSK 1111	0.1631	MVSK 3131	0.1653	MVSK 1111	1.5342
MV 13	0.1586	MV 13	0.1611	MV 13	1.5235
MV 31	0.1434	MV 11	0.1510	MV 31	1.4607
MV 11	0.1413	Omega 10	0.1345	MV 11	1.4511
Omega 10	0.1218	MV 31	0.1303	Omega 10	1.4146
Omega 15	0.0662	Omega 15	0.0813	Omega 15	1.2113

Using Reference Point of 7% annual

Sharpe Ratio		Sortino Ratio		Omega Ratio	
Omega 7	0.0403	Omega 7	0.0447	Omega 7	1.1197
Omega 8	0.0266	Omega 8	0.0288	Omega 8	1.0803
Omega 10	−0.0270	Omega 10	−0.0300	Omega 10	0.9226
Omega 15	−0.0306	Omega 15	−0.0379	Omega 15	0.9135
Omega 1	−0.0387	Omega 1	−0.0400	Omega 1	0.8965
MVS 131	−0.0806	MVS 131	−0.0853	MVS 131	0.7977
MVSK 1313	−0.1122	MVSK 1313	−0.1143	MVSK 1313	0.7273
MV 13	−0.1122	MV 13	−0.1152	MV 13	0.7249
MVS 313	−0.1388	MVS 313	−0.1428	MVS 313	0.6743
MVS 111	−0.1438	MVS 111	−0.1552	MVS 111	0.6631
MVSK 3131	−0.1617	MV 31	−0.1611	MVSK 3131	0.6288
MVSK 1111	−0.1625	MVSK 3131	−0.1682	MVSK 1111	0.6273
MV 11	−0.1711	MVSK 1111	−0.1768	MV 11	0.6110
MV 31	−0.1722	MV 11	−0.1836	MV 31	0.6084

Using Reference Point of 8% annual

Sharpe Ratio		Sortino Ratio		Omega Ratio	
Omega 7	−0.0101	Omega 7	−0.0112	Omega 7	0.9717
Omega 8	−0.0198	Omega 8	−0.0214	Omega 8	0.9437
Omega 15	−0.0543	Omega 15	−0.0671	Omega 15	0.8511
Omega 10	−0.0634	Omega 10	−0.0705	Omega 10	0.8256
Omega 1	−0.0991	Omega 1	−0.1026	Omega 1	0.7528
MVS 131	−0.1500	MV 13	−0.1840	MVS 131	0.6499
MV 13	−0.1785	MVSK 1313	−0.1897	MV 13	0.5921
MVSK 1313	−0.1849	MVS 131	−0.1967	MVSK 1313	0.5840
MVS 313	−0.2171	MV 31	−0.2346	MVS 313	0.5335
MVS 111	−0.2219	MVSK 3131	−0.2525	MVS 111	0.5238
MVSK 3131	−0.2415	MVSK 1111	−0.2645	MVSK 3131	0.4936
MVSK 1111	−0.2421	MV 11	−0.2661	MVSK 1111	0.4923
MV 11	−0.2476	MVS 313	−0.2667	MV 11	0.4827
MV 31	−0.2494	MVS 111	−0.2827	MV 31	0.4794

Table 4 General statistics of weights across time

General statistics across time for in sample optimal weights

Mean Var allocation with $\gamma_1 = 1$ and $\gamma_2 = 1$

	mean	std	Q_1	Q_2	Q_3	min	max
Hedge Fund Index	0.0103	0.0423	0	0	0.2320	0	0.2223
Conv Arbi	0.1081	0.1313	0	0.0255	0.1257	0	0.3885
Dedi Short Bias	0.1101	0.0281	0.0932	0.1029	0	0.0543	0.1780
Emer Mark	0.0180	0.0353	0	0	0	0	0.1234
Event Driven	0.0073	0.0434	0	0.0372	0.0902	0	0.3154
Event Driven Distres	0.0925	0.1362	0.0032	0	0	0	0.4846
Event Driven Multi-Strat	0.0096	0.0338	0	0.3709	0.4840	0	0.1776
Event Driven Risk Arbit	0.3878	0.2432	0.2573	0.0326	0.2721	0	0.8592
Fixed Income Arbit	0.1417	0.1728	0	0	0	0	0.4979
Global Macro	0.0267	0.0598	0	0	0.0734	0	0.2508
Long/Short Equity	0.0365	0.0589	0	0.0218	0.0877	0	0.1999
Managed Futures	0.0516	0.0600	0			0	0.1915

Mean Var Skew allocation with $\gamma_1 = 1$, $\gamma_2 = 1$ and $\gamma_3 = 1$

	mean	std	Q_1	Q_2	Q_3	min	max
Hedge Fund Index	0.0052	0.0265	0	0	0.2324	0	0.1777
Conv Arbi	0.1089	0.1309	0	0.0240	0.1213	0	0.3875
Dedi Short Bias	0.1051	0.0313	0.0889	0.0995	0.0043	0	0.1782
Emer Mark	0.0184	0.0358	0	0	0	0	0.1243
Event Driven	0.0072	0.0436	0	0.0236	0.0613	0	0.3167
Event Driven Distres	0.0719	0.1229	0	0	0	0	0.4819
Event Driven Multi-Strat	0.0104	0.0342	0	0.3712	0.5232	0	0.1785
Event Driven Risk Arbit	0.4014	0.2540	0.2586	0.0533	0.2651	0	0.8736
Fixed Income Arbit	0.1446	0.1708	0	0	0.0060	0	0.4970
Global Macro	0.0297	0.0678	0	0	0.0756	0	0.2767
Long/Short Equity	0.0386	0.0612	0	0.0524	0.0978	0	0.2001
Managed Futures	0.0585	0.0693	0			0	0.3037

Mean Var Skew Kurt allocation with $\gamma_1 = 1$, $\gamma_2 = 1$, $\gamma_3 = 1$ and $\gamma_4 = 1$

	mean	std	Q_1	Q_2	Q_3	min	max
Hedge Fund Index	0.0236	0.0774	0	0	0.1861	0	0.4227
Conv Arbi	0.0823	0.1087	0	0.0054	0.1186	0	0.3369
Dedi Short Bias	0.1080	0.0278	0.0954	0.1049	0	0	0.1816
Emer Mark	0.0150	0.0329	0	0	0.0349	0	0.1107
Event Driven	0.0066	0.0407	0	0	0	0	0.2951
Event Driven Distres	0.0483	0.0933	0	0.4169	0.5396	0	0.3624
Event Driven Multi-Strat	0.0113	0.0328	0	0.1093	0.2442	0	0.1652
Event Driven Risk Arbit	0.4323	0.2315	0.3017	0	0	0.0163	0.8174
Fixed Income Arbit	0.1487	0.1490	0	0	0.0530	0	0.4441
Global Macro	0.0178	0.0481	0	0.0836	0.1319	0	0.2090
Long/Short Equity	0.0323	0.0550	0			0	0.1987
Managed Futures	0.0738	0.0715	0			0	0.2700

Mean Var allocation with $\gamma_1 = 1$ and $\gamma_2 = 3$

	mean	std	Q_1	Q_2	Q_3	min	max
Hedge Fund Index	0	0	0	0	0.3355	0	0
Conv Arbi	0.1436	0.1877	0	0	0.1157	0	0.5573
Dedi Short Bias	0.0947	0.0486	0.0586	0.0830	0.0013	0.0286	0.2092
Emer Mark	0.0134	0.0255	0	0	0	0	0.0850
Event Driven	0.0147	0.0574	0	0.0329	0.2102	0	0.2868
Event Driven Distres	0.1322	0.2090	0	0	0	0	0.6997
Event Driven Multi-Strat	0.0198	0.0498	0.1363	0.2044	0.3524	0	0.2578
Event Driven Risk Arbit	0.2858	0.2587	0	0.0758	0.2009	0	0.8427
Fixed Income Arbit	0.0914	0.1415	0	0	0.1457	0	0.3986
Global Macro	0.0966	0.1145	0	0	0.1009	0	0.4414
Long/Short Equity	0.0608	0.1046	0	0.0080	0.0856	0	0.3545
Managed Futures	0.0471	0.0576	0			0	0.1793

Mean Var Skew allocation with $\gamma_1 = 1$, $\gamma_2 = 3$ and $\gamma_3 = 3$

	mean	std	Q_1	Q_2	Q_3	min	max
Hedge Fund Index	0.1430	0.1844	0	0.0081	0.3339	0	0.5510
Conv Arbi	0.0863	0.0520	0.0543	0.0778	0.1085	0	0.2121
Dedi Short Bias	0.0160	0.0283	0	0	0.0331	0	0.0854
Emer Mark	0.0146	0.0619	0	0	0	0	0.3569
Event Driven							
Event Driven Distres	0.0905	0.1843	0	0	0.0597	0	0.6962
Event Driven Multi-Strat	0.0196	0.0549	0	0	0	0	0.2706
Event Driven Risk Arbit	0.3313	0.2991	0.1421	0.2307	0.6021	0	0.9090
Fixed Income Arbit	0.1010	0.1399	0	0	0.1930	0	0.3960
Global Macro	0.0716	0.1195	0	0	0.0839	0	0.4066
Long/Short Equity	0.0614	0.1060	0	0	0.1006	0	0.3616
Managed Futures	0.0647	0.0716	0.0267	0	0.1278	0	0.2222

Mean Var Skew Kurt allocation with $\gamma_1 = 1$, $\gamma_2 = 3$, $\gamma_3 = 1$ and $\gamma_4 = 1$

	mean	std	Q_1	Q_2	Q_3	min	max
Hedge Fund Index	0.0123	0.0553	0	0	0	0	0.3668
Conv Arbi	0.1064	0.1340	0	0.0072	0.2274	0	0.4079
Dedi Short Bias	0.0960	0.0453	0.0745	0.0957	0.1113	0	0.2129
Emer Mark	0.0135	0.0275	0	0	0	0	0.1007
Event Driven	0.0149	0.0513	0	0	0	0	0.2778
Event Driven Distres	0.0691	0.1392	0	0	0.0785	0	0.5524
Event Driven Multi-Strat	0.0211	0.0566	0	0	0	0	0.2672
Event Driven Risk Arbit	0.3843	0.2677	0.2165	0.3384	0.5688	0	0.8753
Fixed Income Arbit	0.1043	0.1375	0	0	0.1907	0	0.4393
Global Macro	0.0481	0.0854	0	0	0.0611	0	0.3098
Long/Short Equity	0.0525	0.0891	0	0	0.0866	0	0.3159
Managed Futures	0.0775	0.0762	0	0.0689	0.1432	0	0.2183

Mean Var allocation with $\gamma_1 = 3$ and $\gamma_2 = 1$

	mean	std	Q_1	Q_2	Q_3	min	max
Hedge Fund Index	0.0119	0.0488	0	0	0	0	0.1690
Conv Arbi	0.0980	0.1212	0	0.0255	0.2178	0	0.3330
Dedi Short Bias	0.1112	0.0279	0.0933	0.1095	0.1266	0.0574	0.1776
Emer Mark	0.0182	0.0355	0	0	0	0	0.1307
Event Driven	0.0067	0.0413	0	0	0	0	0.3081
Event Driven Distres	0.0940	0.1288	0.0023	0.0500	0.0966	0	0.4506
Event Driven Multi-Strat	0.0089	0.0325	0	0	0	0	0.1607
Event Driven Risk Arbit	0.3895	0.2358	0.2578	0.3684	0.4847	0	0.8609
Fixed Income Arbit	0.1511	0.1737	0	0.0725	0.2977	0	0.5027
Global Macro	0.0251	0.0590	0	0	0	0	0.3086
Long/Short Equity	0.0346	0.0576	0	0	0.0585	0	0.1952
Managed Futures	0.0507	0.0594	0	0.0209	0.0787	0	0.2714

Mean Var Skew allocation with $\gamma_1 = 3$, $\gamma_2 = 1$ and $\gamma_3 = 3$

	mean	std	Q_1	Q_2	Q_3	min	max
Hedge Fund Index	0.0037	0.0229	0	0	0	0	
Conv Arbi	0.0990	0.1208	0	0.0232	0.2200	0	
Dedi Short Bias	0.1009	0.0321	0.0845	0.0964	0.1180	0	
Emer Mark	0.0198	0.0370	0	0	0.0177	0	
Event Driven	0.0061	0.0416	0	0	0	0	
Event Driven Distres	0.0561	0.1148	0	0	0.0465	0	
Event Driven Multi-Strat	0.0101	0.0327	0	0	0	0	
Event Driven Risk Arbit	0.4161	0.2586	0.261299573	0.3740	0.6652	0	
Fixed Income Arbit	0.1545	0.1707	0	0.0748	0.2913	0	
Global Macro	0.0343	0.0740	0	0	0.0227	0	
Long/Short Equity	0.0381	0.0602	0	0	0.0704	0	
Managed Futures	0.0614	0.0683	0	0.0520	0.1043	0	

Mean Var Skew Kurt allocation with $\gamma_1 = 3$, $\gamma_2 = 1$, $\gamma_3 = 3$ and $\gamma_4 = 1$

	mean	std	Q_1	Q_2	Q_3	min	max
Hedge Fund Index	0.0238	0.0774	0	0	0	0	0.4158
Conv Arbi	0.0792	0.1059	0	0.0063	0.1736	0	0.3268
Dedi Short Bias	0.1060	0.0280	0.0946	0.1032	0.1172	0	0.1813
Emer Mark	0.0150	0.0326	0	0	0	0	0.1106
Event Driven	0.0066	0.0403	0	0	0.0351	0	0.2916
Event Driven Distres	0.0449	0.0894	0	0	0	0	0.3520
Event Driven Multi-Strat	0.0111	0.0329	0	0	0	0	0.1727
Event Driven Risk Arbit	0.4352	0.2310	0.3007	0.4224	0.5871	0.0255	0.8118
Fixed Income Arbit	0.1523	0.1493	0	0.1196	0.2626	0	0.4442
Global Macro	0.0191	0.0515	0	0	0	0	0.2083
Long/Short Equity	0.0319	0.0549	0	0	0.0491	0	0.1978
Managed Futures	0.0749	0.0721	0	0.0838	0.1339	0	0.2619

Table 4 (continued)

Portfolio allocation using Omega with Reference point 1% annual

	mean	std	Q_1	Q_2	Q_3	min	max
Hedge Fund Index	0	0	0	0	0	0	0
Conv Arbi	0.1243	0.1752	0.0282	0	0.3518	0	0.4127
Dedi Short Bias	0.0718	0.0649	0	0.0632	0.0887	0	0.2677
Emer Mark	0.0248	0.0436	0	0	0.0412	0	0.1332
Event Driven	0.0069	0.0295	0	0	0	0	0.1675
Event Driven Distres	0.1626	0.2698	0	0	0.2546	0	0.9092
Event Driven Multi-Strat	0.0202	0.0489	0	0	0	0	0.2763
Event Driven Risk Arbit	0.2087	0.2463	0	0.1206	0.3289	0	0.8366
Fixed Income Arbit	0.0592	0.1215	0	0	0.0464	0	0.5301
Global Macro	0.1996	0.2093	0	0.1204	0.3854	0	0.6023
Long/Short Equity	0.0686	0.1298	0	0	0.1254	0	0.5128
Managed Futures	0.0533	0.0841	0	0	0.0991	0	0.2278

Portfolio allocation using Omega with Reference point 3% annual

	mean	std	Q_1	Q_2	Q_3	min	max
Hedge Fund Index	0	0	0	0	0	0	0
Conv Arbi	0.1417	0.2050	0.0017	0	0.3987	0	0.5228
Dedi Short Bias	0.0662	0.0658	0	0.0587	0.0806	0	0.2446
Emer Mark	0.0234	0.0439	0	0	0.0346	0	0.1701
Event Driven	0.0031	0.0203	0	0	0	0	0.1491
Event Driven Distres	0.1821	0.2610	0	0.0056	0.2543	0	0.8606
Event Driven Multi-Strat	0.0118	0.0443	0	0	0	0	0.2709
Event Driven Risk Arbit	0.1226	0.1878	0	0.0117	0.1717	0	0.6137
Fixed Income Arbit	0.0560	0.1207	0	0	0	0	0.4334
Global Macro	0.2639	0.2837	0	0.1898	0.4881	0	1
Long/Short Equity	0.0828	0.1454	0	0	0.1431	0	0.5071
Managed Futures	0.0463	0.0801	0	0	0.1029	0	0.2126

Portfolio allocation using Omega with Reference point 7% annual

	mean	std	Q_1	Q_2	Q_3	min	max
Hedge Fund Index	0	0	0	0	0	0	0
Conv Arbi	0.1953	0.2962	0	0	0.5141	0	0.8001
Dedi Short Bias	0.0423	0.0591	0	0.0061	0.0713	0	0.2047
Emer Mark	0.0152	0.0351	0	0	0.0040	0	0.1806
Event Driven	0	0	0	0	0	0	0
Event Driven Distres	0.2165	0.3048	0	0	0.3387	0	0.9939
Event Driven Multi-Strat	0.0059	0.0354	0	0	0	0	0.2544
Event Driven Risk Arbit	0.0175	0.0686	0	0	0	0	0.3961
Fixed Income Arbit	0.0196	0.0796	0	0	0	0	0.5110
Global Macro	0.3897	0.4118	0	0.2464	0.7819	0	1
Long/Short Equity	0.0846	0.1456	0	0	0.1493	0	0.4802
Managed Futures	0.0136	0.0317	0	0	0	0	0.1201

Portfolio allocation using Omega with Reference point 8% annual

	mean	std	Q_1	Q_2	Q_3	min	max
Hedge Fund Index	0	0	0	0	0	0	0
Conv Arbi	0.2030	0.3165	0	0	0.5311	0	0.8371
Dedi Short Bias	0.0356	0.0538	0	0	0.0648	0	0.1880
Emer Mark	0.0203	0.0731	0	0	0	0	0.4890
Event Driven	0	0	0	0	0	0	0
Event Driven Distres	0.2213	0.3109	0	0	0.4008	0	0.9841
Event Driven Multi-Strat	0.0095	0.0508	0	0	0	0	0.3258
Event Driven Risk Arbit	0.0052	0.0299	0	0	0	0	0.2092
Fixed Income Arbit	0.0073	0.0388	0	0	0	0	0.2391
Global Macro	0.4071	0.4172	0	0.3178	0.8287	0	1
Long/Short Equity	0.0833	0.1487	0	0	0.1257	0	0.4679
Managed Futures	0.0074	0.0213	0	0	0	0	0.0827

Portfolio allocation using Omega with Reference point 10% annual

	mean	std	Q_1	Q_2	Q_3	min	max
Hedge Fund Index	0	0	0	0	0	0	0
Conv Arbi	0.2087	0.3340	0	0	0.5765	0	0.9207
Dedi Short Bias	0.0037	0.0139	0	0	0	0	0.0784
Emer Mark	0.0390	0.1148	0	0	0	0	0.4596
Event Driven	0	0	0	0	0	0	0
Event Driven Distres	0.2277	0.3471	0	0	0.5142	0	1
Event Driven Multi-Strat	0.0064	0.0334	0	0	0	0	0.1915
Event Driven Risk Arbit	0.0007	0.0051	0	0	0	0	0.0378
Fixed Income Arbit	0	0	0	0	0	0	0
Global Macro	0.3659	0.4106	0	0.1267	0.7801	0	1
Long/Short Equity	0.0934	0.1731	0	0	0.1563	0	0.7787
Managed Futures	0.0545	0.2292	0	0	0	0	1

Portfolio allocation using Omega with Reference point 15% annual

	mean	std	Q_1	Q_2	Q_3	min	max
Hedge Fund Index	0	0	0	0	0	0	0
Conv Arbi	0.0182	0.1348	0	0	0	0	1
Dedi Short Bias	0	0	0	0	0	0	1
Emer Mark	0.3054	0.4507	0	0	1	0	1
Event Driven	0	0	0	0	0	0	0
Event Driven Distres	0.0563	0.1685	0	0	0	0	0.7114
Event Driven Multi-Strat	0	0	0	0	0	0	0
Event Driven Risk Arbit	0	0	0	0	0	0	0
Fixed Income Arbit	0	0	0	0	0	0	0
Global Macro	0.1895	0.3729	0	0	0.0348	0	1
Long/Short Equity	0.2669	0.4412	0	0	0.9056	0	1
Managed Futures	0.1636	0.3734	0	0	0	0	1

References

1. Athayde, G., Flores, R.G.: The portfolio frontier with higher moments: The undiscovered country. Computing in Economics and Finance 2002 209, Society for Computational Economics (2002)
2. Bernardo, A., Ledoit, O., Brennan, M.,Grinblatt, M., Roll, R., Santa-clara, P., Vila, J.: Gain, loss and asset pricing (1996)
3. Davies, R., Harry, M.K., Sa, L.: Fund of hedge funds portfolio selection: A multiple-objective approach. Journal of derivatives & Hadge Funds **15**, 91–115 (2009)
4. Dowd, K.: Adjusting for risk: An improved sharpe ratio. International Review of Economics & Finance **9**(3), 209–222 (2000)
5. Farinelli, S., Tibiletti, L.: Sharpe thinking in asset ranking with one-sided measures. European Journal of Operational Research **185**(3), 1542–1547 (2008)
6. Gilli, M., Schumann, E., Tollo, G., Cabej, G.: Constructing long/short portfolios with the omega ratio. Swiss Finance Institute Research Paper Series 08-34, Swiss Finance Institute
7. Hitaj, A., Mercuri, L.: Portfolio allocation using multivariate variance gamma. Financial Markets and Portfolio Management **27**(1), 65–99 (2013)
8. Jensen, M.C.: The performance of mutual funds in the period 1945–1964. Journal of Finance **23**, 389–416 (1968)
9. Jondeau, E., Poon, S., Rockinger, M.: Financial modeling under non-gaussian distributions. Springer Finance. Springer (2007)
10. Kane, S.J., Bartholomew, M.C., Cross, M., Dewar, M.: Optimizing omega. Journal of Global Optimization, 153–167 (2009)
11. Kaplan, P.D, Knowles, J.A.: Kappa: A generalized downside risk-adjusted performance measure. January (2004)
12. Keating, C., Shadwick, W.F.: A universal performance measure. Technical report, The Finance development center, London, May (2002)
13. Kendall, M.G., Hill, A., Bradford: The analysis of economic Time-Series part i: Prices. Journal of the Royal Statistical Society A **116**(1), 11–34 (1953)
14. Martellini, L., Ziemann, V.: Improved estimates of higher-order comoments and implications for portfolio selection. Review of Financial Studies **23**(4), 1467–1502 (2010)
15. Rachev, S.T., Stoyanov, S., Fabozzi, F.J.: Advanced Stochastic Models, Risk Assessment, and Portfolio. John Wiley (2007)
16. Sharpe, W.: Mutual fund performance. The Journal of Business **39**, 119 (1965)
17. Sortino, F.A., Price, L.N.: Performance measurement in a downside risk framework. Journal of Investing **3**(3), 59–64 (1994)
18. Tibiletti, L., Farinelli, S.: Upside and downside risk with a benchmark. Atlantic Economic Journal **31**(4), 387–387 (2003)
19. Zakamouline, V., Koekebakker, S.: Portfolio performance evaluation with generalized sharpe ratios: Beyond the mean and variance. Journal of Banking & Finance **33**(7), 1242–1254 (2009)

Investment Rankings via an Objective Measure of Riskiness: A Case Study

Maria Erminia Marina and Marina Resta

Abstract We introduce a measure for riskiness which is able, at the same time, to let the investor accepting/rejecting gambles in a way as objective as possible (i.e. depending only on the probabilistic features of the gamble), and to take his own risk posture into account, that is by considering the risk attitude expressed by his utility function. We will briefly recall and discuss some theoretical properties, and we will give proof of our results by ranking 30 largest-growth mutual funds; finally, we will compare the results with those of other indexes.

1 Background

Since the publication of Markowitz [18, 19] and Tobin [29] seminal papers, ranking investments via a risk/reward approach has become an appealing and promising instrument to both researcher and practitioners. However, whereas there is a general agreement to assess the reward of a financial position by using its average value, there is still an intense debate concerning how to measure related risk.

As a consequence, it is not surprising that the literature on risk measures is so voluminous: contributions span from dispersion [28] to deviations measures [22]; the Expected Utility approach [1, 21] has been exploited, too; indeed behavioral measures [12] as well as measures that incorporate investor's psychological motivation [13, 14] have been extensively discussed; VaR [11], CVaR [20] and coherent measures of risk [2] undoubtely dominated the scene over the last fifteen years.

M.E. Marina
Department of Economics, University of Genova, via Vivaldi 5, 16126 Genoa, Italy
e-mail: marina@economia.unige.it

M. Resta (✉)
Department of Economics, University of Genova, via Vivaldi 5, 16126 Genoa, Italy
e-mail: resta@economia.unige.it

M. Corazza, C. Pizzi (eds.), *Mathematical and Statistical Methods for Actuarial Sciences and Finance*, DOI 10.1007/978-3-319-02499-8_18, © Springer International Publishing Switzerland 2014

Recently a new promising research stream has been opened by Aumann and Serrano [3] who introduced an objective measure of riskiness: provided the relevance for our work, what remains of this section will be devoted to give some further insights on it.

Let us imagine an individual is asked to judge whether accept or not an investment proposal (gamble) g. From a conceptual point of view, the investor will take his decision depending on two basic factors: (i) how risky the gamble is, and (ii) how averse the individual is to risk. From a formal point of view, the first factor depends only on the probabilistic features of the gamble, while the second factor is related to the investor's risk posture, and hence to his utility function u. Aumann and Serrano [3] say that an agent with utility function u accepts a gamble g at wealth w if: $E[u(w+g)] > u(w)$. Besides, the agent i is uniformly no less risk–averse than agent j if whenever i accepts a gamble at some wealth, j accepts that gamble at any wealth. In order to compare gambles [3] define an index Q to be a mapping that assigns a positive real number to each gamble g: given an index Q, [3] says that *the gamble g is riskier than h if:* $Q(g) > Q(h)$. Moreover [3] proposed two axioms that an index should satisfy. The first is duality: given two gambles g and h and two agents i and j, such that i is uniformly no less risk averse than j, if i accepts a gamble g at wealth w and $Q(g) > Q(h)$, then j accepts the gamble h at wealth w. The second is homogeneity: for any positive real number t: $Q(t g) = t Q(g)$; [3] demonstrate that there is an index R_{AS} that satisfies both duality and homogeneity; for every gamble g, $R_{AS}(g)$ is the unique positive solution of the equation: $E[e^{-g/R_{AS}(g)}] = 1$. Besides, R_{AS} has a number of very appealing features, for example, it is monotonic with respect to first order stochastic dominance. Under such profile R_{AS} is a very inspiring index, like testified by the number of both theoretical and practical contributions that began to bloom on the topic. Moving towards a theoretical direction, [24] provided some general characterizations for the class of risk measures which respect comparative risk aversion. A second research vein concerns the practical application of R_{AS}: [9] analyzed the problem with an eye to actuarial applications, while [10] focused on the evaluation of funds performance, and introduced an index that incorporates R_{AS}. Besides, a comparison between CVaR and R_{AS} in the performance evaluation of Tel Aviv market quoted funds is discussed in [25]. Finally, [7, 8] suggested an operational measure of risk that can be used to monitor the critical wealth of the investor.

Our contribution intends to make a bridge among those streams of research. In particular, we are aimed to extend the original index of Aumann and Serrano: [3] define the R_{AS} index over the set of gambles with positive expected value (equivalently: with expected value of losses lower than the one of gains.) In practical applications, however, the investor could be more demanding, asking to consider for a fixed level $\theta \in (0,1)$ the gambles in the subset $\mathscr{A}_\theta$ such that the ratio between the expected value of losses and the expected value of gains is lower than such threshold value θ. On following we are going to show that for every $\theta \in (0,1]$ it is possible to build up an index of riskiness I_θ over the set $\mathscr{A}_\theta$, with $I_1 = R_{AS}$, satisfying both homogeneity and duality properties. We provide an empirical illustration using mutual funds investments, and we compare the rankings induced by our index I_θ and by other indexes.

The remainder of the paper is as follows. In Sect. 2 we will introduce the value V which lets the investor to accept/reject gambles taking his own risk posture into account ; next we will derive the index of riskiness I_θ. In Sect. 3 we will present some computational results, ranking 30 largest-growth mutual funds. Section 4 will conclude.will describe it and determine proper conditions under which the index can aid the investor to skim among a number of alternatives.

2 An Objective Measure of Riskiness and Its Main Features

We consider a probability space (Ω, M, P), where P is a probability measure on M (the field of measurable subsets of Ω); we assume a gamble g is a random variable defined on Ω, bounded from below, and following [3] with positive expected value and positive probability of negative values, i.e.:

$$E[g] > 0; \; P[g < 0] > 0. \tag{1}$$

We focus on risk averse agents, i.e. those investors whose utility function $u : \mathbb{R} \to \mathbb{R}$ is strictly increasing and concave. We denote by $\mathscr{U}$ the set of all those functions u. For every gamble g we set $g_+(x) = \max\{g(x), 0\}$, and we give the following definition.

Definition 1 For every gamble g and every utility function u it is:

$$V(u, g, w) = \frac{E[u(w + g)] - u(w)}{E\{[u(w + g_+)]\} - u(w)} \tag{2}$$

where w is the investor's wealth.

In practice, we consider a gamble g that is acceptable at the wealth w in the Aumann and Serrano's sense:

$$E[u(w + g)] - u(w) > 0, \tag{3}$$

and in addition V examines the relation existing between the increment in the expected utility value computed over all the possible outcomes of g, and the increment in the expected utility due to all positive values of g, respectively.

Besides, we have discovered [16, 17] that V has some interesting features that we are now going to enumerate. First of all we have:

$$V(u, g, w) < \frac{E[g]}{E[g_+]} < 1. \tag{4}$$

Moreover:

Theorem 1 *Let us assume that g, h are two gambles. If g first order stochastically dominates h, then:*

$$V(u, g, w) > V(u, h, w), \tag{5}$$

for all wealth levels w, and for all u.

Another interesting feature concerns the behaviour of V under concave transformations of the investor's utility function.

Theorem 2 *Let u_1, u_2 being two utility functions. Moreover, assume $u_1 = \phi \circ u_2$, where ϕ is a (strictly) concave function. Then:*

$$V(u_1, g, w)(<) \leq V(u_2, g, w) \tag{6}$$

for all $w \geq 0$, and all the gambles g.

From Theorem 2 it is possible to draw down a notable result under the expected utility framework. To this aim, it aids to remember that for every utility function u twice continuous differentiable, $\rho_u(x) = -\frac{u''(x)}{u'(x)}$ is the Arrow–Pratt [1, 21] coefficient of absolute risk aversion. Given two utility functions u_1, u_2, corresponding to investors i_1 and i_2 respectively, by the Arrow–Pratt theorem we get: $\rho_{u_1}(x)(>) \geq \rho_{u_2}(x)$ iff u_1 is a (strictly) concave transformation of u_2.

Now we are moving to provide an extension of Aumann–Serrano's results. However, whereas in the former case the acceptability of gamble g is stated by (3), here we are strenghtening such concept, since we additionally require that the value V exceeds a fixed threshold. In this spirit, we introduce the following definition.

Definition 2 Consider an agent with utility function u and wealth $w \geq 0$. The agent accepts the gamble g at the threshold value $\theta \in (0, 1]$ if:

$$V(u, g, w) > 1 - \theta.$$

Now we assume an agent with utility function $u_\gamma(x) = -e^{-\gamma x}$, $\gamma > 0$, i.e. a *CARA* agent who has constant absolute risk adversion $\rho_u(x) = \gamma$. For every gamble g let:

$$f_g(\gamma) = \frac{E[1 - e^{-\gamma g}]}{E[1 - e^{-\gamma g_+}]}. \tag{7}$$

It is possible to prove [17] that for every gamble g the continuous function f_g is strictly decreasing in $(0, +\infty)$. Moreover we can show that:

$$\lim_{\gamma \to 0^+} f_g(\gamma) = \frac{E[g]}{E[g_+]},$$

and:

$$\lim_{\gamma \to +\infty} f_g(\gamma) = -\infty.$$

We observe that for every gamble g and for every $w \geq 0$ it is:

$$f_g(\gamma) = V(u_\gamma, g, w)$$

and hence for every $\theta \in (0, 1]$, the equation:

$$f_g(1/\alpha) = 1 - \theta, \tag{8}$$

will admit a unique positive solution if:

$$1 - \theta < \frac{E[g]}{E[g_+]}. \tag{9}$$

Now, let $\theta \in (0,1]$. We define the set:

$$\mathscr{A}_\theta = \left\{ g : \frac{E[g]}{E[g_+]} > 1 - \theta \right\}. \tag{10}$$

Then, for all $g \in \mathscr{A}_\theta$ we denote by $I_\theta(g)$ the unique solution of (8).

As the function f_g is strictly decreasing, it follows that a CARA agent with constant absolute risk aversion $\gamma < \frac{1}{I_\theta(g)}$ accepts the gamble g at the threshold value θ; moreover, if h is another gamble such that $I_\theta(g) > I_\theta(h)$, he also accepts the gamble h at the same threshold value θ, so that I_θ would deem g more "risky"than h.

Note that the index I_θ provides an objective way (that is: independent on the specific agent) to measure the risk at threshold value θ, and from Theorem 2 is monotonic with respect to first order stochastic dominance (i.e.: increasing the gains and decreasing the losses lowers the riskiness). Moreover, by its definition (see (8)) it comes that I_θ has homogeneity properties, i.e. for any positive real number t: $I_\theta(t\,g) = t\,I_\theta(g)$. Finally, one can note that for $\theta = 1$ we have $I_1 = R_{AS}$. To conclude, the following theorem holds.

Theorem 3 *Given two utility functions u and v such that:*

$$\rho_u(x) \geq \bar{\rho} \geq \rho_v(x), \forall x, \tag{11}$$

if $g, h \in \mathscr{A}_\theta$ are two gambles with $I_\theta(g) > I_\theta(h)$, and if $V(u,g,w) > 1 - \theta$, then:

$$V(v,h,w) > 1 - \theta.$$

We observe that the index I_θ posits a duality between riskiness (at the fixed threshold θ) and risk aversion. In fact, if u and v are the utility functions of two agents i and j, the condition stated in (11) is equivalent to say that the agent i is uniformly no less risk averse than j (see our introduction as well as [3]).

3 An Application

We consider a modified version of the datasets employed in the works of [4] and [10]: we examine monthly excess returns (computed from monthly fund returns) of 30 mutual funds investments:

$$R(i) = r_i - r_f, (i = 1, \ldots, 30), \tag{12}$$

where r_i is the monthly return of the i-th investment, and r_f is the one-month US Treasury bill rate. The time horizon under observation spans from January 1990 to January 2011, resulting in 252 overall observations. The data were employed to compute the values of our index I_θ and rank the funds accordingly. We compare our results to those of the reciprocal of the Sharpe Ratio (SR) [26, 27]:

$$SR(R_i) = \frac{E[R(i)]}{\sigma(R(i))}, \ (i = 1, \ldots, 30), \tag{13}$$

where σ is the standard deviation. In addition, we examine the results obtained by using the reciprocal of the Economic Performance Measure (EPM) discussed in [10], being:

$$EPM(R(i)) = \frac{E[R(i)]}{R_{AS}(R(i))}, \ (i = 1, \ldots, 30), \tag{14}$$

where R_{AS} is the index of Aumann and Serrano. Both SR and EPM are performance measures, so moving to their reciprocal, they may straightforwardly considered two indexes of riskiness. Note that, in contrast to SR, both I_θ and EPM are monotonic with respect to first order stochastic dominance. Besides, likewise the R_{AS} index, we can refer to I_θ as an economic index of riskiness. The most important feature of R_{AS}, in fact, relates to the duality axiom which essentially requires that if a gamble is accepted by an investor, less risk averse investors accept less risky gambles. In this light, this is an economically motivated axiom, and therefore rightly Aumann and Serrano claim R_{AS} index to be an economic index of riskiness. Finally, it sticks out immediately that wheres both R_{AS} and EPM are defined over the set $\mathscr{A}_1$, on the other hand, if $\theta \in (0, 1)$, I_θ leaves out the funds that do not belong to the set $\mathscr{A}_\theta$, thus resulting in an additional reduction in the overall number of funds the investor should discard by default.

Moving to the results, Table 1 reports values for the aforementioned indexes, with the rankings they generated given in brackets. The complete list of the funds under examination is provided in Appendix. Here, for sake of readability we adopted the labels: F1, F2, ..., F30, to denote the funds, and we put them in Column 1. Columns 2–4, on the other hand, report the results obtained by R_{AS} and the reciprocal of SR and EPM, respectively. For a better understanding of the discussion, we have reported in Column 5 the values $E[R(i)]/E[R(i)_+]$ for every fund $i = 1, \ldots, 30$. For what is concerning the index I_θ, by (9), the ranking of all 30 funds is possible only if $\theta > 1 - \min\{E[R(i)]/E[R(i)_+], i = 1, \ldots, 30\} = 1 - 0.332316 = 0.667684$. Note that the value 0.332316 corresponds to the fund F29 (Vanguard U.S. Growth in Appendix). Finally, Columns 6-8 show the values and rankings obtained for $\theta = 0.6677$, $\theta = 0.9870$, and $\theta = 0.4000$.

Besides, we analyzed the way the results provided by our index are related to those of the other indexes. This was done by way of the Kendall's τ [15] statistics: it is well known that the Kendall's τ equals one, if two rankings perfectly agree, zero if they are independent, and -1 if they perfectly disagree. Results are provided in Table 2.

Note that the rankings induced by R_{AS}, SR, EPM and I_θ, for $\theta = 0.6677$ and $\theta = 0.987$ involve all thirty funds under observation. However, when the investor is more demanding, asking to consider the value $\theta = 0.40$ at which some funds

Table 1 Results obtained from different risk/performance measures. SR is the acronym for Sharpe Ratio, R_{AS} is the index of Aumann and Serrano, and EPM is the Economic Performance Measure. $E[R(i)]/E[R(i)]_+$ is the ratio opposing the expected value of the funds returns to the expected value of positive returns. Finally I_θ is the abbreviation for our index. Note that $R_{AS} = I_1$

	R_{AS}	$1/SR$	$1/EPM$	$E[R(i)]/$ $E[R(i)]_+$	$I_{\theta=0.6677}$	$I_{\theta=0.987}$	$I_{\theta=0.4}$
F01	2.836 (10)	38.185 (22)	526.316 (19)	0.43554	3.6161 (10)	2.8599 (10)	3.8309 (8)
F02	3.533 (24)	56.167 (24)	1063.83 (25)	0.34134	4.4777 (23)	3.5622 (24)	–
F03	3.358 (21)	34.235 (20)	1000 (23)	0.48082	4.3504 (21)	3.3883 (21)	4.629 (17)
F04	3.046 (18)	27.576 (15)	384.615 (16)	0.38425	3.789 (14)	3.0693 (18)	–
F05	3.556 (25)	34.732 (21)	526.316 (20)	0.38612	4.5583 (24)	3.5869 (25)	–
F06	2.691 (5)	19.945 (5)	140.845 (3)	0.43888	3.3651 (5)	2.712 (5)	3.5463 (4)
F07	2.875 (12)	26.957 (13)	416.667 (17)	0.50545	3.7231 (12)	2.9007 (12)	3.9613 (11)
F08	2.564 (3)	18.882 (2)	75.188 (2)	0.56638	3.2661 (3)	2.5859 (3)	3.4591 (2)
F09	2.857 (11)	29.245 (17)	303.03 (12)	0.42639	3.6266 (11)	2.8813 (11)	3.8371 (9)
F10	2.597 (4)	22.477 (7)	178.571 (5)	0.48731	3.3193 (4)	2.6197 (4)	3.5185 (3)
F11	2.725 (6)	19.435 (4)	270.27 (10)	0.42979	3.4733 (6)	2.7483 (6)	3.679 (5)
F12	2.907 (13)	26.089 (12)	256.41 (9)	0.45311	3.7383 (13)	2.9328 (13)	3.9694 (12)
F13	2.765 (8)	23.419 (8)	303.03 (13)	0.40167	3.5882 (8)	2.7902 (8)	3.8198 (7)
F14	2.755 (7)	18.245 (1)	175.439 (4)	0.44219	3.5707 (7)	2.7802 (7)	3.7998 (6)
F15	3.489 (22)	24.351 (11)	181.818 (6)	0.38274	4.4598 (22)	3.5188 (22)	–
F16	3.813 (28)	22.171 (6)	222.222 (7)	0.48905	4.8579 (27)	3.8455 (28)	5.1451 (20)
F17	2.996 (15)	76.719 (28)	1408.451 (26)	0.34831	3.863 (18)	3.0227 (15)	–
F18	3.012 (17)	60.041 (25)	1538.462 (28)	0.44243	3.842 (16)	3.0379 (17)	4.0705 (14)
F19	2.774 (9)	43.82 (23)	769.231 (22)	0.43682	3.6142 (9)	2.7995 (9)	3.8518 (10)
F20	2.412 (1)	23.578 (9)	344.828 (15)	0.44980	3.1833 (2)	2.4352 (1)	3.4051 (1)
F21	3.124 (19)	33 (19)	588 (21)	0.43306	4.1588 (19)	3.1544 (19)	4.4599 (16)
F22	4.033 (30)	1835.451 (30)	100000 (30)	0.43304	5.2226 (30)	4.0693 (30)	5.5565 (22)
F23	2.967 (14)	65.081 (27)	1428.571 (27)	0.43138	3.8481 (17)	2.9939 (14)	4.0959 (15)
F24	3.785 (27)	77.17 (29)	1666.667 (29)	0.42324	4.8597 (28)	3.8178 (27)	5.1582 (21)
F25	2.432 (2)	19.189 (3)	238.095 (8)	0.33761	2.9413 (1)	2.4486 (2)	–
F26	3.677 (26)	26.968 (14)	285.714 (11)	0.42663	4.6633 (26)	3.7081 (26)	4.9328 (19)
F27	3.518 (23)	30.587 (18)	416.667 (18)	0.44355	4.5637 (25)	3.5497 (23)	4.8579 (18)
F28	3.861 (29)	28.431 (16)	322.581 (14)	0.38134	5.0211 (29)	3.8959 (29)	–
F29	3.263 (20)	61.818 (26)	1000 (24)	**0.33232**	4.1973 (20)	3.2917 (20)	–
F30	2.998 (16)	24.342 (10)	24.342 (1)	0.46603	3.8166 (15)	3.0233 (16)	4.0414 (13)

Table 2 Kendall's τ for the rank correlation among the rankings based on R_{AS}, $1/SR$, $1/EPM$ and, I_θ, for $\theta = 0.6677$, $\theta = 0.987$, and $\theta = 0.4$

	R_{AS}	$1/SR$	$1/EPM$	$I_{\theta=0.6677}$	$I_{\theta=0.987}$	$I_{\theta=0.4}$
R_{AS}	1	0.3885	0.2650	0.9436	1	0.9567
$1/SR$	0.3885	1	0.6060	0.4143	0.3885	0.4545
$1/EPM$	0.2650	0.6060	1	0.2814	0.2650	0.3957
$I_{\theta=0.6677}$	0.9436	0.4143	0.2814	1	0.9436	0.9827
$I_{\theta=0.987}$	1	0.3885	0.2650	0.9436	1	0.9567
$I_{\theta=0.4}$	0.9567	0.4545	0.3957	0.9827	0.9567	1

(gambles) do not satisfy the requirements stated by (10), I_θ cannot be computed for all thirty funds, but only for twenty-two of them. Obviously, in this latter case Kendall's τ compares the twenty-two funds ranked by $I_{0.40}$.

A closer look to Table 2 shows that all indexes are positively correlated. Moreover, in the case of $\theta = 0.987$ the rankings induced by I_θ and R_{AS} are identical, but varying the threshold level θ could modify the order of riskiness. This is, for instance, what happens if instead of $\theta = 0.987$ we consider $\theta = 0.6677$. Finally, as observed in [3], SR and EPM rank normal gambles in the same way, but matters are different for non–normal gambles, as in our discussion case.

4 Conclusions

We introduced and discussed a new index of riskiness I_θ which is able, at the same time, to let the investor accepting/rejecting gambles in way as objective as possible (i.e. depending only on the probabilistic features of the gamble), and to take his own risk posture into account; in particular, given two investors, if the one *more* risk averse accepts at threshold value θ the riskier of two gambles, then the other less adverse investor accepts the less risky gamble at the same threshold value θ (duality property). Our index is inspired by Aumann and Serrano's R_{AS}, and when $\theta = 1$ it results: $I_1 = R_{AS}$. Moreover, our index also verifies properties such as homogeneity, monotonicity, and consistency with respect to first order stochastic dominance. We tested the effectiveness of our index in a practical application. We ranked 30 largest-growth mutual funds; the time horizon under observation spans from January 1990 to January 2011, resulting in 252 overall observations. We computed I_θ for different values of the threshold θ ($\theta = 0.4$, $\theta = 0.6677$, $\theta = 0.987$ and $\theta = 1$).

We then compared the results with those provided by the reciprocals of the Sharpe Ratio (SR), and of the Economic Performance Measure (EPM) recently discussed by Homm and Pigorsch.

By comparison of the rankings we found out that our index is undoubtedly *more demanding* than the others under examination, as in our example in the case $\theta = 0.4000$ the investor has to discard some funds from the ranking. In practice, we tried to offer a tool assuring the investor a greater safety: more precisely, once fixed a level θ, our index takes under consideration a gamble (an investment) g only if the ratio between the expected value of its losses and the expected value of its gains is less then such threshold value θ.

Appendix

Table 3 reports the complete name of the funds we examined, coupled to the labels we employed throughout the paper to denote them.

Table 3 Funds employed in our study: the complete list. The first column reports the labels used to indicate each fund throughout the paper, the second column shows the fund name

ID	Fund name
F01	AIM Value A
F02	AIM Weingarten A
F03	Amcap
F04	American Cent-20thC Growth
F05	American Cent-20thC Select
F06	Brandywine
F07	Davis NY Venture A
F08	Fidelity Contrafund
F09	Fidelity Destiny I
F10	Fidelity Destiny II
F11	Fidelity Growth Company
F12	Fidelity Magellan
F13	Fidelity OTC
F14	Fidelity Retirement Growth
F15	Fidelity Trend
F16	Fidelity Value
F17	IDS Growth A
F18	IDS New Dimensions A
F19	Janus
F20	Janus Twenty
F21	Legg Mason Value Prim
F22	Neuberger&Berman Part
F23	New Economy
F24	Nicholas
F25	PBHG Growth PBHG
F26	Prudential Equity B
F27	T. Rowe Price Growth Stock
F28	Van Kampen Am Cap Pace A
F29	Vanguard U.S. Growth
F30	Vanguard/Primecap

References

1. Arrow, K.J.: Alternative approaches to the theory of choice in risk–taking situations. Econom. **19**, 404–37 (1951)
2. Artzner, P., Delbaen, F., Eber, J.M., Heath, D.: Coherent Measures of Risk. Math. Fin. **9**(3) 203–228 (1999)
3. Aumann, R., Serrano, R.: An economic index of riskiness. J. Polit. Ec. **116**(5), 810–836 (2008)
4. Bao, Y.: Estimation risk-adjusted Sharpe ratio and fund performance ranking under a general return distribution. J Fin. Econom. **7**, 152–173 (2009)

5. Barndorff-Nielsen, O.E.: Normal inverse Gaussian distributions and stochastic volatility modelling. Scandin J Stat. **24**, 1–13 (1997)
6. Bollerslev, T., Kretschmer, U., Pigorsch, C., Tauchen, G.: A discrete-time model for daily S&P 500 returns and realized variations: jumps and leverage effects. J Econom. **150**, 151–166 (2009)
7. Foster, D.P., Hart, S.: An Operational Measure of Riskiness. J. Polit. Ec. **117**(5), 785–814 (2009)
8. Hart, S.: Comparing Risks by Acceptance and Rejection. J. Polit. Ec. **119**(4), 617–638 (2011)
9. Homm, U., Pigorsch, C.: An operational interpretation and existence of the Aumann-Serrano index of riskiness. Ec Lett **114**(3), 265–267 (2012)
10. Homm, U., Pigorsch, C.: Beyond the Sharpe Ratio: An Application of the Aumann-Serrano Index to Performance Measurement. J. Bank Finance (2012). http://dx.doi.org/10.1016/j.jbankfin.2012.04.005
11. Morgan, J.P.: RiskMetrics Technical Document. Morgan Guarantee trust Company (1996)
12. Jia, J., Dyer, J.S., Butler, J.C.: Measures of Perceived Risk. Manag Sc **45**, 519–532 (1999)
13. Luce, R.D.: Several possible measures of risk. Theor. Dec. **12**, 217–228 (1980)
14. Luce, R.D.: Several possible measures of risk. Correction. Theor. Dec. **13**, 381(1981)
15. Kendall, M.: A New Measure of Rank Correlation Biometrika **30**(1–2), 81–89 (1938)
16. Marina, M.E., Resta, M.: On some measures for riskiness. Proc XVIII Conf on Risk Theory, Campobasso (5 September 2011)
17. Marina, M.E., Resta, M.: A generalization of the Aumann–Serrano index of riskiness: theoretical results and some applications, submitted (2012)
18. Markowitz, H.M.: Portfolio Selection. J. Fin. **7**, 77–91 (1952)
19. Markowitz, H.M.: Portfolio Selection: Efficient Diversification of Investment. Yale University Press. New Haven, USA (1959)
20. Pflug, G.C.: Some remarks on the value-at-risk and the conditional value-at-risk. In: Uryasev, S.P. (ed.) Probabilistic Constrainted Optimization: Methodology and Applications, MA, pp. 278–287. Kluwer, Norwell (2000)
21. Pratt, J.: Risk Aversion in the Small and in the Large. Econom. **32**(1, 2), 122–136 (1964)
22. Rockafellar, R.T., Uryasev, S., Zabarankin, M.: Generalized deviations in risk analysis. Fin. Stoch. **10**, 51–74 (2006)
23. Schultze, K., Existence and Computation of the Aumann-Serrano Index of Riskiness. Working Paper. McMaster University (2011)
24. Schultze, K.: Risk Measures Respecting Comparative Risk Aversion. Working Paper. McMaster University (2011)
25. Shalit, H.: Measuring Risk in Israeli Mutual Funds: Conditional Value-at-Risk vs. Aumann-Serrano Riskiness Index. Working paper Ben-Gurion. University of the Negev (2011)
26. Sharpe, W.F.: Mutual fund performance. J Busin **39**, 119–138 (1966)
27. Sharpe, W.F.: The Sharpe ratio. J. Port. Man. **21**, 49–58 (1994)
28. Stone, B.K.: A General Class of Three-Parameter Risk Measures. J. Fin. **28**, 675–685 (1973)
29. Tobin, J.: Liquidity Preference as a Behavior Toward Risk. Rev. Ec. Stud. **25**, 65–86 (1958)

A Squared Rank Assessment of the Difference Between US and European Firm Valuation Ratios

Marco Marozzi

Abstract Financial ratios are useful in determining the financial strengths and weaknesses of firms. The most commonly used methods for comparing firms through financial ratios are multivariate analysis of variance and multiple discriminant analysis. These methods have been very often used for inferential purposes when the underlying assumptions (e.g. random sampling, normality, homogeneity of variances...) are not met. A method for comparing firm financial ratios is proposed, it is based on squared ranks and does not require any particular assumptions. The proposed method is devised to explicitly consider the possible difference in variances and to take also into account the dependence among the financial ratios. It is robust against skewness and heavy tailness. This aspect is very important because usually financial ratios, even after removing outliers, are highly skewed and heavy tailed. An application for studying the difference between US and European firms is discussed.

1 Introduction

In the financial literature, comparison of financial characteristics of groups of firms through financial ratios is a popular research. Financial ratios are useful in determining financial strengths and weaknesses of firms. Liquidity, profitability, leverage, solvency and activity ratios are generally considered. In the literature, opinions on the order of their importance differ widely [11]. For a very simple method to select financial ratios according to their importance in ranking firms see [22] and [23]. The most commonly used methods for comparing firms through financial ratios are multivariate analysis of variance (MANOVA) and multiple discriminant analysis

M. Marozzi (✉)
Department of Economics, Statistics and Finance, University of Calabria, Via Bucci 0C, 87036 Rende (CS), Italy
e-mail: marco.marozzi@unical.it

M. Corazza, C. Pizzi (eds.), *Mathematical and Statistical Methods for Actuarial Sciences and Finance*, DOI 10.1007/978-3-319-02499-8_19, © Springer International Publishing Switzerland 2014

(MDA). These methods, as discussed in the next section, have been very often used for inferential purposes when the underlying assumptions (e.g. random sampling, normality, homogeneity of variances ...) are not met. It is very important to note that when the underlying assumptions of MANOVA and MDA are not met the conclusions can be only merely descriptive rather than inferential as (wrongly) claimed in many papers. In this paper we address the comparison of financial ratios from the descriptive point of view by proposing a method based on squared ranks that can be applied to financial data even when the assumptions required by the traditional methods are not met.

In Sect. 2 we discuss the shortcomings of traditional methods for comparing firms through financial ratios. Our method is presented in Sect. 3 and it is applied in Sect. 4 for studying the difference between US and European firms. Section 5 summarizes the findings and conclusions of the study.

2 Traditional Methods for Comparing the Financial Characteristics of Firms

The most commonly used methods for comparing firms through financial ratios are MANOVA and MDA. They are very often used for inferential purposes. In this case, the methods require stringent assumptions. The assumptions needed for MANOVA are [8]:

- units (firms) are randomly sampled from the populations of interest;
- observations are independent of one another;
- the dependent variables (financial ratios) have a multivariate normal distribution within each group (in our case US and European firm groups);
- the K groups (in our case $K = 2$) have a common within-group population variance/covariance matrix. This assumption is twofold: the homogeneity of variance assumption should be met for each financial ratio; the correlation between any two financial ratios must be the same in the two groups.

It is important to note that in practice it is unlikely that all assumptions are met. MANOVA is relatively robust against violations of assumptions in many situations. MANOVA is not robust to violations of one or both of the first two assumptions. In the absence of outliers, MANOVA is quite robust to violations of the normal assumption. MANOVA is not robust to violations of the variance/covariance homogeneity assumption in particular when the sample sizes are unequal. It is important to limit the number of financial ratios because the power of MANOVA tests tends to decrease as the number of variables increases (unless the sample sizes increase as well). [27] used MANOVA for comparing the financial characteristics of US and Japanese electric and electrical equipment manufacturing firms. They used MANOVA for inferential purposes even thought the data are not random samples, the financial ratios have non normal distribution and the homogeneity of variances has not been checked. Very similar comments apply also to [25].

The major assumptions of MDA are [19]:

- the group of firms are random samples;
- each financial ratio is normally distributed;
- US and European group sizes should not be grossly different and should be at least five times the number of financial ratios;
- the variance/covariance matrices for the US and European groups should be approximately equal.

The very well known paper of [1] uses MDA to analyze a set of financial ratios for the purposes of bankruptcy prediction. Inferential conclusions are drawn from the data even thought data are not random samples, both normal distribution assumption and homogeneity of variance/covariance assumption have not been checked. For these reasons, conclusions should be considered descriptive rather than inferential (as wrongly done by the author). The majority of financial papers on MDA draw inferences from the data (for example computing p-values) even though one, some or all of the assumptions are not fulfilled in the data, among other see [5, 12, 15, 26, 32, 34, 36, 37].

In finance, the analysis of data that do not fulfill the requirements of MANOVA nor MDA is not an issue we should wave aside. The typical data analyzed in finance are taken from a database of publicly traded firms. Data are not random samples because all the firms that have no missing data are considered. Moreover, many financial ratios are highly skewed and heavy tailed. Most of financial ratios are restricted from taking on values below zero but can be very large positive values [11], as a consequence they are not normally distributed. Non normality due to skewness and heavy tails was noted by many early empirical studies, see [3, 6, 7, 13, 17, 24, 29, 33]. Data transformations using square root or logarithm and procedures for removing outliers may help to ease the problem of non normality [13]. If it is not possible to use random samples of firms, a solution is to work within the permutation/resampling framework. The permutation framework is justified because we may assume that under the null hypothesis of no difference due to grouping, the observed datum may be indifferently assigned to either group 1 or group 2 (i.e. the exchangeability assumption under the null hypothesis is met) and therefore conditional (on the observed data) inference can be drawn [31]. Rarely in practice you have random sampling and therefore unconditional inferences associated with parametric tests, being based on random sampling, often cannot be drawn in practice. Financial literature show a lack of attention on the formal assumptions required by statistical methods like MANOVA and MDA especially to draw inferential conclusions from the data.

3 A New Method for Assessing the Differences Between Financial Ratios

Financial ratios are commonly used for comparing performance, financial status, solvency position and borrowing power of firms. They provide shareholders, managers, creditors, potential investors and bankers valuable information in assessing the liquidity, profitability and debt position of a firm. In this section we propose a method to compare firm financial ratios which does not require any particular assumptions because we follow a descriptive point of view. Contrary to MANOVA and MDA our method does not require random sampling nor normality nor homogeneity of variance/covariance. It is devised to explicitly consider the possible difference in variances as well as the dependence among the financial ratios.

Let $\{_lX_{ij}; i = 1,2; j = 1,\ldots,n_i; l = 1,\ldots,L\}$ be the data set, where $_lX_{ij}$ denotes the value of financial ratio l for firm j of group i, $n = n_1 + n_2$. We say that the two groups are not different so far as $_lX$ is concerned if both means $M(_lX_i)$ and variances $VAR(_lX_i)$ of $_lX$ in the two groups are equal. To grade the difference between groups when $M(_lX_1) \neq M(_lX_2)$ and/or $VAR(_lX_1) \neq VAR(_lX_2)$ we compute the following statistic

$$_lC = C(_l\underline{X}_1, _l\underline{X}_2) = C(_lU, _lV) = _lU^2 + _lV^2 - 2\rho_l U_l V, \tag{1}$$

where $_l\underline{X}_i$ is the $_lX$ i-th sample,

$$_lU = U(_l\underline{R}_1) = \frac{6\sum_{i=1}^{n_1} {}_lR_{1i}^2 - n_1(n+1)(2n+1)}{\sqrt{n_1 n_2(n+1)(2n+1)(8n+11)/5}},$$

$$_lV = V(_l\underline{R}_1) = \frac{6\sum_{i=1}^{n_1} (n+1 - {}_lR_{1i})^2 - n_1(n+1)(2n+1)}{\sqrt{n_1 n_2(n+1)(2n+1)(8n+11)/5}},$$

$_l\underline{R}_1 = (_lR_{11}, \ldots, _lR_{1n_1})$, $_lR_{1i}$ denotes the rank of $_lX_{1i}$ in the pooled sample

$$_l\underline{X} = (_l\underline{X}_1, _l\underline{X}_2) = (_lX_{11}, \ldots, _lX_{1n_1}, _lX_{21}, \ldots, _lX_{2n_2}) = (_lX_1, \ldots, _lX_{n_1}, _lX_{n_1+1}, \ldots, _lX_n)$$

and

$$\rho = corr(_lU, _lV) = \frac{2(n^2 - 4)}{(2n+1)(8n+11)} - 1.$$

The $_lU$ and $_lV$ statistics are respectively the standardized sum of squared ranks and squared contrary ranks of the first group. Note that the $_lC$ statistic is a combination of $_lU$ and $_lV$ taking into account their negative correlation ρ. When there is no difference between the groups so far as $_lX$ is concerned $M(_lU) = M(_lV) = 0$, $VAR(_lU) = VAR(_lV) = 1$ and $(_lU, _lV)$ is centered on $(0,0)$, whereas it is not when the two groups are different in means and/or variances of $_lX$, see [10,21]. Therefore $_lC$ increases as the difference between groups increases.

For the purpose of comparing the grade of difference of various financial ratios, the $_lC$ statistic should be normalized to lay between 0 and 1. The steps of the normalization procedure are:

1. randomly permute $_l\underline{X}$ obtaining $_l^1\underline{X}^* = (_l^1\underline{X}_1^*, _l^1\underline{X}_2^*) = (_lX_{u_1^*}, \ldots, _lX_{u_n^*}) = (_lX_1^*, \ldots, _lX_n^*)$ where $(u_1^*, \ldots, u_n^*)$ is a random permutation of $(1, \ldots, n)$;
2. compute

$$_l^1C^* = C(_l^1\underline{X}_1^*, _l^1\underline{X}_2^*) = C(_l^1U^*, _l^1V^*) = (_l^1U^*)^2 + (_l^1V^*)^2 - 2\rho_l^1U^*_l^1V^*$$

where $_l^1 U^* = U(_l^1 \underline{R}_1^*)$, $_l^1 V^* = V(_l^1 \underline{R}_1^*)$ and $_l^1 \underline{R}_1^*$ contains the ranks of the $_l^1 \underline{X}_1^*$ elements;

3. repeat step 1 and step 2 for $B - 1$ times, where $B = \frac{n!}{n_1! n_2!}$ in order to consider all the possible random permutations of $(1,\ldots,n)$. Note that B is not $n!$ because the C statistic does not depend on the order of the firms within the groups;

4. compute

$$_l \tilde{C} = \frac{B - \sum_{b=1}^{B} I(_l^b C^* \geq {_l^0 C})}{B}$$

where $_l^0 C = C(_l \underline{X}_1, {_l}\underline{X}_2)$;

5. repeat steps 1 to 4 for every $l = 1,\ldots,L$. Note that the B permutations of $(1,\ldots,n)$ must be considered in the same order for every $_l X$.

Note that when there is no difference in $_l X$ between the groups of firms $_l \tilde{C}$ is 0, whereas when the difference is the largest one among all the possible resampling/permutation values of the $_l C$ statistic $_l \tilde{C}$ is 1, in general it is $0 \leq {_l}\tilde{C} \leq 1$. Therefore you can compare $_1 \tilde{C},\ldots,{_L}\tilde{C}$ to grade the difference in each financial ratio and find out which are the most (or least) different financial ratios. $_l \tilde{C}$ can be estimated by considering a random sample of the permutations when it is computationally impractical to consider all the possible B permutations. It is important to emphasize that $_l \tilde{C}$ may be seen as the complement to 1 of the permutation p-value of the test for the location-scale problem based on the $_l C$ statistic. Therefore it is a normalized measure of how far the groups of firms are from the situation of no difference in means and variances of $_l X$. The $_l C$ statistic is a monotone function of a statistic proposed by [10] for jointly testing for location and scale differences. The corresponding test has been further studied in [21] by computing for the very first time the table of exact critical values; by showing that the test maintains its size very close to the nominal significance level and is more powerful than the most familiar test for the location-scale problem due to Lepage. Moreover it has been shown that the Cucconi test is very robust against highly skewness and heavy tailness and that should be preferred to tests like the Kolmogorov-Smirnov and Cramer-Von Mises when the distributions under comparison may be different in shape other than location and/or scale. These characteristics of the test make the $_l C$ statistic particularly suitable for analyzing financial data.

Univariate analysis of financial ratios may be misleading, for example a fictitious firm with a poor profitability ratio might be regarded as potential distress if one does not look to its good liquidity ratio. Therefore several financial ratios should be combined for a complete picture of the firm. The need for a combination of the various financial ratio measures of difference naturally arises. The steps of the combination procedure of $_l C$, $l = 1,\ldots,L$ are:

1. compute

$$^0 MC = \sum_{l=1}^{L} \ln\left(\frac{1}{1 - {_l^0}\tilde{C}}\right)$$

where $_l^0 \tilde{C} = {_l}\tilde{C}$;

2. compute

$$^1MC^* = \sum_{l=1}^{L} \ln \left(\frac{1}{1 - {}_l^1\tilde{C}^*} \right)$$

where $ {}_l^1\tilde{C}^* = \frac{B - \sum_{b=1}^{B} I({}_l^b C^* \geq {}_l^1 C^*)}{B} $;

3. repeat step 2 for $b = 2, \ldots, B$ where the permutations of $(1, \ldots, n)$ are considered in the same order as in the normalization procedure;

4. compute $\widetilde{MC} = \frac{B - \sum_{b=1}^{B} I({}^b MC^* \geq {}^0 MC)}{B}$.

The procedure is an adaptation of the nonparametric combination of dependent tests [31] which is a flexible and very useful procedure of combination because it assesses nonparametrically the dependence structure among the partial aspects to be combined without any parametric assumption on it. Note that since permutation of individual firm data have been considered then all the underlying dependence relations between the financial ratios are preserved. $\widetilde{MC}$ is the normalized measure of difference between the groups of firms which simultaneously considers all the financial ratios. Note that when there is no difference in $_lX$, $l = 1, \ldots, L$, between the groups of firms $\widetilde{MC}$ is 0, when the difference is the largest one among all the possible resampling/permutation values of the MC statistic $\widetilde{MC}$ is 1, in general it is $0 \leq \widetilde{MC} \leq 1$. $\widetilde{MC}$ can be used also to measure how different are the various industry sectors if one classifies the firms of the two groups according to the industry sector they belong to and compute $\widetilde{MC}$ for every sector of interest. Note that the algorithm of the combined procedure uses the Fisher omnibus combining function.

4 Studying the Difference Between US and European Firms

In this section, we use the method presented in the previous one to assess the difference between US and European firm financial ratios. In general, ratios measuring profitability, liquidity, solvency, leverage and activity are considered in financial ratio analysis to assess the financial strength (or weakness) of a firm. [9] and [18] reviewed the literature finding 41 and 48 ratios, respectively, to be used in practice. Unfortunately, there is no clear indications on which are the most important financial ratios. Here, we are interested in listed firm valuation and then following the suggestions of [11] we consider the following ratios, very popular in valuation:

- $_1X = P/E =$ price to earnings ratio $= \frac{\text{market capitalization}}{\text{net income}}$;

- $_2X = P/B =$ price to book equity ratio $= \frac{\text{market capitalization}}{\text{current book value of equity}}$;

- $_3X = P/S =$ price to sales ratio $= \frac{\text{market capitalization}}{\text{revenues}}$;

- $_4X = EV/EBITDA =$ enterprise value to $EBITDA$ ratio $= \frac{\text{enterprise value}}{EBITDA}$, where the enterprise value is the market value of debt and equity of a firm net of cash and $EBITDA$ stands for earnings before interest, taxes, depreciation and amortization;

- $_5X = EV/C =$ enterprise value to capital ratio $= \frac{\text{enterprise value}}{\text{current invested capital}}$;

- $_6X = EV/S =$ enterprise value to sales ratio $= \frac{\text{enterprise value}}{\text{revenues}}$.

P/E, P/B and P/S represent the market capitalization of a firm as a multiple of its net income, book value of equity and revenues respectively. $EV/EBITDA$, EV/C and EV/S represent the enterprise value of a firm as a multiple of its $EBITDA$, invested capital and revenues respectively. It is well known that these financial ratios can vary substantially across industry sectors and therefore we compare US and European firms belonging to the same industry.

Financial ratio data about US and European firms have been downloaded from Damodaran Online website at pages.stern.nyu.edu/~adamodar. The data are updated on January 1, 2011 and refer to 5928 publicly traded US firms and to 4816 publicly traded European firms. Since financial ratios may be extreme, usually due to very small denominators, and many firms have missing or negative values, the initial data have been selected. General practice (see e.g. [4, 16, 20]) suggests to exclude: any firm with missing or negative financial data necessary to compute the financial ratios of interest; with a capitalization less than 100 millions USD; with data necessary to compute the financial ratios of interest not lying within the 1st and 99th percentile of data distribution. The firms that survived this selection are 1784 US and 1465 European firms. The aim of the selection is to obtain a data set of "regular" firms in the sense that the firms are suitable to be analyzed with valuation ratios.

We consider the ten most numerous industry sectors, see Table 1. Industry sectors have been classified according to the SIC (Standard Industrial Classification) code, a four digit code used by the United States Security and Exchange Commission for classify industries.

It is important to note that the usual inferential statistical techniques cannot be used because the groups of firms are not random samples since the firms have not been randomly selected from the population of US and European firms under issue. The groups are made up of the firms that survived the initial selection. Moreover, as shown in Table 2, the data are highly skewed, heavy tailed and there is no homo-

Table 1 The industry sectors of interest

Industry sector	SIC code	US firms	Eur. firms
Computer Software Services	3579	71	77
Machinery	3500	61	66
Retail (Special Lines)	5600	57	35
Food Processing	2000	48	61
Oilfield Services Equipment	3533	47	33
Electronics	3670	42	35
Chemical (Specialty)	2820	36	26
Electrical Equipment	3600	34	29
Telecom Services	4890	28	25
Apparel	2300	22	31

Table 2 Financial ratio summary statistics for US and European firms

SIC code	statistic	US firms						European firms					
		P/E	P/B	P/S	EV/EBITDA	EV/C	EV/S	P/E	P/B	P/S	EV/EBITDA	EV/C	EV/S
3579	mean	41.5	3.8	3.4	12.1	5.4	3.2	25.3	2.9	1.8	10.6	9.1	1.8
	st. dev.	37.7	2.5	2.2	6.8	4.6	2.2	18.5	1.8	1.9	5.5	19.5	1.8
	kurtosis	5.3	2.8	0.0	2.9	2.7	0.1	12.5	1.7	5.6	4.5	54.6	5.4
	skewness	2.4	1.7	0.8	1.5	1.8	0.9	3.0	1.4	2.2	1.8	7.0	2.2
3500	mean	44.0	3.4	1.7	10.2	3.3	1.7	54.0	3.5	1.6	12.4	5.4	1.7
	st. dev.	52.0	2.2	1.1	4.2	2.8	1.1	98.7	3.1	1.2	5.3	10.4	1.2
	kurtosis	13.0	2.7	6.7	4.0	19.4	4.2	31.0	22.2	1.5	0.1	46.9	1.7
	skewness	3.5	1.7	2.1	1.3	3.8	1.7	5.1	4.0	1.4	0.7	6.5	1.4
5600	mean	27.3	3.5	1.2	7.2	4.9	1.2	18.7	3.5	1.3	9.4	4.7	1.3
	st. dev.	18.9	2.3	0.9	3.3	4.7	0.9	16.2	2.5	1.5	5.8	4.4	1.4
	kurtosis	3.4	3.1	5.0	0.8	4.7	5.6	14.7	0.8	7.9	7.8	2.0	8.3
	skewness	1.9	1.8	2.1	1.1	2.2	2.2	3.4	1.2	2.6	2.3	1.5	2.6
2000	mean	25.7	3.7	1.4	9.2	2.9	1.6	19.4	2.0	0.8	8.0	2.6	1.0
	st. dev.	38.3	2.8	1.3	5.5	2.0	1.5	22.5	1.1	0.7	3.2	2.6	0.7
	kurtosis	40.3	1.7	20.4	8.9	5.4	26.4	28.9	0.5	4.2	1.3	11.2	3.2
	skewness	6.1	1.5	3.9	2.7	2.2	4.5	5.0	1.0	1.9	0.9	3.2	1.7
3533	mean	32.6	2.5	2.7	8.5	2.5	3.0	27.6	2.9	1.8	14.5	4.4	2.6
	st. dev.	33.7	1.6	1.9	4.4	1.8	2.2	37.9	3.4	1.3	18.2	7.6	1.8
	kurtosis	16.2	5.6	1.9	1.7	4.9	3.4	15.0	10.3	1.0	27.9	8.2	0.5
	skewness	3.6	2.0	1.5	1.1	2.1	1.9	3.6	3.2	1.0	5.1	3.0	0.8
3670	mean	40.2	2.8	2.2	9.9	4.0	1.9	36.6	3.4	2.0	13.5	4.9	2.1
	st. dev.	48.5	1.9	2.2	6.2	3.7	1.9	35.2	1.6	1.2	6.9	2.9	1.2
	kurtosis	15.2	15.9	3.5	3.6	5.6	6.7	9.2	3.1	1.3	0.1	1.1	0.9
	skewness	3.6	3.4	1.9	1.9	2.2	2.4	2.9	1.6	1.3	0.9	1.2	1.2
2820	mean	30.3	3.6	1.8	8.5	3.2	1.8	38.9	4.1	2.0	13.7	4.4	2.1
	st. dev.	20.4	2.6	1.7	3.5	2.7	1.6	31.8	4.1	1.7	5.7	4.0	1.6
	kurtosis	3.2	6.9	6.8	2.1	15.0	7.2	0.9	11.2	1.2	2.4	6.4	0.5
	skewness	1.9	2.4	2.3	1.2	3.5	2.4	1.4	3.2	1.4	1.4	2.4	1.2
3600	mean	42.1	3.0	2.2	10.3	3.5	2.1	30.5	2.7	1.2	12.0	4.0	1.3
	st. dev.	45.8	1.2	1.3	3.9	2.1	1.2	36.5	2.4	0.6	7.8	4.3	0.6
	kurtosis	4.2	1.9	0.5	2.8	5.6	0.7	15.1	14.2	-1.2	3.7	11.9	-1.1
	skewness	2.2	1.1	0.1	1.4	2.1	1.1	3.7	3.5	0.1	1.9	3.1	-0.1
4890	mean	27.8	2.7	1.9	5.5	2.7	2.1	87.7	3.7	1.4	6.5	3.2	1.8
	st. dev.	40.5	1.7	1.1	3.7	2.2	1.0	343.3	3.7	0.9	3.2	2.2	1.2
	kurtosis	13.9	2.3	3.0	13.7	5.8	0.6	24.9	14.0	6.6	5.4	9.5	5.4
	skewness	3.6	1.7	1.4	3.3	2.3	0.7	5.0	3.4	2.2	2.0	2.7	2.1
2300	mean	22.7	3.3	1.7	8.6	3.9	1.8	34.0	4.4	2.5	14.3	5.3	2.6
	st. dev.	10.7	2.0	1.5	3.1	3.6	1.8	31.8	4.2	2.5	11.8	6.0	2.5
	kurtosis	5.0	1.8	2.8	3.7	6.9	7.8	15.8	4.5	4.8	15.1	13.8	6.3
	skewness	1.9	1.4	1.7	1.6	2.5	2.6	3.5	2.1	2.1	3.5	3.4	2.4

geneity of variances between groups. Similarly to our method, neural networks (NN) do not require restrictive assumptions on the data set, see [28]. In finance they are used to separate a set a firms in different groups, generally the healthy firm group and the unhealthy firm group, see [2]. Thus NN are an alternative to MDA rather than to our method. Note that according to [35] in the literature there is no evidence that NN are superior to MDA and that [2] concluded that NN are not dominant compared to MDA. NN are widely used in finance also for prediction, see [14, 30]. Our aim is different: to measure the difference between US and European firms as far as several financial ratios are considered and therefore NN are not suitable to pursue our aim. Among the methods traditionally used in finance, a method that in principle is similar to the method proposed in the paper is a descriptive (not inferential since the groups are not random sample) MANOVA. Unfortunately, MANOVA cannot be applied to our data because as shown in Table 2 the data are highly skewed, heavy tailed and there is no homogeneity of variances between groups, see [8]. On the contrary, the method proposed in Sect. 3 is particularly suitable in this situation. It is a descriptive method and then does not require random sampling. It is based on a statistic that is robust against skewness and heavy tailness in the data and takes explicitly into account the difference in variability as well as in central tendency between groups. The method considers also the dependence relations among financial ratios that are strong and then should be taken into account. The dependence among the financial ratios can be assessed also when the correlations are different in the two groups, note that this is very difficult to be assessed within the MANOVA setting (traditional MANOVA cannot assess it).

The industry sectors of Table 1 have been compared using the normalized measure of difference proposed in the previous section. Table 3 displays the results obtained using a random sample of 20,000 permutations in the algorithms for normalizing C and MC. By computing $_l\tilde{C}$ $l = 1,\ldots,L$ the comparison between US and European firms is addressed by the univariate point of view and we can grade the difference in each financial ratio and find out which are the most (or least) different financial ratios for what concerns central tendency and variability. Table 4 shows that the overall most different financial ratios are the P/S ratio and the $EV/EBITDA$ ratio. In particular, the P/S ratio is the most different financial ratio for computer, retail, electrical equipment and telecom services industry sector. The $EV/EBITDA$ ratio is the most different financial ratio for machinery, oilfield services/equipment, chemical and apparel industry sector. On the other side, the overall least different financial ratio is the P/B ratio. In particular, the P/B ratio is the least different financial ratio for machinery, retail, chemical and telecom services industry sector.

By computing $\widetilde{MC}$ we compare the US and European firms by simultaneously considering the six financial ratios and then we address the problem from the multivariate point of view and we get a complete picture of valuation difference between US and European firms. A very important feature of $\widetilde{MC}$ is in fact the consideration of the dependence relations among the financial ratios. A second, but not less important, utilization of $\widetilde{MC}$ is to find out what are the most (or least) different industry sectors as far as the six ratios are simultaneously considered. Table 3 shows that the industry sectors ranked in decreasing order of difference at the basis of all the financial ratios are the computer, food processing, electronics, electrical equipment, oilfield services/equipment, retail (special lines), chemical (specialty), apparel, telecom services and machinery. Since the least different industry sector has a difference of .815 we conclude that US and European firms are very different as far as valuation ratios are considered.

Even if the data sets have been cleaned by excluding firms with extreme ratios, several financial ratios are still highly skewed and heavy tailed. Therefore we log transform the data to reduce skewness and kurtosis. By computing the $_l\tilde{C}$ and $\widetilde{MC}$

Table 3 Normalized difference between US and European firms

Ratio	SIC code				
	3579	3500	5600	2000	3533
P/E	0.999	0.889	0.989	0.985	0.914
P/B	0.983	0.019	0.557	0.999	0.817
P/S	1.000	0.398	0.992	0.999	0.900
EV/EBITDA	0.759	0.975	0.884	0.243	0.990
EV/C	0.860	0.826	0.702	0.921	0.962
EV/S	1.000	0.386	0.859	0.995	0.755
Combined	1.000	0.815	0.975	0.998	0.985

Ratio	SIC code				
	3670	2820	3600	4890	2300
P/E	0.130	0.611	0.674	0.786	0.959
P/B	0.957	0.051	0.964	0.575	0.652
P/S	0.989	0.089	0.995	0.923	0.371
EV/EBITDA	0.982	1.000	0.953	0.887	0.992
EV/C	0.999	0.652	0.950	0.589	0.639
EV/S	0.983	0.313	0.976	0.777	0.610
Combined	0.997	0.944	0.994	0.880	0.903

measures we obtain practically the same results as with non transformed data and we find evidence that our difference measure is robust against skewness and heavy tailness. This result confirms the findings of [21] found in the inferential context. Note that the method is robust because the C statistic is based on ranks.

This application has two limitations. The first limitation is common to all the studies that compared US and European firms and refers to the differences in the accounting practices of the countries that may partly distort the comparability of US and European financial statement data. Unfortunately we have at our disposal financial ratios computed for a single year. This is the second limitation because it is preferable to consider several year average financial ratios because single year financial ratios may be influenced by some temporary or extraordinary circumstances.

5 Conclusions

A method for comparing firms through financial ratios has been proposed. The method is based on squared ranks and does not require any particular assumptions because it is a descriptive method. This aspect is very important because very often financial data sets do not meet the assumptions (e.g. random sampling, normality, homogeneity of variances ...) required by traditional methods like MANOVA and MDA. The proposed method is devised to explicitly consider the possible difference in variances. It is a sort of measure of difference between groups of firms which takes also into account the dependence among the financial ratios. It is robust against skewness and heavy tailness. This second aspect is also very important because usually financial ratios, even after removing outliers, are highly skewed and heavy tailed. From the univariate point of view, the method can be used to grade the difference in each financial ratio and find out which are the most (or least) different financial ratios for what concerns central tendency and variability. From the multivariate point of view, you assess the comparison by simultaneously considering all the financial ratios and get a complete picture of the difference between the groups of firms. If the firms have been classified according to industry sectors, you can find out what are the most (or least) different industry sectors.

An application for studying the difference between US and European firms has been discussed. Six valuation ratios and ten industry sectors have been considered. We used our method because the usual inferential statistical techniques cannot be used because the samples are not random samples, the data are not symmetric and are generally heavy-tailed (even after removing the outliers), there is no homogeneity of variances between groups. We found that the overall most different financial ratios are the P/S ratio and the $EV/EBITDA$ ratio, while the overall least different one is the P/B ratio. The most different industry sectors are computer, food processing and electronics while the lest different one is the machinery. We suggest managers, auditors, shareholders, lenders and potential investors to use our method for comparing firm financial ratios because it does not require any particular assumptions, it is devised to explicitly consider the possible difference in variances, it takes into account the dependence relations among the financial ratios and it is very robust against skewness and heavy tailness.

References

1. Altman, E.I.: Financial ratios, discriminant analysis and the prediction of corporate bankruptcy. J. Finan. **23**, 589–609 (1968)
2. Altman, E.I., Marco, G., Varetto, F.C.: Corporate distress diagnosis: comparisons using linear discriminant analysis and neural networks (the Italian experience). J. Bank. Financ. **18**, 505–529 (1994)
3. Bedingfield, J.P., Reckers, P.M.J., Stagliano, A.J.: Distributions of financial ratios in the commercial banking industry. J. Financ. Res. **8**, 77–81 (1985)
4. Bhojraj, S., Lee, C.M.C.: Who is my peer? A valuation-based approach to the selection of comparable firms. J. Acc. Res. **40**, 407–439 (2002)
5. Bhunia, A., Sarkar, R.: A study of financial distress based on MDA. J. Manag. Res. **3**, 1–11 (2011)
6. Bird, R.G., McHugh, A.J.: Financial ratios - an empirical study. J. Bus. Finan. Account. **4**, 29–45 (1977)
7. Boughen, P.D., Drury, J.C.: U.K. statistical distribution of financial ratios. J. Bus. Finan. Account. **7**, 39–47 (1980)
8. Bray, J.H., Maxwell, S.E.: Multivariate analysis of variance. Sage Publications. Thousand Oaks, CA (1985)

9. Chen, K.H., Shimerda, T.A.: An empirical analysis of useful financial ratios. Finan. Manage. **10**, 51–60 (1981)

10. Cucconi, O.: Un nuovo test non parametrico per il confronto tra due gruppi campionari. Giornale Economisti **27**, 225–248 (1968)

11. Damodaran, A.: Damodaran on valuation 2nd ed.. Wiley, New York (2006)

12. Deakin, E.B.: A discriminant analysis of predictors of business failure. J. Acc. Res. **10**, 167–179 (1972)

13. Deakin, E.B.: Distributions of financial accounting ratios: some empirical evidence. Account. Rev. **51**, 90–96 (1976)

14. Dietrich, M.: Using simple neural networks to analyse firm activity. Sheffield Econ. Res. Papers **14** (2005)

15. Edmister, R.O.: An empirical test of financial ratio analysis for small business failure prediction. J. Financ. Quant. Anal. **7**, 1477–1493 (1972)

16. Henschke, S., Homburg, C.: Equity valuation using multiples: controlling for differences between firms. Unpublished Manuscript (2009)

17. Horrigan, J.O.: A short history of financial ratio analysis. Account. Rev. **43**, 284–294 (1968)

18. Hossari, G., Rahman, S.: A comprehensive formal ranking of the popularity of financial ratios in multivariate modeling of corporate collapse. J. Am. Acad. Bus. **6**, 321–327 (2005)

19. Klecka, W.R.: Discriminant analysis. Sage Publications. Beverly Hills, CA (1980)

20. MacKay, P., Phillips, G.M.: How does industry affect firm financial structure? Rev. Financ. Stud. **18**, 1431–1466 (2005)

21. Marozzi, M.: Some notes on the location-scale Cucconi test. J. Nonparametr. Stat. **21**, 629–647 (2009)

22. Marozzi, M., Santamaria, L.: A dimension reduction procedure for corporate finance indicators. In: Corazza, M., Pizzi, C. (eds.) Mathematical and Statistical Methods for Actuarial Sciences and Finance, pp. 205–213. Springer, Milano (2010)

23. Marozzi, M.: Composite indicators: a sectorial perspective. In: Perna, C., Sibillo, M. (eds.) Mathematical and Statistical Methods for Actuarial Sciences and Finance, pp. 287–294. Springer, Milano (2012)

24. Mecimore, C.D.: Some empirical distributions of financial ratios. Manag. Account. **50**, 13–16 (1968)

25. Meric, G., Leveen, S.S., Meric, I.: The financial characteristics of commercial banks involved in interstate acquisitions. Finan. Rev. **26**, 75–90 (1991)

26. Meric, I., Lentz, C., Meric, G.: A comparison of the financial characteristics of French German and UK firms in the electronic and electrical equipment manufacturing industry: 2001–2005. Eur. J. Econ. Financ. Admin. Sci. **7**, 87–93 (2007)

27. Meric, I., McCall, C.W., Meric, G.: U.S. and Japanese electronic and electrical equipment manufacturing firms: a comparison. Internat. Bull. Bus. Admin. **3**, 1–5 (2008)

28. Nilsson, J.N.: Introduction to machine learning. Unpublished book (2010)

29. O'Connor, M.C.: On the usefulness of financial ratios to investors in common stock. Account. Rev. **48**, 339–352 (1968)

30. Paliwal, M., Kumar, U.A.: Neural networks and statistical techniques: A review of applications. Expert. Syst. Appl. **36**, 2–17 (2009)

31. Pesarin, F., Salmaso, L.: Permutation tests for complex data. Wiley, Chichester (2010)

32. Ray, B.: Assessing corporate financial distress in automobile industry of India: an application of Altman's model. Res. J. Finan. Account. **2**, 2222–2847 (2011)

33. Ricketts, D., Stover, R.: An examination of commercial bank financial ratios. J. Bank. Res. **9**, 121–124 (1978)
34. Salmi, T., Dahlstedt, R., Luoma, M., Laakkonen, A.: Financial Ratios as Predictors of Firms' Industry Branch. Finnish J. Bus. Econ. **37**, 263–277 (1988)
35. Smith, M.: Research methods in accounting, 2nd ed.. Sage Publications, London (2011)
36. Stevens, D.L.: Financial characteristics of merged firms: a multivariate analysis. J. Financ. Quant. Anal. **8**, 149–158 (1973)
37. Zhang, L., Altman, E., Yen, J.: Corporate financial distress diagnosis model and application in credit rating for listing firms in China. Frontiers Comp. Sc. China **4**, 220–236 (2010)

A Behavioural Approach to the Pricing of European Options

Martina Nardon and Paolo Pianca

Abstract Empirical studies on quoted options highlight deviations from the theoretical model of Black and Scholes; this is due to different causes, such as assumptions regarding the price dynamics, markets frictions and investors' attitude toward risk. In this contribution, we focus on this latter issue and study how to value options within the *continuous cumulative prospect theory*. According to prospect theory, individuals do not always take their decisions consistently with the maximization of expected utility. Decision makers have biased probability estimates; they tend to underweight high probabilities and overweight low probabilities. Risk attitude, loss aversion and subjective probabilities are described by two functions: a value function and a weighting function, respectively. As in Versluis *et al.* [15], we evaluate European options; we consider the pricing problem both from the writer's and holder's perspective, and extend the model to the put option. We also use alternative probability weighting functions.

1 Introduction

Black and Scholes [3] model is considered as a milestone in the option pricing literature, it is widely applied in financial markets, and has been developed in many directions. Nevertheless, empirical studies on quoted options highlight deviations from the theoretical model; this is due to different causes, such as assumptions regarding

M. Nardon (✉)
Ca' Foscari University of Venice, Department of Economics, Center for Quantitative Economics, Cannaregio 873, 30121 Venice, Italy
e-mail: mnardon@unive.it

P. Pianca
Ca' Foscari University of Venice, Department of Economics, Center for Quantitative Economics, Cannaregio 873, 30121 Venice, Italy
e-mail: pianca@unive.it

M. Corazza, C. Pizzi (eds.), *Mathematical and Statistical Methods for Actuarial Sciences and Finance*, DOI 10.1007/978-3-319-02499-8_20, © Springer International Publishing Switzerland 2014

the price dynamics, and volatility in particular, the presence of markets frictions, information imperfections, and investors' attitude toward risk. With reference to this latter issue, we study how to value options within the *cumulative prospect theory* developed by Kahneman and Tversky [7, 14].

According to the prospect theory, individuals do not always take their decisions consistently with the maximization of expected utility. Decision makers evaluate their choices based on potential gains and losses relative to a *reference point*, rather than in terms of final wealth. They are risk-averse when considering gains and risk-seeking with respect to losses. Individuals are loss averse: they are much more sensitive to losses than they are to gains of comparable magnitude. Risk attitude and loss aversion are described by a value function, which is typically concave (risk-aversion) for gains and convex (risk-seeking) for losses, and steeper for losses.

A value function is not able to capture the full complexity of observed behaviours: the degree of risk aversion or risk seeking appears to be dependent not only on the value of the outcomes but also the probability and ranking of outcomes. Empirical studies suggest that individuals have also biased probability perceptions, modeled by a weighting function: small probabilities are overweighted, whereas individuals tend to underestimate large probabilities[1]. This turns out in a typical *inverse-S shaped* weighting function: the function is initially concave (probabilistic risk seeking or *optimism*) and then convex (probabilistic risk aversion or *pessimism*)[2]. An inverse-S shaped form for the weighting function combines the increased sensitivity with concavity for small probabilities and convexity for medium and large probabilities. In particular, such a function captures the fact that individuals are extremely sensitive to changes in (cumulative) probabilities which approach to 0 and 1.

The literature on behavioural finance (see e.g. [5] for a survey) and prospect theory is huge, whereas a few studies in this field focus on financial options. A first contribution which applies prospect theory to options valuation is the work of Shefrin and Statman [10], who consider covered call options in a one period binomial model. A list of papers on this topic includes: Poteshman and Serbin [8], Abbink and Rockenbach [1], Breuer and Perst [4], and more recently Versluis *et al.* [15]. As in [15], we evaluate European plain vanilla options. In particular, we use alternative probability weighting functions (see Prelec [9]). We extend the model to the European put option. We perform some numerical examples.

The rest of the paper is organized as follows. Section 2 synthesizes the main features of prospect theory and introduces the value and the weighting functions; Sect. 3 focuses on the application of continuous cumulative prospect theory to European option pricing; in Sect. 5 numerical results are provided and discussed. Finally we present some conclusions.

---

[1] Kahneman and Tversky [7] provide empirical evidence of such behaviours.

[2] Abdellaoui et al. [2] discuss how optimism and pessimism are possible sources of increased probability sensitivity.

2 Prospect Theory

Prospect theory (PT), in its formulation proposed by Kahneman and Tversky [7], is based on the subjective evaluation of *prospects*. With a finite set[3] of potential future states of nature $S = \{s_1, s_2, \ldots, s_n\}$, a prospect

$$((\Delta x_1, p_1), (\Delta x_2, p_2), \ldots, (\Delta x_n, p_n))$$

is a collection of pairs $(\Delta x_i, p_i) \in \mathbb{R} \times [0,1]$, for $i = 1, 2, \ldots, n$, where Δx_i is an outcome and p_i its probability. Outcome Δx_i is defined relative to a certain reference point x^*; being x_i the absolute outcome, we have $\Delta x_i = x_i - x^*$. Assume $\Delta x_i \leq \Delta x_j$ for $i < j$, $i, j = 1, 2, \ldots, n$; prospects assign to any possible ordered outcome a probability p_i.

2.1 Cumulative Prospect Theory

According to PT, subjective values $v(\Delta x_i)$ are not multiplied by probabilities p_i, but using *decision weights* $\pi_i = w(p_i)$. It is relevant to separate gains from losses as in PT negative and positive outcomes may be evaluated differently: the function v is typically convex and steeper in the range of losses and concave in the range of gains, and subjective probabilities may be evaluated through a weighting function w^- for losses and w^+ for gains, respectively. In order to emphasize this fact, let us denote with Δx_i, for $-m \leq i < 0$ (strictly) negative outcomes and with Δx_i, for $0 < i \leq n$ (strictly) positive outcomes, with $\Delta x_i \leq \Delta j$ for $i < j$. Subjective value of the prospect is displayed as follows:

$$V = \sum_{i=-m}^{n} \pi_i \cdot v(\Delta x_i), \tag{1}$$

with decision weights π_i and values $v(\Delta x_i)$. In the case of expected utility, the weights are $\pi_i = p_i$ and the utility function in not based on relative outcomes.

Cumulative prospect theory (CPT) developed by Tversky and Kahneman [14] overcomes some drawbacks (such as violation of stochastic dominance) of the original prospect theory. In the CPT, decision weights π_i are differences in transformed cumulative probabilities (of gains or losses):

$$\pi_i = \begin{cases} w^-(p_{-m}) & i = -m \\ w^-\left(\sum_{j=-m}^{i} p_j\right) - w^-\left(\sum_{j=-m}^{i-1} p_j\right) & i = -m+1, \ldots, -1 \\ w^+\left(\sum_{j=i}^{n} p_j\right) - w^+\left(\sum_{j=i+1}^{n} p_j\right) & i = 0, \ldots, n-1 \\ w^+(p_n) & i = n. \end{cases} \tag{2}$$

It is worth noting that the subjective probabilities not necessarily sum up to one, due to the fact that different weighting functions w^+ and w^- can be used.

The shape of the value function and the weighting function becomes significant in describing actual choice patterns. Both functions will be discussed later.

---

[3] Infinitely many outcomes may also be considered.

2.2 Framing and Mental Accounting

PT postulates that decision makers evaluate outcomes with respect to deviations from a reference point rather than with respect to net final wealth. The definition of such a reference point is crucial due to the fact that individuals evaluate results through a value function which gives more weight to losses than to gains of comparative magnitude. Individual's framing of decisions around a reference point is of great importance in PT.

People tend to segregate outcomes into separate *mental accounts*, these are then evaluated separately for gains and losses. Thaler [12] argues that, when combining such accounts to obtain overall result, typically individuals do not simply sum up all monetary outcomes, but use *hedonic frame*, such that the combination of the outcomes appears best possible.

Consider a combination of two sure (positive) outcomes Δx and Δy, the hedonic frame can be described as follows (see [13]):

$$V = \max\{v(\Delta x + \Delta y),\ v(\Delta x) + v(\Delta y)\}. \tag{3}$$

Outcomes Δx and Δy are aggregated, and in such a case we have $v(\Delta x + \Delta y)$, or segregated $v(\Delta x) + v(\Delta y)$, depending on what yields the highest prospect value.

An extension of the hedonic frame rule is (see also [4]):

$$V = \sum_{i=1}^{n} \pi_i \cdot \max\{v(\Delta x_i + \Delta y),\ v(\Delta x_i) + v(\Delta y)\} +$$
$$+ \left(1 - \sum_{i=1}^{n} \pi_i\right) \cdot \max\{v(0 + \Delta y),\ v(0) + v(\Delta y)\}, \tag{4}$$

where Δx_i are possible results (for $i = 1, \ldots, n$) with subjective probabilities π_i and Δy is a sure result.

Regarding the valuation of financial options, different aggregation or segregation of the results are possible. One can consider a single option position (*narrow framing*) or a portfolio, a naked or a covered position. It is also possible to segregate results across time: e.g. one can evaluate separately the premium paid for the option and its final payoff.

2.3 Continuous Cumulative Prospect Theory

In order to apply CPT to option valuation, one has to deal with continuous results. Davis and Satchell [6] provide the continuous cumulative prospect value V:

$$V = \int_{-\infty}^{0} \Psi^-[F(x)]\, f(x)\, v^-(x)\, dx + \int_{0}^{+\infty} \Psi^+[1 - F(x)]\, f(x)\, v^+(x)\, dx, \tag{5}$$

where $\Psi = \frac{dw(p)}{dp}$ is the derivative of the weighting function w with respect to the probability variable, F is the cumulative distribution function and f is the probability density function of the outcomes x; v^- and v^+ denote the value function for losses and gains, respectively.

Specific parametric forms have been suggested in the literature for the value function. The value function v is continuous, strictly increasing, it displays concavity in the domain of gains and convexity in the domain of losses (it is not required to be differentiable, which is not generally the case in correspondence of the reference point), and is steeper for losses. A function which is used in many empirical studies is the following value function

$$v(x) = \begin{cases} -\lambda(-x)^b & x < 0 \\ x^a & x \geq 0, \end{cases} \qquad (6)$$

with positive parameters which control risk attitude $0 < a < 1$ and $0 < b < 1$, and loss aversion, $\lambda > 1$. Of course, in the limit case $a = b = 1$ and $\lambda = 1$ one recovers the risk neutral value function. Function (6) has zero as reference point and it satisfies the above mentioned properties. As in equation (5), we will denote with v^+ and v^- in order to emphasize the value function calculated for gains and losses. Figure 1 shows an example of the value function defined by (6).

Prosect theory involves also a probability weighting function which models probabilistic risk behaviour. A weighting function w is uniquely determined and maps the probability interval $[0, 1]$ into $[0, 1]$. The function w is strictly increasing, with $w(0) = 0$ and $w(1) = 1$. In this work we will assume continuity of w on $[0, 1]$, even thought in the literature discontinuous weighting functions are also considered.

As weighting function one can consider the function suggested by Tversky and Kahneman [14]:

$$w(p) = \frac{p^\gamma}{(p^\gamma + (1-p)^\gamma)^{1/\gamma}}, \qquad (7)$$

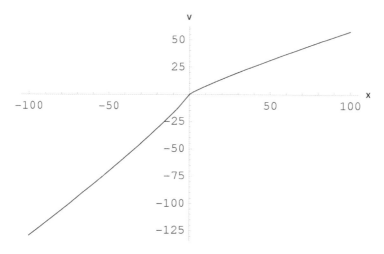

Fig. 1 Value function (6) with parameters $\lambda = 2.25$ and $a = b = 0.88$

where γ is a positive constant (with some constraint in order to have an increasing function). Note that (7) satisfies $w(0) = 0$ and $w(1) = 1$.

The parameter γ captures the the degree of sensitivity to changes in probabilities from impossibility (zero probability) to certainty; the lower the parameter, the higher is the curvature of the function. When $\gamma < 1$, one obtains the typical inverse-S shaped form, where low probabilities are overweighted (which results in heavier tails of the distribution) and medium and high probabilities are underweighted. In the applications, we consider two distinct functions w^+ and w^-, with parameters γ^+ and γ^- respectively, for probabilities associated to gains and losses.

As an alternative, Prelec [9] suggests a two parameters function of the form

$$w(p) = e^{-\delta(-\ln p)^\gamma} \qquad p \in (0,1), \tag{8}$$

with $w(0) = 0$ and $w(1) = 1$. The parameter δ (with $0 < \delta < 1$) governs elevation of the weighting function relative to the $45°$ line, while γ (with $\gamma > 0$) governs curvature and the degree of sensitivity to extreme results relative to medium probability outcomes. When $\gamma < 1$, one obtains the inverse-S shaped function.

In the applications, we consider Prelec's weighting function with a single parameter

$$w(p) = e^{-(-\ln p)^\gamma} \qquad p \in (0,1). \tag{9}$$

Note that in this case, the unique solution of equation $w(p) = p$ for $p \in (0,1)$ is $p = 1/e \simeq 0.367879$ and does not depend on the parameter γ.

The weighting function w may be one of the main causes of the options' mispricing through its effect to the prospect value (see [11]). Figures 2 and 3 show some examples of weighting functions defined by (7) and (9) for different values of the parameters. As the parameters tend to the value 1, the weight tends to the objective probability and the function w tends to the identity function.

3 European Options Valuation

Let S_t be the price at time t (with $t \in [0,T]$) of the underlying asset of a European option with maturity T; in a Black-Scholes setting, the underlying price dynamics is driven by a geometric Brownian motion. As a result, the probability density function (pdf) of the underlying price at maturity S_T is

$$f(x) = \frac{1}{x\sigma\sqrt{2\pi T}} \exp\left(\frac{-\left[\ln(x/S_0) - (\mu - \sigma^2/2)T\right]^2}{2\sigma^2 T} \right), \tag{10}$$

where μ and $\sigma > 0$ are constants, and the cumulative distribution function (cdf) is

$$F(x) = \Phi\left(\frac{\ln(x/S_0) - (\mu - \sigma^2/2)T}{\sigma\sqrt{T}} \right), \tag{11}$$

where $\Phi(\cdot)$ is the cdf of a standard Gaussian random variable.

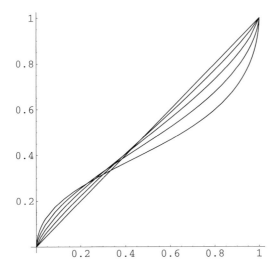

Fig. 2 Weighting function (7) for different values of the parameter γ. As γ approaches the value 1, the w tends to the identity function

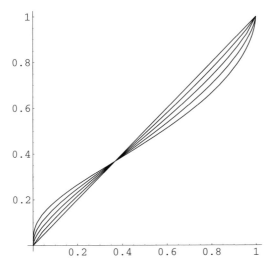

Fig. 3 Prelec's weighting function (9) for different values of the parameter γ. As γ approaches the value 1, the w tends to the identity function

Versluis *et al.* [15] provide the prospect value of writing L call options. Without loss of generality, we will consider $L = 1$. Let c be the option premium with strike price X. At time $t = 0$, the option's writer receives c and can invest the premium at the risk-free rate r until maturity, obtaining $c\,e^{rT}$, when he has to pay the amount $S_T - X$ if the option expires in-the-money.

Considering zero as a reference point (*status quo*), the prospect value of the writer's position in the *time segregated* case is

$$V_s = v^+ \left(c\, e^{rT} \right) + \int_X^{+\infty} \Psi^- \left(1 - F(x) \right) f(x) v^- \left(X - x \right) dx, \tag{12}$$

with f and F being the pdf and the cdf defined in (10) and (11) of the future underlying price S_T, and v is defined as in (6).

In equilibrium, we equate V_s at zero and solve for the price c:

$$c = e^{-rT} \left(\lambda \int_X^{+\infty} \Psi^- \left(1 - F(x) \right) f(x) \left(x - X \right)^b dx \right)^{1/a}, \tag{13}$$

which requires numerical approximation of the integral.

When considering the *time aggregated* prospect value, one obtains

$$\begin{aligned}
V_a = {}& w^+ \left(F(X) \right) v^+ \left(c\, e^{rT} \right) + \\
& + \int_X^{X + c\, \exp(rT)} \Psi^+ \left(F(x) \right) f(x) v^+ \left(c \exp(rT) - (x - X) \right) dx + \\
& + \int_{X + c\, \exp(rT)}^{+\infty} \Psi^- \left(1 - F(x) \right) f(x) v^- \left(c \exp(rT) - (x - X) \right) dx.
\end{aligned} \tag{14}$$

In this latter case, the option price in equilibrium has to be determined numerically.

In the case of a put option, one can not use put-call parity arguments. The prospect value of the writer's position in the time segregated case is

$$V_s = v^+ \left(p\, e^{rT} \right) + \int_0^X \Psi^- \left(F(x) \right) f(x) v^- \left(x - X \right) dx, \tag{15}$$

and one obtains

$$p = e^{-rT} \left(\lambda \int_0^X \Psi^- \left(F(x) \right) f(x) \left(X - x \right)^b dx \right)^{1/a}. \tag{16}$$

In the time aggregated case the put option value is defined by the following expression

$$\begin{aligned}
V_a = {}& \int_0^{X - p\, e^{rT}} \Psi^- \left[F(x) \right] f(x) v^- \left[p\, e^{rT} - (X - x) \right] dx + \\
& + \int_{X - p\, e^{rT}}^X \Psi^+ \left[1 - F(x) \right] f(x) v^+ \left[p\, e^{rT} - (X - x) \right] dx + \\
& + w^+ \left[1 - F(X) \right] v^+ \left[p\, e^{rT} \right].
\end{aligned} \tag{17}$$

Equation $V_a = 0$ has to be solved numerically for p.

3.1 Option Valuation from Holder's Perspective

When one considers the problem from the holder's viewpoint, the prospect values both in the time segregated and aggregated cases change. Holding zero as a reference point, the prospect value of the holder's position in the time segregated case is

$$V_s^h = v^- \left(-c\,e^{rT}\right) + \int_X^{+\infty} \Psi^+ \left(1 - F(x)\right) f(x) v^+ \left(x - X\right) dx. \qquad (18)$$

We equate V_s^h at zero and solve for the price c, obtaining

$$c^h = e^{-rT} \left(\frac{1}{\lambda} \int_X^{+\infty} \Psi^+ \left(1 - F(x)\right) f(x) \left(x - X\right)^a dx\right)^{1/b}. \qquad (19)$$

In the time aggregated case, the prospect value has the following integral representation

$$V_a^h = w^- \left(F(X)\right) v^- \left(-c\,e^{rT}\right) +$$
$$+ \int_X^{X+c\,\exp(rT)} \Psi^- \left(F(x)\right) f(x) v^- \left((x - X) - c\,\exp(rT)\right) dx +$$
$$+ \int_{X+c\,\exp(rT)}^{+\infty} \Psi^+ \left(1 - F(x)\right) f(x) v^+ \left((x - X) - c\,\exp(rT)\right) dx. \qquad (20)$$

In order to obtain the call option price in equilibrium, one has to solve numerically the problem.

In an analogous way one can derive the holder's prospect values for the put option.

4 Numerical Results

We have calculated the options prices both in the time segregated and aggregated case. Tables 1 and 2 report the results for the European calls and puts, respectively,

Table 1 Call option prices in the Black-Scholes model and in the segregated and aggregated prospects, from the writer's viewpoint. The parameters are: $S_0 = 100$, $\mu = 0.05$, $r = 0.05$, $\sigma = 0.2$, $T = 1$, and $L = 1$. In the BS model: $a = b = 1$, $\lambda = 1$, and $\gamma = 1$. The Tversky-Kahneman parameters are: $a = b = 0.88$, $\lambda = 2.25$, $\gamma^+ = 0.61$, and $\gamma^- = 0.69$. The moderate sentiment parameters are: $a = b = 0.988$, $\lambda = 1.125$, $\gamma^+ = 0.961$, and $\gamma^- = 0.969$. The last two columns report the results when Prelec's weighting function is used

		TK sentiment		Moderate sentiment		Moderate sentiment Prelec's w function	
X	c_{BS}	c_S	c_a	c_S	c_a	c_S	c_a
70	33.5401	79.7250	45.6392	37.8210	34.5707	37.7386	34.6245
80	24.5888	59.1367	37.0964	27.7655	25.6391	27.6832	25.7038
90	16.6994	42.2814	29.3001	18.9376	17.7360	18.8640	17.7956
100	10.4506	29.2077	22.3341	11.9501	11.3814	11.9028	11.4247
110	6.0401	19.4366	16.2796	6.9907	6.7702	6.9824	6.8049
120	3.2475	12.3977	11.2361	3.8140	3.7446	3.8394	3.7830
130	1.6395	7.5530	7.2823	1.9564	1.9387	1.9982	1.9819

Table 2 Put option prices in the Black-Scholes model and in the segregated and aggregated prospects, from the writer's viewpoint. The parameters are: $S_0 = 100$, $\mu = 0.05$, $r = 0.05$, $\sigma = 0.2$, $T = 1$, and $L = 1$. In the BS model: $a = b = 1$, $\lambda = 1$, and $\gamma = 1$. The Tversky-Kahneman parameters are: $a = b = 0.88$, $\lambda = 2.25$, $\gamma^+ = 0.61$, and $\gamma^- = 0.69$. The moderate sentiment parameters are: $a = b = 0.988$, $\lambda = 1.125$, $\gamma^+ = 0.961$, and $\gamma^- = 0.969$. The last two columns report the results when Prelec's weighting function is used

		TK sentiment		Moderate sentiment		Moderate sentiment Prelec's w function	
X	p_{BS}	p_s	p_a	p_s	p_a	p_s	p_a
70	0.1262	0.8496	0.8747	0.1555	0.1553	0.1670	0.1666
80	0.6872	3.1572	3.0876	0.8204	0.8145	0.8379	0.8321
90	2.3101	7.9546	7.0728	2.6922	2.6362	2.6932	2.6485
100	5.5735	15.6905	12.5269	6.3848	6.1385	6.3488	6.1483
110	10.6753	26.4833	19.0565	12.0983	11.4250	12.0262	11.4463
120	17.3950	40.2732	26.3766	19.5999	18.2390	19.5062	18.2748
130	25.2994	56.8373	34.2999	28.4329	26.1692	28.3307	26.2097

from the writer's viewpoint, for different strikes and values of the parameters. When we set $\mu = r$, $a = b = 1$, $\lambda = 1$, and $\gamma = 1$, we obtain the same results as in the Black-Scholes model (BS prices are reported in the second column). We compare BS premia with prices obtained considering the parameters used in [14] by Tversky and Kahneman. We then used the *moderate sentiment* parameters as in [15] and compare the prices obtained considering different weighting functions: in particular, we applied (7) and Prelec's function (9); results are reported in the last columns of the tables.

Table 3 reports the results for the call option evaluated by the holder. The prices are below the writer's results. What we obtain is an interval for the option in which BS value lies.

It is worth noting that hypothesis on the segregation of the results are also important: in particular, results obtained in the time aggregated case with moderate sentiment do not deviate too far from BS prices. The segregated prospect model, combined with TK sentiment, provide too high writer's option prices and to low holder's option prices to be used in practice. This is also true, but somehow mitigated in the aggregated model with TK sentiment.

5 Concluding Remarks and Future Research

Prospect theory has recently begun to attract attention in the literature on financial options valuation; when applied to option pricing in its continuous cumulative version, it seems a promising alternative to other models proposed in the literature, for its potential to explain option mispricing with respect to Black and Scholes model. In particular, the weighting function may be one of the main causes of the mispric-

Table 3 Call option prices in the Black-Scholes model and in the segregated and aggregated prospects from the holder's viewpoint. The parameters are: $S_0 = 100$, $\mu = 0.05$, $r = 0.05$, $\sigma = 0.2$, $T = 1$, and $L = 1$. In the BS model: $a = b = 1$, $\lambda = 1$, and $\gamma = 1$. The Tversky-Kahneman parameters are: $a = b = 0.88$, $\lambda = 2.25$, $\gamma^+ = 0.61$, and $\gamma^- = 0.69$. The moderate sentiment parameters are: $a = b = 0.988$, $\lambda = 1.125$, $\gamma^+ = 0.961$, and $\gamma^- = 0.969$. The last two columns report the results when Prelec's weighting function is used

		TK sentiment		Moderate sentiment		Moderate sentiment Prelec's w function	
X	c_{BS}	c_S	c_a	c_S	c_a	c_S	c_a
70	33.5401	12.1241	26.6775	29.8193	32.7509	29.7427	32.7826
80	24.5888	9.0924	18.9273	21.9088	23.8714	21.8318	23.9137
90	16.6994	6.6805	12.7900	14.9708	16.1214	14.9014	16.1617
100	10.4506	4.8156	8.3010	9.4774	10.0519	9.4332	10.0828
110	6.0401	3.3871	5.2167	5.5693	5.8103	5.5628	5.8402
120	3.2475	2.3074	3.1864	3.0552	3.1403	3.0816	3.1787
130	1.6395	1.5131	1.8997	1.5767	1.6024	1.6193	1.6473

ing; in this work we compared the results obtained with two alternative weighting functions.

We evaluated options from both the writer's and holder's perspective, obtaining an interval for the option price. Such an interval may be affected by the functional form of the weighting function and the values of its parameters. It will be interesting to use other weighting functions among those proposed in the literature.

We also obtained prices for the call and the put options. As we cannot use put-call parity arguments, another problem which will be interesting to study more in depth is if there exist relations between put and call options prices.

Finally, option prices are sensitive to the choice of the values of the parameters. Calibrating model parameters to market data in order to obtain an estimate of the market sentiment is an important issue which requires further investigation.

References

1. Abbink, K., Rockenbach, B.: Option pricing by students and professional traders: A behavioral investigation. Managerial and Decision Economics **27**(6), 497–510 (2006)
2. Abdellaoui, M., L'Haridon, O., Zank, H.: Separating curvature and elevation: A parametric probability weighting function. Journal of Risk and Uncertainty **41**, 39–65 (2010)
3. Black, F., Scholes, M.: The pricing of options and corporate liabilities. Journal of Political Economy **81**(3), 637–654 (1973)
4. Breuer, W., Perst, A.: Retail banking and behavioral financial engineering: The case of structured products. Journal of Banking and Finance **31**(3), 827–844 (2007)
5. Barberis, N., Thaler, R.: A survey of behavioral finance. In: Constantinides, G.M., Harris, M., Stulz, R. (eds) Handbook of Economics and Finance, Elsevier Science, pp. 1052–1121 (2003)

6. Davies, G.B., Satchell, S.E.: The behavioural components of risk aversion. Journal of Mathematical Psychology **51**(1), 1–13 (2007)
7. Kahneman, D., Tversky, A.: Prospect theory: An analysis of decision under risk. Econometrica **47**(2), 263–292 (1979)
8. Poteshman, A.M., Serbin, V.: Clearly irrational financial market behavior: Evidence from the early exercise of exchange traded stock options. Journal of Finance **58**(1), 37–70 (2003)
9. Prelec, D.: The probability weighting function. Econometrica **66**(3), 497–527 (1998)
10. Shefrin, H., Statman, M.: Behavioral aspects of the design and marketing of financial products. Financial Management **22**(2), 123–134 (1993)
11. Shiller, R.J.: Human behavior and the efficiency of the financial system. In: Taylor, J.B., Woodford, M. (eds.) Handbook of Macroeconomics, vol. **1C**, pp. 1305–1340. Elsevier, Amsterdam (1999)
12. Thaler, R.H.: Mental accounting and consumer choice. Marketing Science **4**, 199–214 (1985)
13. Thaler, R.H.: Mental accounting matters. Journal of Behavioral Decision Making **12**, 183–206 (1999)
14. Tversky, A., Kahneman, D.: Advances in prospect theory: cumulative representation of the uncertainty. Journal of risk and uncertainty **5**, 297–323 (1992)
15. Versluis, C., Lehnert, T., Wolff, C.C.P.: A cumulative prospect theory approach to option pricing. Working paper (2010). Available at SSRN: http://ssrn.com/abstract=1717015

Threshold Structures in Economic and Financial Time Series

Marcella Niglio and Cosimo Damiano Vitale

Abstract In this paper we present some nonlinear autoregressive moving average (NARMA) models proposed in the literature focusing then the attention on the Threshold ARMA (TARMA) model with exogenous threshold variable. The main features of this stochastic structure are shortly discussed and the forecasts generation is presented.
It is widely known that in presence of most economic time series the nonlinearity of the data generating process can be well caught by the threshold model under analysis even if, at the same time, the forecast accuracy is not always equally encouraging. Starting from this statement we evaluate how the forecast accuracy of the US Consumer Price Index can be improved when a TARMA model with exogenous threshold variable is fitted to the data. We give empirical evidence that predictors based on a squared loss function can be more accurate when the spread between US Treasury Bonds and US Treasury Bills is selected as threshold variable.

1 Introduction

The asymmetry of most economic and financial time series has been widely investigated in the literature. The abrupt declines that often characterize short periods of time and the smooth increases, usually longer than the pervious phases, are ascribed as the main causes of largely recognized asymmetric effects.

M. Niglio (✉)
Di.S.E.S., Univerisità degli Studi di Salerno, Via Giovanni Paolo II 132, 84084 Fisciano (SA), Italy
e-mail: mniglio@unisa.it

C.D. Vitale
Di.S.E.S., Univerisità degli Studi di Salerno, Via Giovanni Paolo II 132, 84084 Fisciano (SA), Italy
e-mail: cvitale@unisa.it

M. Corazza, C. Pizzi (eds.), *Mathematical and Statistical Methods for Actuarial Sciences and Finance*, DOI 10.1007/978-3-319-02499-8_21, © Springer International Publishing Switzerland 2014

To catch these features of data, a significant number of models has been proposed in the literature and some of them can be considered as direct generalization of the linear autoregressive moving average (ARMA) model, widely presented in [5].

In fact, after evaluating the inability of ARMA models to take into account the mentioned asymmetry, a number of variants has been given to this stochastic structures. Among them, [24] proposes the so called Self Exciting Threshold ARMA model (SETARMA) given as:

$$X_t = \sum_{k=1}^{\ell} \left[\phi_0^{(k)} + \sum_{i=1}^{p_k} \phi_i^{(k)} X_{t-i} + e_t - \sum_{j=1}^{q_k} \theta_j^{(k)} e_{t-j} \right] I(X_{t-d} \in R_k), \qquad (1)$$

where d is the *threshold delay*, $R_k = (r_{k-1}, r_k]$ with $-\infty = r_0 < r_1 < \ldots < r_{\ell-1} < r_\ell = \infty$ and $\{e_t\}$ is a sequence of independent and identically distributed (*iid*) random variables with $E(e_t) = 0$ and $E(e_t^2) = \sigma^2 < \infty$, whereas p_k and q_k are the autoregressive and moving average order of regime k respectively. New results have been recently given in [13] and [14] for model (1) that, among the others, has a number of theoretical and empirical interesting features: for example it allows to take advantage of some results (properly revised) given in the linear ARMA context; its structure can lead to easy interpretations when applied in empirical domains where the observed phenomena are characterized by regimes changes; it catches asymmetric effects in the data that, in presence of economic time series, are often due to business cycles.

[18] proposes a variant of model (1) where X_t is given by:

$$X_t = \mu_{I(t)} + \sum_{j=1}^{p} \phi_j^{(i)} \left(X_{t-j} - \mu_{I(t-j)} \right) + e_t - \sum_{k=1}^{q} \theta_k^{(i)} e_{t-k}, \qquad r_{i-1} \le Y_t < r_i, \quad (2)$$

for $i = 1, 2, \ldots \ell$, where $I(t) = i$ if $r_{i-1} \le Y_t < r_i$, $\mu_{I(t-j)}$ is the mean value of X_{t-j}, $\mu_{I(t)}$ changes according to the switching among regimes induced by the threshold variable Y_t and the error sequence is defined as:

$$e_t = \sigma_i \eta_t, \qquad \text{if} \quad r_{i-1} \le Y_t < r_i,$$

with η_t a sequence of Gaussian white noises.

The main differences that can be appreciated from the comparison of the models (1) and (2) are that in the autoregressive part of the latter model the difference $(X_{t-j} - \mu_{I(t-j)})$ is obtained subtracting to X_{t-j} the mean value of the regime that has generated X_{t-j} itself (in model (2) each regime has its own mean) and further the threshold variable in model (2) is exogenous without threshold delay d.

Following the spirit that has inspired the two previous models, [6] propose the Autoregressive Asymmetric Moving Average (ARasMA) model defined as:

$$X_t = \phi_1 X_{t-1} + \ldots + \phi_p X_{t-p} + e_t + \beta_1^+ e_{t-1}^+ + \ldots + \beta_q^+ e_{t-q}^+ + \beta_1^- e_{t-1}^- + \ldots + \beta_q^- e_{t-q}^-$$

$$(3)$$

where e_t is a Gaussian white noise sequence with mean zero and finite variance σ^2, $e_{t-i}^+ = \max\{e_{t-i}, 0\}$ and $e_{t-i}^- = \min\{e_{t-i}, 0\}$, for $i = 1, 2, \ldots, q$. In practice the moving average component of model (3) has two separate moving average filters that characterize the asymmetry of X_t whereas the autoregressive component is left unchanged for $t = 1, 2, \ldots, T$.

[10] propose another variant of regime switching ARMA model:

$$X_t = m_0(s_t) + \sum_{i=1}^{p} a_{0i}(s_t) \left[X_{t-i} - m_0(s_{t-i}) \right] = e_t + \sum_{i=1}^{q} b_{0i}(s_t) e_{t-i} \tag{4}$$

where $\{e_t\}$ is an independent sequence of centered random variables whose variance can be time dependent when $e_t = \sigma_0(s_t) \eta_t$ (with $\{\eta_t\}$ a sequence of *iid* random variables), and s_t is the realization of the stochastic process S_t defined on $\{1, 2, \ldots, d\}$. Note that when $d = \ell$ model (4) generalizes model (2). In fact this last model can be seen as particular case of model (4) when S_t degenerates to a deterministic process defined over $\{0, 1\}$ where the switching between the two states depends from an exogenous variable Y_t.

More recently another example of model obtained introducing a threshold effect in the corresponding ARMA model has been given in [12]. In is mainly motivated from the fact that long memory effects and structural changes of data can be easily confused in empirical analysis. So they propose a threshold ARFIMA model having form:

$$\left(X_t^- - \phi_1^- X_{t-1}^- \ldots - \phi_p^- X_{t-p}^- \right) (1 - B)^{d^-}$$
$$+ \left(X_t^+ - \phi_1^+ X_{t-1}^+ \ldots - \phi_p^+ X_{t-p}^+ \right) (1 - B)^{d^+} = \Theta(B) e_t \tag{5}$$

with $\{e_t\}$ a sequence of *iid* errors with mean zero and finite variance $\sigma^2 > 0$, $\Theta(B) = 1 + \theta_1 B + \theta_2 B^2 + \ldots + \theta_q B^q$, $X_t^- = (X_t - \mu^-) I(Z_{t-1} \leq r)$, $X_t^+ = (X_t - \mu^+) I(Z_{t-1} > r)$, μ^- and μ^+ are the mean values of the first and the second regime respectively, the threshold variable $Z_{t-1} = f(e_{t-1}, \ldots, e_{t-p})$ with $f(\cdot)$ known, and where the parameter of fractional integration becomes d^- if $Z_{t-1} \leq r$ and d^+ if $Z_{t-1} > r$.

Starting from model (5), [12] propose two further models: the short memory threshold ARFIMA obtained when $d^- = d^+$ and the long memory threshold ARFIMA where the autoregressive parameters of model (5) become $\phi_i^+ = \phi_i^- = \phi_i$, for $i = 1, 2, \ldots, p$. They further note that in the short memory case the threshold variable Z_{t-1} can be substituted by X_{t-1} itself (or even an exogenous variable). In this last case we further remark that if $\mu^- = \mu^+$ and $d = 0$ the model degenerates to a particular threshold ARMA model whose properties have been examined in [7].

In this wide class of models that generalize the ARMA stochastic structure, the attention will be given to a variant of model (1). More precisely in Sect. 2 we discuss some feature of the proposed model specification and we focus the attention on the forecasts generation whereas in Sect. 3 we apply this variant to evaluate its forecast accuracy to predict the US Consumer Price Index. Some concluding remarks are given at the end.

2 TARMA Model with Exogenous Threshold Variable

The dynamic structure of model (1) is characterized by a threshold variable given by X_s, with $s = t - d$. The feedback of X_s on X_t, with $s < t$, heavy impacts the statistical properties (such as stationarity and ergodicity) of the SETARMA process leading to very narrow stationarity region over the parametric space (see [2, 15, 16]).

A variant of model (1) has been recently discussed in [19] where a TARMA model with exogenous threshold variable has been proposed and the conditions for its strong and weak stationarity are provided. In more detail, let X_t be a TARMA($\ell; p_1, \ldots, p_\ell; q_1, \ldots, q_\ell$), its form is given by:

$$X_t = \sum_{k=1}^{\ell} \left[\phi_0^{(k)} + \sum_{i=1}^{p_k} \phi_i^{(k)} X_{t-i} + e_t - \sum_{j=1}^{q_k} \theta_j^{(k)} e_{t-j} \right] I(Y_{t-d} \in R_k), \tag{6}$$

with $\{e_t\}$ a sequence of $iid(0, \sigma^2)$ random variables and Y_t a strictly stationary process independent from e_t. The exogenous threshold variable Y_{t-d} heavy impacts the dynamic structure of X_t in model (6) whose switching among regimes is not due to the process X_t itself as in model (1).

The main advantages obtained from model (6) with respect to the models shortly presented in Sect. 1 are various: the presence of an exogenous threshold variable allows to define a switching structure for X_t that is related to a delayed variable Y_{t-d} which influences its stochastic dynamic. This property of model (6) has interesting implications in empirical domain (for example in economic domain where a change of state is due to other phenomena observed in the past) and makes the use of the TARMA model more flexible with respect to model (1) and (2). A further difference of model (6) when compared to model (3) and (5), is that the threshold dynamic involves both the autoregressive and the moving average components. It increases the ability of the model to catch asymmetric effects and business cycles in the data. Finally the main advantage of model (6) with respect to model (4) is related to its more easy interpretation which is greatly appreciated from practitioners (and not only from them).

In the following, to simplify the notation, we assume that $p_1 = p_2 = \ldots = p_\ell = p$ and $q_1 = q_2 = \ldots = q_\ell = q$ (for this purpose null parameters can be included in the model) such that we can shortly say that $X_t \sim TARMA(\ell; p, q)$.

The estimation of the model parameters $\Psi = (\Phi^{(1)}, \ldots, \Phi^{(\ell)}, \Theta^{(1)}, \ldots, \Theta^{(\ell)}, \sigma^2)$, with $\Phi^{(i)} = (\phi_0^{(i)}, \phi_1^{(i)}, \ldots, \phi_p^{(i)})$ and $\Theta^{(i)} = (\theta_1^{(i)}, \ldots, \theta_q^{(i)})$, can be based of the iterative procedure given in Tong (1983) and discussed for the SETARMA model in [19]. The algorithm, that for the sake of simplicity is presented for $\ell = 2$, is based on the following steps:

1. select a set of non negative integer values for the p and q orders, define a set $\mathscr{D} \in N^*$ of possible threshold delays d (with N^* the set of positive integers) and a subset $[R_{\mathscr{L}}, R_{\mathscr{U}}]$ of α-quantiles of Y_t that represent the candidate values of the threshold parameter r; for all combinations of (p, q, d, r) define a grid;

2. select the cells of the grid defined in step 1. such that each regime has an adequate number of data;
3. estimate the $\phi_i^{(k)}$ and $\theta_j^{(k)}$ parameters of model (6) such that

$$\hat{\beta} = \underset{\hat{\beta} \in B}{\arg\min} \sum_{i=s}^{T} e_i^2(\beta),$$

with $\beta = \left(\Phi^{(1)}, \Phi^{(2)}, \Theta^{(1)}, \Theta^{(2)} \right)$, $s = \max(p,q,d)+1$, $e_i(\beta) = X_i - E[X_i|\mathscr{F}_{i-1}]$ and $\mathscr{F}_{i-1}$ a $\sigma-$field generated by $(X_1, \dots, X_{i-1}, Y_1, \dots, Y_{i-1-d})$;
4. use the residuals $\hat{e}_t = X_t - \hat{X}_t$, for $t = s, \dots, T$, obtained from step 3., and estimate the variance $\hat{\sigma}^2 = (T-s)^{-1} \sum_{t=s}^{T} \hat{e}_t^2$.

The estimation of the model parameters is the preliminary step for the forecasts generation which represents one of the major aims of the time series analysis. Here we focus the attention on point forecasts whose accuracy is evaluated in detail.

2.1 Forecast Generation

The generation of forecasts from nonlinear time series models has been differently discussed (for recent reviews see [11] and [23, Chap. 14]). In the present paper we focus the attention on one-steps ahead forecasts (denoted $f_{T,1}$) generated from model (6). As well known the generation of $f_{T,1}$ requires an information set ($\mathscr{F}_T$) that contains all available informations on X_t and on the threshold variable Y_t, up to time T.

After choosing a cost function $C(e_{T,1})$ with $e_{T,1} = X_{T+1} - f_{T,1}$, the predictor $f_{T,1}$ is obtained from the minimization of $C(e_{T,1})$. The selection of the function $C(\cdot)$ can be differently chosen (see among the others [20]). In the following a quadratic cost function is selected for $C(\cdot)$ such that $f_{T,1}$ is obtained from:

$$\underset{f_{T,1}}{\arg\inf} E\left[(X_{T+1} - f_{T,1})^2 \right], \tag{7}$$

where $f_{T,1}$ is unbiased and with serially uncorrelated prediction errors $e_{T,1}$.

Taking advantage of the assumptions on model (6) and after the minimization of (7), the predictor $f_{T,1} = E[X_{T+1}|\mathscr{F}_T]$ is defined as:

$$f_{T,1} = E\left[\sum_{k=1}^{\ell} \left(\phi_0^{(k)} + \sum_{i=1}^{p} \phi_i^{(k)} X_{t+1-i} + e_{t+1} - \sum_{j=1}^{q} \theta_j^{(k)} e_{t+1-j} \right) I(Y_{t+1-d} \in R_k) | \mathscr{F}_T \right]$$

$$= \sum_{k=1}^{\ell} \left[\phi_0^{(k)} + \sum_{i=1}^{p} \phi_i^{(k)} X_{t+1-i} - \sum_{j=1}^{q} \theta_j^{(k)} e_{t+1-j} \right] I(Y_{t+1-d} \in R_k), \tag{8}$$

with $d \geq 1$ and all parameters are estimated following the four-steps procedure illustrated in Sect. 2.

The predictor (8), even called *analytical* in [11], completely changes when h-steps ahead forecasts are generated. In more detail when $h > d$ the threshold variable Y_{t+h-d} has to be predicted making use of its conditional expectation. It requires the knowledge of the dynamic structure of the threshold variable that needs to be investigated. This problem, faced in the SETARMA domain in [1], is beyond the aims of the present paper and is left for future research.

3 U.S. Monthly Consumer Price Index

The behavior of the US Consumer Price Index (CPI) is a central concern in economics and in the last years the attention is further grown on it. A question largely discussed in the literature is why the CPI can be hard to forecast and how the introduction of proper variables can be of help in this difficult task (see among the others [8, 22]). In this context the use of TARMA models with exogenous threshold variable can give some advantage for different reasons: they are able to catch the asymmetry of the CPI; the estimated model has interesting empirical interpretations; the exogenous threshold variable allows to include informations in the model that can improve the forecast accuracy.

Following [8], we have generated for the monthly US CPI one-step ahead forecasts from model (6) using as exogenous threshold variable the Spread (based on monthly data) between the US treasury notes (10-years US treasury rates) and the US treasury bills (3-months US treasury rates). The results in terms of forecast accuracy are described below.

The monthly US CPI and the Spread have been examined over the period 1985:01-2008:07 for a total of 284 observations. The original Consumer Price Index data and $lCPI = \log(CPI_t/CPI_{t-1})$ are presented in Fig. 1 whereas the time plot and the autocorrelation function (ACF) of $lSpread = (Spread_t - Spread_{t-1})$, with $Spread = \log(TNotes_t/TBills_t)$, is shown in Fig. 2. From the comparison of the two figures, it can be noted an increasing variability first observed in $lSpread$ and then in $lCPI$. The marked changes in the $lSpread$ variability, that start just before the terroristic attacks to the twin towers in US, have heavy consequences over all the next 5 years.

To start the presentation of data under analysis, some summary statistics are presented in Table 1 where, as expected, it can be noted the higher variability, skewness and kurtosis of $lSpread$ with respect to $lCPI$.

Table 1 Summary statistics of $lCPI$ and $lSpread$ data (mean, variance, median, skewness, kurtosis, minimum, maximum)

Serie	μ	σ^2	Me	γ_1	γ_2	min	max
$lCPI$	0.0026	$4.86e^{(-06)}$	0.0025	0.2437	3.6490	−0.0055	0.014
$lSpread$	−0.0024	0.0050	−0.0039	0.7455	5.1055	−0.3094	0.3452

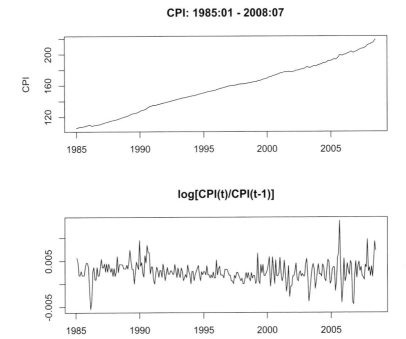

Fig. 1 Top: Time plot of the monthly US CPI over the period 1985:01-2008:07; Bottom: Time plot of $lCPI = \log(CPI_t/CPI_{t-1})$

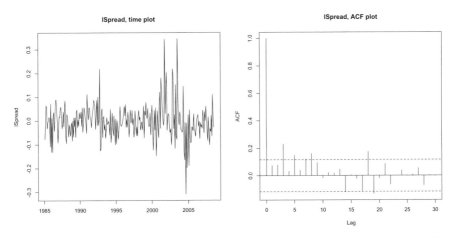

Fig. 2 Time plot and ACF plot of the spread between the US treasury notes (10 years US treasury rates) and the US treasury bills (3 months US treasury rates), where $lSpread = (Spread_t - Spread_{t-1})$ with $Spread = \log(TNotes_t/TBills_t)$

The dynamic structure of the US CPI has been largely studied and the nonlinearity of the underlying data generating process mainly due to asymmetric effects is widely accepted (see among the others [4]). The nonlinearity recognized in the US CPI has lead to fit switching structures to the observed data that in most cases belong to the wide class of nonlinear autoregressive (NLAR) processes (see [3, 4, 17]).

In our analysis we have selected for the *lCPI* the nonlinear autoregressive moving average model (6) where the moving average component allows to select a more parsimonious model with respect to NLAR structures used in [17]. After the identification and estimation of the model based on the four steps procedure presented in Sect. 2, we have generated one-step ahead forecasts over the period 2006:08-2008:07. In more detail let $T = 284$ the total sample size, using the predictor (8) we have generated from $T - h + 1$ to T, $h = 24$ one-step ahead forecasts using an *expanding window algorithm* based on the following steps:

1. define a forecast horizon h;
2. for $i = 0, \ldots, h - 1$:

    (a) select the sample $X_1, \ldots, X_{T-h+i}$ over which to estimate the model parameters;
    (b) generate the one-step ahead forecast $f_{T-h+i+1}$ for $X_{T-h+i+1}$, given the threshold variable Y_{T-h+i}.

This algorithm allows to increase the estimation window as i grows producing, at each iteration of step 2., an out of sample one-step ahead forecast.

The sequence of forecasts $\mathbf{f}_T = (f_{T-h+1}, \ldots, f_T)$ is then compared to the observed data $\mathbf{X}_T = (X_{T-h+1}, \ldots, X_T)$ using a square loss function:

$$E[L_2(\mathbf{X}_T, \mathbf{f}_T)] = \frac{1}{h} \sum_{i=0}^{h-1} (X_{T-h+i+1} - f_{T-h+i+1})^2, \qquad \text{[MSFE]}$$

to evaluate their accuracy.

As discussed in the previous pages, the introduction of an exogenous threshold variable and the use of nonlinear models can improve the accuracy of the forecasts generated for the US CPI. To further remark this point we have generated forecasts for *lCPI* using the following models:

1. TARMA model (6) with *lSpread* as threshold variable;
2. SETARMA model (1);
3. SETAR model, obtained from (1) with $\theta_j^{(k)} = 0$, for $j = 1, \ldots, q_k$ and $k = 1, \ldots, \ell$;
4. ARMA model, given by (1) with $\ell = 1$;
5. Random Walk (RW).

For models 1. to 4. the identification has been based on the Akaike Information Criterion whereas the parameters are estimated at each iteration of step 2. Further, for all five models, the predictors obtained from (7) are considered and their MSFE's are compared.

Table 2 Evaluation, in terms of MSFE, of the SETARMA model (1) with three competitor models

	$d=1$	$d=2$	$d=3$	$d=4$	$d=5$	$d=6$
$\dfrac{MSFE(SETARMA)}{MSFE(SETAR)}$	1.0256	0.9504	0.9887	1.0096	0.9956	0.9899
$\dfrac{MSFE(SETARMA)}{MSFE(ARMA)}$	1.0134	1.0329	1.0444	1.1295	1.0559	1.0311
$\dfrac{MSFE(SETARMA)}{MSFE(RW)}$	0.6783	0.6913	0.6990	0.7560	0.7067	0.6901

Table 3 Evaluation, in terms of MSFE, of the TARMA model (6) with three competitor models

	$d=1$	$d=2$	$d=3$	$d=4$	$d=5$	$d=6$
$\dfrac{MSFE(TARMA)}{MSFE(SETAR)}$	0.9932	0.9181	0.9884	0.9176	0.9597	0.9426
$\dfrac{MSFE(TARMA)}{MSFE(ARMA)}$	0.9814	0.9977	1.0441	1.0266	1.0179	0.9817
$\dfrac{MSFE(TARMA)}{MSFE(RW)}$	0.6568	0.6678	0.6988	0.6871	0.6812	0.6571

In Table 2 the MSFE of the SETARMA model is compared with the MSFE's of the SETAR, ARMA and RW models whereas in Table 3 the MSFE of the TARMA model is compared with the MSFE's of the same three competitor models.

We have generated for the TARMA (and to make easier the comparison even for the SETARMA) model, forecasts for different values of d ($d = 1, \ldots, 6$) to evaluate if the Spread impacts the CPI prediction in presence of long horizons, as stated in [8].

From Table 2 it can be noted that for $d = 1$ to 6 the SETARMA model does not outperforms the linear ARMA model whereas in four cases some gain, in term of forecast accuracy, is obtained with respect to the SETAR forecasts.

The poor performance of the forecasts generated from nonlinear models with respect to those obtained from linear structures is not new in presence of economic time series (in this very wide literature see [9, 21]). The complexity of the generating process and the number of variables that impacts this kind of phenomena suggest the use of models that are able to catch, at the same time, the nonlinearity of data and the relation with other economic time series.

Following this advice, in Table 3 it can be noted that the introduction of *lSpread* as threshold variable allows to generate more accurate forecasts from the TARMA model when compared with the predictions obtained from the SETAR, ARMA and RW models. When the MSFE's of the TARMA and SETAR model are compared, the better forecast accuracy of the former model with respect to the latter structure can be appreciated for $d = 1, \ldots, 6$. From the comparison TARMA-ARMA forecasts the better performance of the TARMA model is reached for $d = 1$ (which is the threshold delay selected at the identification stage), $d = 2$ and $d = 6$ (this last case confirms what stated in [8] where the author discusses how the Spread can be of help to predict inflation, particularly with long horizons). As expected, in all cases the Random Walk model is characterized by the worst forecast accuracy in both Table 2 and Table 3.

4 Conclusions

In this paper nonlinear autoregressive moving average models often used to analyze economic and financial time series are presented and shortly compared. Among them the attention has been mainly focused on the class of threshold ARMA models and on their forecasts generation and forecasts accuracy. These models have been applied to analyze the monthly US CPI and to investigate if the relation between the US CPI and the Spread (computed in terms of difference between US Treasury Notes and US Treasury Bills rates) can impact the forecast accuracy of the inflation in US. The empirical results show that the introduction of the Spread as threshold variable in (6) impacts the accuracy of the forecasts generated from the model and further that these forecasts always outperform the predictions obtained from model (1) where the threshold variable is endogenous.

Acknowledgements The Authors would like to thanks the Collaborative Research Center 649 of the Center for Applied Statistics and Economics (C.A.S.E.) at the Humboldt-Universität of Berlin (D), for providing the data when the first Author has been guest researcher.

References

1. Amendola, A., Niglio, M., Vitale, C.D.: Least squares predictors for threshold models: properties and forecast evaluation. In: Perna, C., Sibillo, M. (eds.) Mathematical and Statistical Methods in Insurance and Finance, pp. 1–9. Springer (2008)
2. Amendola, A., Niglio, M., Vitale, C.D.: Statistical properties of SETARMA models. Commun. Stat.: Theory and Methods **38**, 2479–2497 (2009)
3. Ball, L., Cecchetti, S.G., Gordon, R.J.: Inflation and uncertainty at short and long horizon, Brookings Pap. Econ. Activity. **1**, 215–254 (1990)
4. Beecheya, M., Österholmb, P.: Revisiting the uncertain unit root in GDP and CPI: testing for nonlinear trend reversion. Econ. Lett. **100**, 221–223 (2008)
5. Box, G.E.P., Jenkins, G.M.: Time series analysis, forecasting and control. Holden-Day, San Francisco (1976)
6. Brännas, K., De Gooijer, J.G.: Asymmetries in conditional mean and variance: modeling stock returns by asMA-asGARCH. J. Forecast. **23**, 155–171 (2004)

7. Brockwell, P.J., Liu, J., Tweedie, R.L.: On the existence of stationary threshold AR-MA process. J. Time Ser. Anal. **13**, 95–107 (1992)
8. Estrella, A.: Why does the yield curve predict output and inflation? Econ. J. **115**, 722–744 (2004)
9. Faust, J., Wright, J.: Forecasting Inflation. Manuscript for Handbook of Forecasting (2012)
10. Franq, C., Gautier, A.: Large sample propserties of parameter least squares estimates for time-varying ARMA models. J. Time Ser. Anal. **25**, 765–783 (2003)
11. Kock, A.B., Teräsvirta, T.: Forecasting with nonlinear time series models. In: Clements M.P., Hendry D.F. (eds.) Oxford Handbook on Economic Forecasting, pp. 61-88. Oxford University Press (2011)
12. Lahiani, A. Scaillet, O.: Testing for threshold effects in ARFIMA models: application to US unemployment data rate. Int. J. Forecast. **25**, 418–428 (2009)
13. Li, G., Li, W.K.: Testing a linear time series model against its threshold extension. Biometrika **98**, 243–250 (2011)
14. Li, D., Li, W.K., Ling, S.: On the least squares estimation of threshold autoregressive and moving-average models. Stat. Interface **4**, 183–196 (2011)
15. Ling, S.: On probability properties of a double threshold ARMA conditional heteroskedasticity model, J. Appl. Probab. **36**, 688–705 (1999)
16. Liu, J., Susko, E., On strict stationarity and ergodicity of a non-linear ARMA model. J. Appl. Probab. **29**, 363–373 (1992)
17. Marcellino, M.: A benchmark for models of growth and inflation. J. Forecast. **27**, 305–340 (2008)
18. Mélard, G., Roy R.: Modèles de séries chronologiques avec seuils. Rev. Stat Appl. **36**, 5–23 (1988)
19. Niglio, M., Vitale, C.D.: Local unit roots and global stationarity of TARMA models, Methodol. Comput. Appl. Probab. **14**, 17–34 (2012)
20. Patton, A., Timmermann A.: Properties of Optimal Forecasts under Asymmetric Loss and Nonlinearity, J. Econom. **140**, 884–918 (2007)
21. Rossi, B., Sekhposyan, T.: Have economic model's forecasting performance for US output growth and inflation changed over time, and when? Int. J. Forecast. **26**, 808–835 (2010)
22. Stock, J.H, Watson, M.W.: Why has U.S. inflation become harder to forecast? J. Money Credit Bank. **39**, 3–33 (2007)
23. Teräsvirta, T., Tjøstheim, D.T., Granger, C.W.J.: Modelling nonlinear economic time series. Oxford University Press, New York (2010)
24. Tong, H.: Threshold models in nonlinear time series analysis. Springer-Verlag, London (1983)

Intelligent Algorithms for Trading the Euro-Dollar in the Foreign Exchange Market*

Danilo Pelusi, Massimo Tivegna and Pierluigi Ippoliti

Abstract In order to improve the profitability of Technical Analysis (here represented by Moving Average Crossover, DMAC), suitable optimization methods are proposed. Artificial Intelligence techniques can increase the profit performance of technical systems. In this paper, two intelligent trading systems are proposed. The first one makes use of fuzzy logic techniques to enhance the power of genetic procedures. The second system attempts to improve the performances of fuzzy system through Neural Networks. The target is to obtain good profits, avoiding drawdown situations, in applications to the DMAC rule for trading the euro-dollar in the foreign exchange market. The results show that the fuzzy system gives good profits over trading periods close to training period length, but the neuro-fuzzy system achieves the best profits in the majority of cases. Both systems show an optimal robustness to drawdown and a remarkable profit performance. In principle, the algorithms, described here, could be programmed on microchips. We use an hourly time series (1999–2012) of the Euro-Dollar exchange rate.

D. Pelusi (✉)
Department of Communication Science, University of Teramo, Teramo, Italy
e-mail: dpelusi@unite.it

M. Tivegna
Department of Theories and Policies for Social Development, University of Teramo, Teramo, Italy
e-mail: mc1223@mclink.it

P. Ippoliti
Department of Theories and Policies for Social Development, University of Teramo, Teramo, Italy
e-mail: ippopippo79@libero.it

  * Supported by the Research and Enterprise Investment Programme 2012/13, administered by the University of Greenwich.

M. Corazza, C. Pizzi (eds.), *Mathematical and Statistical Methods for Actuarial Sciences and Finance*, DOI 10.1007/978-3-319-02499-8_22, © Springer International Publishing Switzerland 2014

1 Introduction

The trading performance of Technical Analysis (TA) can, in general, be improved using intelligent data mining techniques. Among the main procedures, intelligent methods such as Fuzzy Logic and Genetic Algorithms (GA) have frequently achieved good results. In many cases, the combination of these techniques produces relevant findings. Allen and Karjalainen [1] used a genetic algorithm to learn technical trading rules for the S&P 500 index using daily prices. They concluded that the GA application finds little evidence of substantial improvements in economically significant technical trading rules. This happens, most likely, because their pioneering experiment aims at discovering randomly, through GA methods, technical rules not necessarily ever used by a human trader. However, good results are found in [4] where the behaviour of traders tends to adapt the chosen algorithm to market conditions, by dropping trading rules as soon as they become loss-making or when more profitable rules are found. A Genetic Algorithm is applied here to a number of indicators calculated on a set of US Dollar/British Pound exchange rates, selecting rules based on combinations of different indicators at different frequencies and lags. Moreover, strong evidence of economically significant out-of-sample excess returns to technical trading rules are found using genetic programming techniques by [6].

A relatively new approach to genetic procedures application to TA is introduced in [9]. The authors use Genetic Algorithms (GA) to obtain optimal (and constrained optimal) parameters for high profit in a very popular trading rule, the Dual Moving Average Crossover (DMAC). This is a mathematically well-defined rule and GA is able to hit at least a good local optimum there.

In this paper, we design a Neuro-Fuzzy system for trading in the Euro-Dollar foreign exchange market, using a DMAC trading rule (described in Sect. 2). The Fuzzy Logic, together with genetic optimization, is good for computing and re-computing optimal and near-optimal trading parameters of the DMAC, whereas the Neural Networks (NN), being able to learn from the data in repeated experiments, âĂŞ is devoted to give optimal weights to the parameters found by the Fuzzy Logic Controller(FLC). The idea is using the constrained algorithms of [9] to define the Membership Functions (MF) and the fuzzy rules of the trading system (Sect. 3). After that, in order to achieve the best performance of the overall fuzzy logic-GA-enhanced DMAC, the weights of fuzzy rules are computed by a suitable Neural Network, as done by applications in a different field [10–12] (Sect. 4). The target is to achieve optimal profits with DMAC in the euro-dollar exchange rate market, avoiding drawdown situations [9]. The intelligent model is programmed in Matlab with Fuzzy Logic and Neural Network tools. We use an hourly time series (1999–2012) of the Euro-Dollar exchange rate supplied by the US data provider CQG (with an extensive work to filter out errors).

A last thought on the innovative strength of the algorithmic procedures here proposed. Genetic Algorithms, Fuzzy Logic Controllers and the use of Neural Networks are more frequently used separately in finance. Their joint use is new in this field. Moreover, this novelty can have huge implications for the optimal design of unattended trading machines. They can signal trading opportunities or execute trades automatically, within the limits of boundary values supplied by the users of these instruments. In principle, the internal architecture of these machines coul be programmed on microcircuits.

2 The DMAC Rule

Technical Analysis is a forecasting method of price movements for trading. It aims at forecasting future price movements from chart patterns of past values (see [13]), which can be expressed in parametric mathematical form. Parameters can therefore be optimized in a trading environment, where you have a period to evaluate and optimize your trading rule (called the Training Set, TNS) and an unknown future where you plunge your optimized tool for trading and earn money (called Trading Set, TDS).

A technical trading system consists of a set of trading rules which depend on technical parameters. The rule generates trading signals according to its parameter values. The most well-known types of technical trading systems are moving averages, channels and momentum oscillators [8]. Among such technical rules, the DMAC rule is here considered. The Dual Moving Average Method is one of the few technical trading procedures that is mathematically well defined. The DMAC system generates trading signals by identifying when the short-term trend rises above or below the long-term trend.

Let p_t be the hourly foreign exchange rate at time t. We define the Fast Moving Average (FMA) A_f, over n hours, and the Slow Moving Average (SMA) A_s, over m hours: $A_f(t) = \sum_{i=1}^{n} \frac{p_{t-1-i}}{n}$ and $A_s(t) = \sum_{i=1}^{m} \frac{p_{t-1-i}}{m}$, with $t \geq m > n$.

The trading rules are:

- if $A_f(t) > A_s(t)$ than go long (i.e. buy the asset) at p_{t+1};
- if $A_f(t) < A_s(t)$ than go short (i.e. sell the asset) at p_{t+1}.

The FMA computes a moving average over a smaller number of hours than in the SMA. The FMA thus picks up the short-term movements of the rate, whereas the SMA draws the longer term trend of it. Trading signals are obtained by their contemporaneous movements. A trade is opened if one of two trading rules holds. In other words, if the Fast Moving Average curve crosses the Slow Moving Average curve than a trade (long or short) is opened. Such trade is closed when a suitable threshold is reached. For both longs and shorts we have two thresholds: a Take Profit (TP), if the exchange rate went in the direction established by your DMAC, a Stop Loss (SL), if the direction was the opposite.

In the DMAC rule, the technical parameters to be optimized are the Fast Moving Average, the Slow Moving Average, the Take Profit and the Stop Loss. Practitioners choose the values of these parameters according to experience, market trends, volatility and their personal trading styles. They also frequently prefer to base decisions on a mix of technical indicators. The techniques we proposed in [9] and propose here can offer further information to professional and non-professional traders.

3 Design of the Decision Protocol

In order to obtain good profits from the application of DMAC rule, a suitable choice of the technical parameters - FMA, SMA, TP and SL - is necessary. Our methodology starts with the definition and actual indication of two features (among the many possible ones) of the euro-dollar exchange rate behaviour overtime. The first one is the slope of trend (as representing the longer term direction of the exchange rate), whereas the second one is the number of crossings of the exchange rate curve around its time trend (as representing the volatility of the rate). As standard in the evaluation of trading rules (in order to avoid data snooping), the profitability of our trading protocol is analyzed by dividing the available sample into a Training Set (TNS) and a Trading Set (TRS).

A Fuzzy Logic Controller (FLC) characterizes the slope values and the crossings number. In fact, with the help of this technique it is possible to define suitable scaling parameters of the inputs. The proposed FLC has two fuzzy inputs (slope and crossings number) and four fuzzy outputs (TP, SL, FMA and SMA). To define the Membership Functions (MF) number and the relative scaling factors, we consider the results obtained in a Master thesis, written at Teramo by one of our brightest students, Pierluigi Ippoliti.

The first fuzzy input (trend) reflects five states of the forex market. It is therefore represented by five MF: Fast Decreasing (FD), Slow Decreasing (SD), Lateral Trend (LT), Slow Increasing (SI), Fast Increasing (FI). The crossings number and the fuzzy outputs have three MF: Low (L), Medium (M), High (H). To establish the MF scaling factors, the methods, defined in [9] and in the above-mentioned thesis, are used. A constrained algorithm is applied to the DMAC rule, avoiding losses greater than four per cent in one month. This constraint avoids draw-down situations during the trading and reduces losses.

As customary in the literature, the choice of training and trading sets length is a delicate issue [1, 7]. We analyze the profitability of our technical rule considering a TNS length of four months (various ones) between 1999 and 2007. Once the training periods are chosen, slope and crossings number are computed. Over such periods, the DMAC rule is applied. Through genetic procedures, the best values of TP, SL, FMA and SMA are found [9]. This means that the GA searches the parameters which give optimal profits.

Using trial and error procedures, we establish the ranges of FD, SD, LT, SI, FI Membership Functions (see Fig. 1, upper panel) normalized between $[-1, 1]$. Besides, according to the literature, we choose the triangular/trapezoidal shape for our MF. The Membership Functions scaling parameters of fuzzy input crossings number (see Fig. 1, lower panel) and fuzzy outputs are defined according with the genetic procedures results. The MF of output parameters are shown in Fig. 2, where the unit of measurement of TP and SL are the basis points of the Euro-$ exchange rate, the so-called pips in the tradersâAZ jargon. Note that the Take Profit scale is greater than Stop Loss scale by ten factor. This choice comes from the genetic procedures results and – as a matter of fact – it is frequent (on a much smaller scale) among traders.

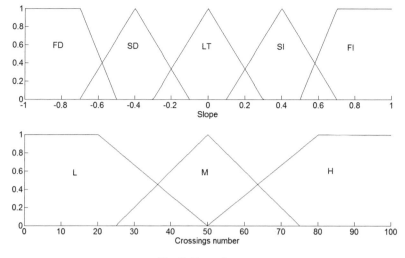

Fig. 1 Fuzzy inputs

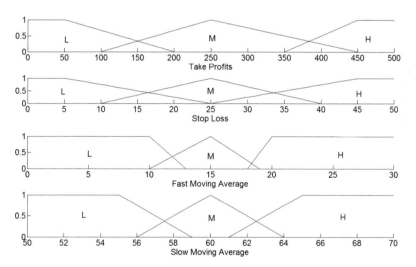

Fig. 2 Fuzzy outputs

The next step to define the fuzzy model is the fuzzy rules definition. Because we have one input with 5 MF and the second one with 3 MF, then $5 \times 3 = 15$ optimality criteria (rules in other terms) can be used. An example of fuzzy rule is: If slope is FD and crossings number is L, then TP is H and SL is L and FMA is L and SMA is M. By exploiting the information of the constrained algorithms, suitably adjusted for our trading aims, the rules base is defined. In other words, the knowledge introduced in the fuzzy system comes from genetic-constrained investigations. Denoting the slope input with S and the crossings number with N_c, the fuzzy rules are shown in Table 1.

Table 1 Fuzzy rules base

$S \setminus N_c$	L	M	H
FD	$H;L;L;M$	$H;M;L;H$	$M;L;H;L$
SD	$M;M;H;H$	$H;L;L;M$	$H;M;M;L$
LT	$H;H;L;M$	$H;M;M;L$	$L;H;M;M$
SI	$L;H;L;M$	$M;H;L;H$	$H;L;L;M$
FI	$H;L;L;L$	$M;L;L;L$	$H;L;L;H$

Notice, at this point, that all the fuzzy rules are defined to have the same weights values in the choice of the parameters of DMAC. To improve the trading decision model, the most profitable rules must be spotted. This means that the fuzzy rules which give good profits must have higher weights values in determining the best DMAC parameters. To solve this problem, we design a suitable Neural Network.

4 Neuro-Fuzzy Algorithm

Once defined the fuzzy sets and the rules base, the next step is the optimization of rules weights. To do this, a Neuro-Fuzzy algorithm is proposed. First of all, we define a training set for the Neural Network. The patterns are characterized by 2 inputs and 15 outputs. The two inputs are the slope S and the crossings number N_c, whereas the outputs are the optimal parameters of the DMAC, as discovered by the FLC, as reported in Table 1. Such training set for NN is drawn by the following steps.

Step 1. Select $m = 100$ training period of four months length. For each period, compute the slope S and the crossings number N_c.
Step 2. For each (S, N_c) pair, generate $n = 15$ random values between 0 and 1 which correspond to fuzzy rules weights w_i ($i = 1, 2, \ldots, 15$), with $w_i \varepsilon [0, 1]$.
Step 3. Compute TP, SL, FMA and SMA according fuzzy inputs S, N_c and weights w_i.
Step 4. Calculate the DMAC profits p_j, $j = 1, 2, \ldots, n$ and compute the max profit.
Step 5. Select the weights w_i, $i = 1, 2, \ldots, 15$ associated with max profit.

In this way, a training set of 100 patterns to train the NN is designed. The patterns are composed by the inputs S and N_c and the weights w_j as outputs. A blocks diagram of training set selection is shown in Fig. 3. The FLC receives the fuzzy inputs S and N_c and the weights w_i from NN. The parameters S and N_c are also inputs of NN. The Fuzzy Logic Controllers gives the outputs TP, SL, FMA and SMA which serve as inputs to the DMAC rule.

The architecture of a Neural Network very much depends on the kind of application under investigation. Because we have two fuzzy inputs, the input layer of NN is composed by 2 neurons, whereas the output layer has 15 neurons, as the number of

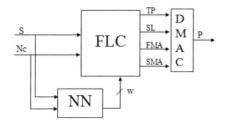

Fig. 3 Blocks diagram of Neuro-fuzzy system

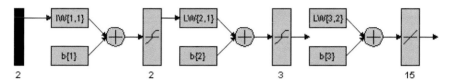

Fig. 4 Neural Network architecture

fuzzy weights. Because there are no fix rules to establish the hidden layer neurons number, trial and error procedures are used. This number sometimes depends on specific application. For instance, some indications can be found in [14]: they propose an algorithm able to obtain the number of necessary hidden neurons of single-hidden-layer feed forward networks for different pattern recognition application tasks. The hidden layer of our NN is made up by 3 neurons (see Fig. 4).

In order to design the NN architecture, the Matlab Neural Network tool is used. Establishing an epochs number of 250 and a goal of 0.02, the net achieves a performance index of 0.00266812.

Once trained the NN, the intelligent system is ready for trading the euro-dollar exchange rate in the foreign exchange market.

5 Experimental Results

Our intelligent algorithm works on hourly foreign exchange rate. The TNSs go from 1999 to 2007, whereas the TDSs go from 2008 to 2012. We consider four-month TNSs length and test the algorithm on monthly, two-months, three-months, four-months, six-month and annual TDSs lengths. Matlab was used for computations.

In all the experiments, the same drawdowns are roughly obtained with the fuzzy controller only and its neuro-fuzzy extension (see, for instance, all the charts in the paper and Table 2, column 3, where only one number is reported). The smaller the relative frequency of draw-down number (N_{dd}, is small), the better is the intelligent system performance. The neuro-fuzzy system, though, improves the cumulative profit in four experimental typologies out of six (compare P_f and P_{nf} in the following table and look at the last column).

Table 2 Draw-down number and cumulative profits for fuzzy and neuro-fuzzy systems.

Period	N_p^*	N_{dd}^*	R^*	P_f^*	P_{nf}^*	$\Delta = P_{nf} - P_f$
Monthly	55	2	0.0364	0.5590	0.7367	0.1777
Two-month	27	2	0.0741	0.6477	0.7529	0.1052
Three-month	18	1	0.0556	0.2702	0.3415	0.0713
Four-month	13	1	0.0769	0.8013	0.6401	−0.1613
Six-month	9	0	0	0.3498	0.2421	−0.1077
Annual	4	0	0	0.3342	0.4032	0.0689

$^*N_p$ is the number of TNSs. N_{dd} is the draw-down number. $R = \frac{N_{dd}}{N_p}$ is the relative frequency of draw-down number. P_f is the fuzzy cumulative profit. P_{nf} is the neuro-fuzzy cumulative profit. They are both expressed in basis points of the Eurodollar exchange rate.

Table 2 shows the cumulative profits achieved with fuzzy and neuro-fuzzy models. It can be noticed that the fuzzy procedures seem to fare better than neuro-fuzzy ones for medium period lengths with respect to TNS length. In fact, fuzzy cumulative profit P_f is greater than neuro-fuzzy cumulative profit P_{nf} over the four-month and six-month periods. Viceversa, the neuro-fuzzy algorithms give better results than fuzzy methods over period lenghts smaller and relatively bigger than TNS length. Remind that the chosen Training Set length is four-month.

We decided - for space limitations - to show only six charts (monthly, bi-monthly and six-month experiments for fuzzy and neuro-fuzzy trades), which give information on the performance overtime of our techniques. All the charts have the same structure and show the profit performance for the period (left axis), the cumulative performance (right axis)and the drawdown horizontal line (referred to the left axis). The time profiles of profit results (single period and cumulative) are not dramatically different for monthly and bi-monthly fuzzy and neuro-fuzzy experiments. More difference is observed in the six-month case for fuzzy and neuro-fuzzy. The cumulative performance line remains always above zero and shows a more subdued profile in the first part of the sample (roughly corresponding to the sub-prime crisis in 2008-09) than in the second. Our algorithms seem to be disturbed more by the subprime crisis (2008-09) than by the European sovereign debt crisis (2010–2012).

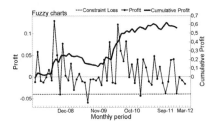 

Fig. 5 Monthly profits and cumulative profits of fuzzy system

Fig. 6 Monthly profits and cumulative profits of neuro-fuzzy system

Fig. 7 Bi-monthly profits and cumulative profits of fuzzy system

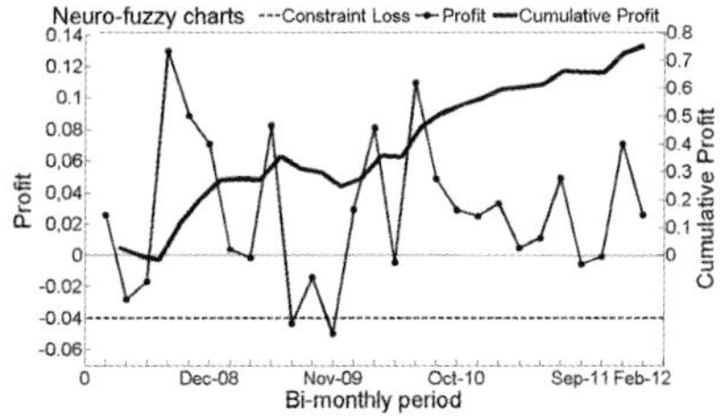

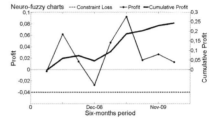

Fig. 8 Bi-monthly profits and cumulative profits of neuro-fuzzy system

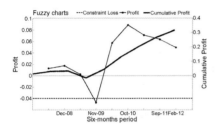

Fig. 9 Semestral profits and cumulative profits of fuzzy system

Fig. 10 Semestral profits and cumulative profits of neuro-fuzzy system

6 Conclusions

The profitability of standard Technical Analysis techniques is a very open issue in foreign exchange market [5, 8]. Technical rules are still "the obstinate passion" for traders in the foreign exchange market [5]. Among these rules, the DMAC is the most popular technical indicator. To simulate the behaviour of market agents, artificial intelligence techniques can be applied. This paper proposes two intelligent systems based on fuzzy and neuro-fuzzy features. The first one is designed with the help of genetic constrained algorithms, whereas the second one is based on a suitable Neural Network architecture. The results show that the fuzzy system is more profitable over trading period lengths close to the four-month training period length. Viceversa, the neuro-fuzzy model improves upon the fuzzy system over periods with length lower or much greater than the training period length. Moreover, both fuzzy system and neuro-fuzzy system give small drawdown numbers. The aggregate profit rates over more than four years (2008–2012), in Table 2 columns 4 and 5, give good numbers with respect to profit rates on a yearly basis.

Going into future planned research on the work in this paper, probably the most useful line of activity for traders is replicating our results with higher frequency data: tick or one-minute. This would indicate how to manage TNS and TDS with a timing much closer to the actual functioning of todayâĂŹs forex market. Another useful step will be the definition of intelligent systems also for others technical rules, such as momentum, oscillators and channel rules. The main future challenge will be the design of neural networks able to reduce to a minimum the drawdown number.

References

1. Allen, F., Karjalainen, R.: Using Genetic Algorithms to Find Technical Trading Rules. Journal of Financial Economics **51**, 245–271 (1999)
2. Cagliesi, G., Tivegna, M.: Rationality, Behavior and Switching Idiosincracies in the Euro-Dollar Exchange Rate. In: Klein L.R. (ed.) Essays in Honor of Albert Ando. E. Elgar (2006)
3. Tivegna, M., Chiofi, G.: News and Exchange rate dynamics. Ashgate, London (2004)
4. Dempster, M., Jones, C.M.: A real time adaptive trading system using genetic programming. Judge Institute of Management. WP **36.** University of Cambridge (2000)
5. Menkhoff, L.,Taylor, M.: The Obstinate Passion of Foreign Exchange Professionals: Technical Analysis. Journal of Economic Litterature, pp. 936–972 (December 2007)
6. Neely, C.J., Weller, P. A., Dittmar, R.: Is Technical Analysis Profitable in the Foreign Exchange Market? A Genetic Programming Approach. Journal of Financial and Quantitative Analysis **32**, 405–426 (1997)
7. Neely, C.J.: Risk-adjusted, ex ante, optimal technical trading rules in equity markets. International Review of Economics and Finance **12**(1), 69–87 (Spring 2003)
8. Park, C., Irwin, S.H.: What do you we know about the profitability of Technical Analysis. Journal of Economic Surveys **21**(4) (September 2007)
9. Pelusi, D., Tivegna, M.: Optimal trading rules at hourly frequency in the foreign exchange markets. Mathematical and Statistical Methods for Actuarial Sciences and Finance, pp. 341–348. Springer (2012)
10. Pelusi, D.: Genetic-Neuro-Fuzzy Controllers for second order control systems. Proceedings of the IEEE European Modelling Symposium on Mathematical Modelling and Computer Simulation. Madrid (2011)
11. Pelusi, D.: On designing optimal control systems through genetic and neuro-fuzzy techniques. Proceedings of the IEEE International Symposium on Signal Processing and Information Technology, pp. 134–139. Bilbao, Spain (2011)
12. Pelusi, D.: Improving Settling and Rise Times of Controllers Via Intelligent Algorithms. Proceedings of the IEEE International Conference on Modelling and Simulation, pp. 187–192. Cambridge, United Kingdom (2012)
13. Pelusi, D.: A pattern recognition algorithm for optimal profits in currency trading. Mathematical and Statistical Methods for Actuarial Sciences and Finance **XII**, pp. 233–242. Springer (2010)
14. Silvestre, M.R., Oikawa, S.M., Vieira, F.H.T., Ling, L.L.: A Clustering Based Method to Stipulate the Number of Hidden Neurons of mlp Neural Networks: Applications in Pattern Recognition. TEMA Tend. Mat. Apl. Comput. **9**(2), 351–361 (2008)

Risk Management and Capital Allocation for Non-Life Insurance Companies

Marco Pirra, Salvatore Forte and Matteo Ialenti

Abstract This paper deals with some key issues related to risk and capital allocation in non life insurance. There is an increasing concern about quantifying the capital requirements for financial entities. Actually, the major part of the insurance regulation aims to provide a clear guideline for insurers to manage with this issue. For example, the new European insurance regulatory system Solvency II focuses in this direction. In this risk based capital system, the required capital is based on two building blocks: the minimum capital requirement and the solvency capital requirement very close to the theoretical concept of economic capital. How much capital should an insurance firm hold? And what rate of return must the firm achieve on this capital? How should the required capital be allocated to different lines of business and products sold? While these questions are of critical importance to the firm, external forces in the operating environment often dictate the answers. For example, regulators and rating agencies greatly influence the amount of capital the firm must hold; in addition, investors influence both the amount of capital the firm holds and the required rate of return on this capital. Therefore, the issues of the amount of capital and the required rate of return on capital are often ultimate beyond the decision-making power of the company; rather, they are demands that the operating environment imposes upon the firm.

M. Pirra (✉)
Department of Economics, Statistics and Finance, University of Calabria, Ponte Bucci cubo 3 C, 87030 Rende (CS), Italy
e-mail: Marco.Pirra@unical.it

S. Forte
Department of Statistics, Sapienza University of Roma, Viale Regina Elena 295, Palazzina G, 00161 Roma, Italy
e-mail: Salvatore.Forte@uniroma1.it

M. Ialenti
Department of Statistics, Sapienza University of Roma, Viale Regina Elena 295, Palazzina G, 00161 Roma, Italy
e-mail: Matteo.Ialenti@uniroma1.it

M. Corazza, C. Pizzi (eds.), *Mathematical and Statistical Methods for Actuarial Sciences and Finance*, DOI 10.1007/978-3-319-02499-8_23, © Springer International Publishing Switzerland 2014

1 Introduction

The identification and quantification of the risks that most companies face are relatively straight forward. This is in part due to the recent supervisory developments such as the Swiss Solvency Test (SST), the Individual Capital Assessment Standards (ICAS) used in the UK, the C3 Phase II standards in use in the US, and the harmonised Solvency II currently under development for the EU. The issues associated with the aggregation of risks and capital allocation are the next area of focus in the economic capital modelling practices of insurance companies. The current focus is on calculating economic capital. However, business decisions need to be made based on risk budget and risk/return optimization. Economic capital plays a central role in prudential supervision, product pricing, risk assessment, risk management and hedging, capital allocation . Available capital is defined as financial resources available as risk-bearing funds to absorb adverse experience. This capital is held as a buffer to meet policyholder claims during adverse climates. It is given by the difference between assets and technical reserves, that is to say $(AC = Assets - Liabilities)$ where the liabilities can be represented approximately by the expected value of X, being X the random variable risk beared by the insurance firm according to the underwritten contracts. (Required) economic capital is calculated based on a risk measure, from which there are many to choose, that is to say that economic capital = f(X), being f an opportune functional or risk measure (i.e. VaR, Tail Var, etc.) of the random variable X. Different risk measures satisfy different purposes of capital determination. Economic capital is aggregated across products, lines of business, business units, geographic and regulatory areas in order to calculate capital requirements at different levels of the organisation. Aggregation generally allows for some diversification benefits between the risks being aggregated, thus resulting in the aggregated capital being less than the sum of the parts. Capital requirements are calculated at the lowest level first (for example, per line of business). Then the aim is to aggregate the capital requirements up to higher levels (for example, at business unit level) to arrive at capital requirements and risk measurement that take into account the interactions between the risks being aggregated (for example, the interaction between two lines of business, say annuities and mortality products). Ultimately, all the capital requirements are aggregated to arrive at holding-level total capital requirements. Total capital requirements (for example, at group level) are therefore smaller than the sum of the capital requirements (for example, at product level). For a range of purposes (for example, pricing or performance measurement), the total capital needs to be allocated back to the lower levels. That is, the diversification benefit achieved by aggregating the risks need to be allocated back to the individual risks. Again, there are a range approaches for doing this, depending upon the intended purpose. The allocation of capital is essential for pricing insurance products and is an important part of the planning and control cycle (risk budgeting and return measurement). This paper examines different techniques for aggregation and allocation of capital; the focus of the paper is on non-life insurance. Section 2 describes the reserve risk and the approach followed to assess it, Sections 3 and 4 explain possible capital aggregation techniques and go through different methods and applications of capital allocation. Section 5 discusses some operational implementation issues given an economic capital framework.

2 Reserve Risk

Loss reserving is a classic actuarial reserving problem encountered extensively in property and casualty as well as health insurance. Typically, losses are arranged in a triangular fashion as they develop over time and as different obligations are incurred from year to year. This triangular format emphasizes the longitudinal and censored nature of the data. The primary goal of loss reserving is to set an adequate reserve to fund losses that have been incurred but not yet developed. For a single line of business written by an insurance company, there is an extensive actuarial literature describing alternative approaches for determining loss reserves. See, for example, [10, 26]. However, almost every major insurer has more than one line of business. One can view losses from a line of business as a financial risk; it is intuitively appealing to think about these risks as being related to one another. It is well-known that if risks are associated, then the distribution of their sum depends on the association. For example, if two risks are positively correlated, then the variability of the sum of risks exceeds the sum of variabilities from each risk. Should an insurer use the loss reserve from the sum of two lines of business or the sum of loss reserves, each determined from a line of business? This problem of additivity was put forth by Ajne, [2], who notes that the most common approach in actuarial practice is the silo method. Here, an insurer divides its portfolio into several subportfolios (silos). A subportfolio can be a single line of business or can consist of several lines with homogeneous development pattern. The claim reserve and risk capital are then calculated for each silo and added up for the portfolio. The most important critique of this method is that the simple aggregation ignores the dependencies among the subportfolios. In loss reserving the evolution of losses over time complicates the determination of dependencies among lines of business. As emphasized in [13] and [23], correlations may appear among losses as they develop over time (within an incurral year) or among losses in different incurral years (within a single development period). Other authors have focussed on correlations over calendar years, thinking of inflationary trends as a common unknown factor inducing correlation. Much of the work on multivariate stochastic reserving methods to date has involved extending the distribution-free method explained in [15]. [3] proposed to estimate the prediction error for a portfolio of correlated loss triangles based on a multivariate chain-ladder method. Similarly, [17] considered the prediction error of another version of the multivariate chain-ladder model proposed in [23], where the dependence structure was incorporated into parameter estimates. Within the theory of linear models, [12] and [16] provided the optimal predictor and the prediction error for the multivariate additive loss reserving method, respectively. Motivated by the fact that not all subportfolios satisfy the same homogeneity assumption, [17] combined chain-ladder and additive loss reserving methods into one single framework. [27] proposed a general multivariate chain-ladder model that introduces correlations among triangles using the seemingly unrelated regression technique. These procedures have desirable properties, focusing on the mean square error of predictions. In this paper, we focus instead on tails of the distribution. Because of this, and the small sample size typically encountered in loss reserving problems, we look more to parametric methods based on distributional families. For example, in a multivariate context, [4] employed a lognormal model for the unpaid losses of each line and a normal copula for the generation of the joint distribution. generalized linear model framework, and demonstrated the calculations of loss reserves and their prediction errors. [6] employed factor analytic techniques to handle several sources of time dependencies (by incurral year, development year, calendar year) as well as correlations among lines of business in a flexible manner. An alternative parametric approach involving Bayesian methods has found applications when studying loss reserves for single lines of business. Some recent work include [5, 18]. The Bayesian methods for multivariate loss reserving problems have rarely been found in the literature. [17] is one example, where the authors considered a bivariate Bayesian model for combining data from the paid and incurred triangles to achieve better prediction. A copula method is employed to associate the claims from multiple run-off triangles. Despite the application of copulas in [4] and [6], both are focused on correlations in a model based on normal distributions. In this paper the reserve risk is evaluated through the Bayesian Fisher Lange method described in [11] and [22]: the reserve risk is valued separetely for each line of business and then aggregated using the approaches described in the next section.

3 Aggregation Techniques

Having chosen a risk measure and calculated the risks, the next step is to aggregate risks across different products, lines, geographic areas, etc. It is generally believed that the aggregated capital should be less than the sum of capital required for each risk being aggregated. However, the recent financial crisis has highlighted the fact that significant interactions can exist between risks. These interactions can have a compounding effect. In such cases aggregating these risks by assuming some diversification effect between them can significantly underestimate the total risk. Bottom-up approaches - that is, calculating capital requirements separately for each risk and then aggregating - can underestimate the total capital required, while a top-down approach - calculating the capital requirements for each contract taking risks into account together - could be a more appropriate way of determining capital requirements. Two bottom-up approaches for aggregating capital are considered in this section: correlations and copulas.

Correlation is a measure of the strength and direction of a linear relationship between random variables. It is is a scale invariant statistic that ranges from -1 to $+1$

and statistically it is determined as follows:

$$Corr(X,Y) = \frac{E((X - E[X])(Y - E[Y]))}{\sigma_X \, \sigma_Y}. \tag{1}$$

Risks are aggregated using the following formula:

$$TotalRisk = (\sum_i \sum_j \rho_{ij} X_i X_j)^{1/2} \qquad i,j = 1,\dots,n, \tag{2}$$

where n is the number of risks being aggregated, ρ_{ij} is the correlation between risks i and j and X_i is the risk measure output (for example, the VaR) of risk i. A correlation matrix is specified for the correlations between risks, and it is used to calculate new totals using (2). In this paper we refer to the correlation matrices proposed by the EIOPA, [9]. The correlation approach assumes that the risks are normally distributed and that the dependence structure can be specified via the margins of a Gaussian distribution. The combined risk distribution is therefore multivariate normal. This assumption may well introduce unacceptable distortion where the risks are not normally distributed. Correlations tend to behave differently in extreme situations. Note, capital is often calculated to provide protection during extreme events: it is this part of the loss distribution that is of greatest interest. It is under stress conditions when the correlation approach of aggregating risks tends to fail.

The idea of the copula comes from Sklar's theorem that can be summarized as follows (for a detailed description of the theory of copulas see [19]). Suppose $M_1(x)$ and $M_2(y)$ are the two marginal distributions of the bivariate distribution $B(x,y)$. Then there exists a function, C, such that

$$B(x,y) = C(M_1(x), M_2(y)). \tag{3}$$

This function, C, is called a copula. All continuous multivariate functions contain a unique copula. When applied to economic capital, M_1 and M_2 can be seen as (a function of) two risk distributions. Under the correlation approach, a risk measure (for example, VaR) would be applied to each of these distributions. The resulting risk measure amounts, say M_1^* and M_2^* would be aggregated using a correlation assumption between risks M_1 and M_2 (for example, a correlation coefficient value of a) in order to calculate the total capital requirement. This is where the correlation approach fails: it assumes the correlation between M_1 and M_2 is constant for all realizations $M_1(x)$ and $M_2(y)$. For example, if M_1 is equity risk and M_2 is lapse risk, there may be a value ρ_a of correlation in normal times. However, if a realization of $M_1(x)$ gave an extremely negative return, the correlation may well be somewhat higher than ρ_a. For economic capital purposes, one is more interested in these more extreme realizations and the corresponding correlations. Copulas solve this problem. The function, C, is a plane (in this case in three dimensions, but can be extended to an n-dimensional plane if additional risks are added) and can be specified such that the interaction between M_1 and M_2 differs at different parts of each of the distributions of M_1 and M_2. Each risk is transformed to a uniform distribution on the interval $[0,1]$. One way of doing this is to use the cumulative distribution function of each risk.

Simulating from the marginal and aggregate risk distributions becomes a relatively easy task after the copula structure is specified.

4 Capital Allocation

Company level capital and risk are allocated down to lower levels such as business units, lines and products for a number of purposes. The initial reason for calculating total capital is often for regulatory reporting; however, insurers are becoming increasingly risk aware and are allocating the capital and risk more actively in order to improve areas such as pricing and performance measurement. This section considers the application of capital allocation. One aim of allocating capital to business units, lines and products is to correctly allow for the cost of the capital in pricing exercises. The cost of capital is usually calculated as a product of the amount of capital allocated to a product and the return-on-capital requirement. Thus, the target price is generally greater where the risk is more concentrated (or less diversified), as more capital is allocated to such a risk. Similarly, risks that are well-diversified are allocated less capital, and hence, they have a lower capital charge in the pricing exercise. A drawback to this is that there is no unique way to allocate capital. Consequently, a line written by two different insurers attracting the same amount of risk may well be allocated different amounts of capital by each insurer. This would result in different premiums being charged by the two insurers for underwriting the same risk. We have to allocate capital to individual lines. The law requires this to some extent, since any insurance company and its subsidiaries will be required to have its own capital. However, a good allocation of capital to individual business lines can be useful for many reasons. It helps evaluate performance, it will allow better pricing of new business, it encourages an individual and quantitative assessment of the lines of business, thus telling central management how each line of business is performing. Also, allocating capital "ex ante" (beforehand) can help avoid moral hazard/agency problems among managers. After aggregating risks in order to take into account the effect of diversification, companies want to allocate the capital back to the lower levels for a range of purposes. In other words, the diversification benefit achieved by aggregating the risks needs to be allocated back to the individual risks. The allocation of capital is an important measure for profitability in relation to risk and is essential for pricing insurance products as well as for the planning and control cycle (risk budgeting and return measurement). There are a range of approaches for allocating capital with different ones being appropriate for different purposes. For a detailed description of the possible approaches see [7, 8, 21, 24]. This paper focuses on the following approaches.

Consider a portfolio of n individual losses $X_1, X_2, \ldots, X_n$ materialising at a fixed future date T. Assume that $(X_1, X_2, \ldots, X_n)$ is a random vector on the probability space $(\Omega, \mathcal{F}, P)$. Throughout the paper, it is always assumed that any loss X_i has a finite mean. The distribution function $P[X_i \leq x]$ of X_i is denoted by $F_{X_i}(x)$. The aggregate loss is defined by the sum

$$AL = \sum_{i=1}^{n} X_i, \tag{4}$$

and it can be interpreted as the total loss of an insurance company with the individual losses corresponding to the losses of the respective business units. We assume that the company has already determined the aggregate level of capital and denote this total risk capital by TRC. The company wishes to allocate this exogenously given total risk capital TRC across its various business units, that is, to determine non-negative real numbers $RC_1, RC_2, \ldots, RC_n$ satisfying the full allocation requirement:

$$\sum_{i=1}^{n} RC_i = TRC. \tag{5}$$

This allocation is in some sense a notional exercise; it does not mean that capital is physically shifted across the various units, as the company's assets and liabilities continue to be pooled. The allocation exercise could be made in order to rank the business units according to levels of profitability. This task is performed by determining at time T, the respective returns on the respective allocated capital $\frac{RC_i - X_i}{RC_i} - 1$, $i = 1, \ldots, n$. For a given probability level $p \in (0, 1)$, we denote the Value-at-Risk (VaR) or quantile of the loss random variable X by $F_X^{-1}(p)$. As usual, it is defined by

$$F_X^{-1}(p) = \inf \{x \in R | F_X(x) \geq p\}, \qquad p \in [0, 1]. \tag{6}$$

The Haircut Allocation Principle A straightforward allocation method consists of allocating the capital $RC_i = \alpha F_{X_i}^{-1}(p)$, $i = 1, \ldots, n$, to business unit i, where the factor α is chosen such that the full allocation requirement (5) is satisfied. This gives rise to the haircut allocation principle:

$$RC_i = \frac{TRC}{\sum_{j=1}^{n} F_{X_j}^{-1}(p)} F_{X_i}^{-1}(p), \qquad i = 1, \ldots, n. \tag{7}$$

For an exogenously given value for TRC, this principle leads to an allocation that is not influenced by the dependence structure between the losses X_i of the different business units. In this sense, one can say that the allocation method is independent of the portfolio context within which the individual losses X_i are embedded. It is a common industry practice, driven by banking and insurance regulations to measure stand-alone losses by a VaR, for a given probability level p. Therefore, let us assume that $TRC = F_{AL}^{-1}(p)$. In addition, we assume that in case business unit i was a stand-alone unit, then its loss would be measured by $F_{X_i}^{-1}(p)$. It is well-known that the quantile risk measure is not always subadditive. Consequently, using the p-quantile as stand-alone risk measure will not necessarily imply that the subportfolios will benefit from a pooling effect. This means that it may happen that the allocated capitals RC_i exceed the respective stand-alone capitals $F_{X_i}^{-1}(p)$.

The Covariance Allocation Principle The covariance allocation principle proposed in [20] is given by the following expression:

$$RC_i = \frac{TRC}{Var[AL]} Cov[X_i, AL], \qquad i = 1, \ldots, n \tag{8}$$

where $Cov[X_i, AL]$ is the covariance between the individual loss X_i and the aggregate loss AL and $Var[AL]$ is the variance of the aggregate loss AL. Because clearly the sum of these individual covariances is equal to the variance of the aggregate loss, the full allocation requirement is automatically satisfied in this case. The covariance allocation rule, unlike the haircut and the quantile allocation principles, explicitly takes into account the dependence structure of the random losses $(X_1, X_2, \ldots, X_n)$. Business units with a loss that is more correlated with the aggregate portfolio loss AL are penalised by requiring them to hold a larger amount of capital than those which are less correlated.

The CTE Allocation Principle For a given probability level $p \in (0, 1)$, the Conditional Tail Expectation (CTE) of the aggregate loss AL is defined as follows:

$$CTE_p[AL] = E[AL|AL > F_{AL}^{-1}(p)]. \tag{9}$$

At a fixed level p, it gives the average of the top $(1..p)\%$ losses. In general, the CTE as a risk measure does not necessarily satisfy the subadditivity property. However, it is known to be a coherent risk measure in case we restrict to random variables with continuous distribution function (see [1, 7]). The CTE allocation principle, for some fixed probability level $p \in (0, 1)$, has the form:

$$RC_i = \frac{TRC}{CTE_p[AL]} E[X_i|AL > F_{AL}^{-1}(p)], \qquad i = 1, \ldots, n. \tag{10}$$

The CTE allocation rule explicitly takes into account the dependence structure of the random losses $(X_1, X_2, \ldots, X_n)$. Interpreting the event $(AL > F_{AL}^{-1}(p))$ as (the aggregate portfolio loss AL is large), we see from (the formula above) that business units with larger conditional expected loss, given that the aggregate loss AL is large, will be penalised with a larger amount of capital required than those with lesser conditional expected loss.

The Market Driven Allocation Principle Let η_M be a random variable such that market-consistent values of the aggregate portfolio loss AL and the business unit losses X_i are given by the following expressions:

$$\pi[AL] = E[\eta_M AL], \tag{11}$$

and

$$\pi[X_i] = E[\eta_M X_i], \qquad i = 1, \ldots, n \tag{12}$$

respectively. Further suppose that at the aggregate portfolio level, a provision $\pi[AL]$ is set aside to cover future liabilities AL. Apart from the aggregate provision $\pi[AL]$, the aggregate portfolio has an available solvency capital equal to $(TRC - \pi[AL])$.

The solvency ratio of the aggregate portolio is then given by

$$\frac{TRC - \pi[AL]}{\pi[AL]}. \tag{13}$$

This solvency ratio is different from the one defined by the EIOPA in [9]. In order to determine an optimal capital allocation over the different business units, we let in $\eta_i = \eta_M$, $i = 1, \ldots, n$, thus allowing the market to determine which states-of-the-world are to be regarded adverse. This yields to the following expression:

$$RC_i = \pi[X_i] + v_i(TRC - \pi[AL]). \tag{14}$$

If we now use the market-consistent prices as volume measures, after substituting

$$v_i = \frac{\pi[X_i]}{\pi[AL]}, \qquad i = 1, \ldots, n, \tag{15}$$

in (14), we find

$$RC_i = \frac{TRC}{\pi[AL]}\pi[X_i], \qquad i = 1, \ldots, n. \tag{16}$$

Rearranging these expressions leads to

$$\frac{RC_i - \pi[X_i]}{\pi[X_i]} = \frac{TRC - \pi[AL]}{\pi[AL]}, \qquad i = 1, \ldots, n. \tag{17}$$

The quantities $\pi[X_i]$ and $(RC_i - \pi[X_i])$ can be interpreted as the market-consistent provision and the solvency capital attached to business unit i, while $\frac{RC_i - \pi[X_i]}{\pi[X_i]}$ is its corresponding solvency ratio.

5 Case Study

The framework described in previous sections can be evaluated through a sample case study in which the initial data set is represented by the run-off triangles of an insurance company operating in the LoBs Motor, General Insurance and Property. The aim is the analysis of the effects of the dependencies between the LoBs on the Solvency Capital Requirement and its allocation to lower levels. First of all the value of the technical provisions is determined through the Bayesian Fisher Lange method described in [11] and [22] separetely for each line of business and then aggregated using the approaches described in Sect. 3. Table 1 shows the results obtained (Best Estimate (BE), Risk Margin (RM), Technical Provision (TP), Solvency Capital Requirement (SCR)). The Solvency Capital requirement is determined using an internal model based on Value-at-Risk techniques (see [22] for a detailed description of the model); the results are compared to the ones that come out using the standard formula proposed by the EIOPA in [9] and a proxy proposed in the same document. Table 2 shows how the Capital is allocated to each LoB through the principles described in Sect. 4, using formulas (7), (8), (10), (16). The results of the case study presented seem to lead to the following conclusions.

Table 1 Technical provisions (Euros)

LoB	Method	BE	RM	TP	SCR
Motor	Internal Model	2,959,231	56,761	3,015,992	355,074
	Standard Formula *(Proxy)*	2,959,231	131,975	3,091,206	790,115
	Standard Formula	2,959,231	125,434	3,084,664	746,132
General Insurance	Internal Model	1,210,733	23,223	1,233,957	155,551
	Standard Formula *(Proxy)*	1,210,733	53,996	1,264,730	399,542
	Standard Formula	1,210,733	51,320	1,262,053	385,589
Property	Internal Model	135,461	2,598	138,059	34,869
	Standard Formula *(Proxy)*	135,461	6,041	141,502	41,451
	Standard Formula	135,461	5,742	141,203	39,675
Aggregate	Internal Model	4,305,425	82,582	4,388,007	456,374
	Standard Formula *(Proxy)*	4,305,425	192,013	4,497,438	1,061,119
	Standard Formula	4,305,425	182,495	4,487,920	1,008,522

Table 2 Solvency capital requirement allocation (Euros)

Allocation Principle	Method	Motor	General Insurance	Property	Aggregate
Haircut	Internal Model	297,064	130,138	29,172	456,374
	Standard Formula *(Proxy)*	681,017	344,374	35,728	1,061,119
	Standard Formula	642,388	331,975	34,159	1,008,522
Covariance	Internal Model	332,924	112,361	11,089	456,374
	Standard Formula *(Proxy)*	774,084	261,252	25,784	1,061,119
	Standard Formula	735,715	248,302	24,506	1,008,522
CTE	Internal Model	316,102	126,334	13,938	456,374
	Standard Formula *(Proxy)*	734,972	293,739	32,407	1,061,119
	Standard Formula	698,542	279,179	30,801	1,008,522
Market Driven	Internal Model	313,678	128,337	14,359	456,374
	Standard Formula *(Proxy)*	729,335	298,398	33,386	1,061,119
	Standard Formula	693,184	283,608	31,731	1,008,522

Table 3 SCR internal model allocation

Allocation Principle	Motor	General Insurance	Property	Aggregate
Haircut	65.09%	28.52%	6.39%	100.00%
Covariance	72.95%	24.62%	2.43%	100.00%
CTE	69.26%	27.68%	3.05%	100.00%
Market Driven	68.73%	28.12%	3.15%	100.00%

- The *SCR* calculated with the standard formula is significantly higher compared to the one obtained with the internal model, both on single lobs and on aggregated level.

- The capital allocation using the haircut principle is straightforward but does not consider dependencies; furthermore, as said before, it is well-known that the quantile risk measure is not always subadditive. Consequently, using the *p*-quantile as stand-alone risk measure will not necessarily imply that the sub-portfolios will benefit from a pooling effect: this means that it may happen that the allocated capitals RC_i exceed the respective stand-alone capitals $F_{X_i}^{-1}(p)$; the covariance allocation principle, unlike the haircut principle, considers the dependencies: business units with a loss that is more correlated with the aggregate portfolio loss are penalised by requiring them to hold a larger amount of capital than those which are less correlated; the CTE allocation rule explicitly takes into account the dependence structure of the random losses capturing, unlike the covariance principle, the effects on the tails of the distributions (a crucial aspect for solvency requirements): business units with larger conditional expected loss, given that the aggregate loss is large, will be penalised with a larger amount of capital required than those with lesser conditional expected loss; the market driven allocation principle considers the market value of the provisions and implicitly takes into account the dependencies between the LoBs.

- The results of the *RC* allocation on each LoB show that, except from the haircut priciple that gives a higher weight to General Insurance and a lower weight to Property, the proportion $RC_i/RCAggregate$ is similar both with the internal model and with the standard formula.

- The allocation of the exogenously given aggregate capital *TRC* to *n* parts, RC_1, RC_2, ..., RC_n, corresponding to the different subportfolios or business units, can be carried out in an infinite number of ways, some of which were illustrated in this paper. It is clear that different capital allocations must in some sense correspond to different questions that can be asked within the context of risk management: the preference of one method over another has a great relevance as the impact on the final results, as shown in Table 3, can be significant.

The results presented and the conclusions exposed depend significantly on the dataset considered and on the insurance company analyzed; the intention is to apply the methodologies to other insurers and verify the possibility to extend the conclusions to other case studies.

References

1. Acerbi, C., Tasche, D.: On the coherence of expected shortfall. Journal of Banking and Finance **26**(7), 1487–1503 (2002)
2. Ajne, B.: Additivity of chain-ladder projections. ASTIN Bulletin **24**(2), 311–318 (1994)
3. Braun, C.: The prediction error of the chain ladder method applied to correlated run-off triangles. ASTIN Bulletin **34**(2), 399–424 (2004)

4. Brehm, P.: Correlation and the aggregation of unpaid loss distributions. In: CAS Forum, pp. 1–23 (2002)
5. De Alba, E., Nieto-Barajas, L.: Claims reserving: a correlated Bayesian model. Insurance: Mathematics and Economics **43**(3), 368–376 (2008)
6. De Jong, P.: Modeling dependence between loss triangles using copula. Working Paper (2010)
7. Dhaene, J., Vanduffel, S., Tang, Q., Goovaerts, M.J., Kaas, R., Vyncke, S.: Risk measures and comonotonicity: A review. Stochastic Models **22**(4), 573–606 (2006)
8. Dhaene, J., Tsanakas, A., Valdez, E.A., Vanduffel, S.: Optimal Capital Allocation Principles. Journal of Risk and Insurance (2009)
9. EIOPA, CEIOPS: QIS5 Technical Specifications (2010). Available on www.ceiops.org
10. England, P., Verrall, R.: Stochastic claims reserving in general insurance. British Actuarial Journal **8**(3), 443–518 (2002)
11. Forte, S., Ialenti, M., Pirra, M.: Bayesian Internal Models for the Reserve Risk Assessment. Giornale dell'Istituto Italiano degli Attuari **LXXI**(1), 39–58 (2008)
12. Hess, K., Schmidt, K., Zocher, M.: Multivariate loss prediction in the multivariate additive model. Insurance: Mathematics and Economics **39**(2), 185–191 (2006)
13. Holmberg, R.D.: Correlation and the measurement of loss reserve variability. In: CAS Forum 1–247 (1994)
14. Kirschner, G., Kerley, C., Isaacs, B.: Two approaches to calculating correlated reserve indications across multiple lines of business. Variance **2**(1), 15–38 (2008)
15. Mack, T.: Distribution-free calculation of the standard error of chain ladder reserve estimates. ASTIN Bulletin **23**(2), 213-225 (1993)
16. Merz, M., Wüthrich M.: Combining chain-ladder and additive loss reserving methods for dependent lines of business. Variance **3**(2), 270–291 (2009)
17. Merz, M., Wüthrich, M.: Prediction error of the multivariate additive loss reserving method for dependent lines of business. Variance **3**(1), 131–151 (2009)
18. Meyers, G.: Stochastic loss reserving with the collective risk model. Variance **3**(2), 239–269 (2009)
19. Nelsen, R.B.: An Introduction to Copulas. Springer Series in Statistics (2006)
20. Overbeck, L.: Allocation of economic capital in loan portfolios. In: Franke, J., Haerdle W., Stahl, G. (eds.) Measuring Risk in Complex Systems. Springer (2000)
21. Panjer, H.: Measurement of risk, solvency requirements and allocation of capital within financial conglomerates (2002)
22. Pirra, M., Forte, S., Ialenti, M.: Implementing a Solvency II internal model: Bayesian stochastic reserving and parameter estimation. Astin Colloquium, Madrid (2011)
23. Schmidt, K.: Optimal and additive loss reserving for dependent lines of business. In: CAS Forum, pp. 319–351 (2006)
24. Tasche, D.: Capital Allocation to Business Units and Sub-Portfolios: the Euler Principle, Risk Books, pp. 423–453 (2008)
25. Taylor, G., McGuire, G.: A synchronous bootstrap to account for dependencies between lines of business in the estimation of loss reserve prediction error. North American Actuarial Journal **11**(3), 70 (2007)
26. Wüthrich, M., Merz, M.: Stochastic Claims Reserving Methods in Insurance. John Wiley and Sons (2008)
27. Zhang, Y.: A general multivariate chain ladder model. Insurance: Mathematics and Economics **46**(3), 588–599 (2010)

Modelling Asymmetric Behaviour in Time Series: Identification Through PSO

Claudio Pizzi and Francesca Parpinel

Abstract In this work we propose an estimation procedure of a specific TAR model in which the actual regime changes depending on both the past value and the specific past regime of the series. In particular we consider a system that switches between two regimes, each of which is a linear autoregressive of order p. The switching rule, which drives the process from one regime to another one, depends on the value assumed by a delayed variable compared with only one threshold, with the peculiarity that even the thresholds change according to the regime in which the system lies at time $t-d$. This allows the model to take into account the possible asymmetric behaviour typical of some financial time series. The identification procedure is based on the Particle Swarm Optimization technique.

1 Introduction

In time series analysis the problem of identification and estimation of a model, as best approximating the real data generating process (in short DGP), is often conducted by iterative procedures. In the classic "Box-Jenkins" framework, the choice of the best model is obtained by a three phases procedure based on the iterations of identification, estimation and validation steps. Let's recall that at each iteration the order of the model are fixed and the parameters are estimated. The use of an Akaike-like criterion enables us to stop the iterative procedure and to select the appropriate model. When we relax the linearity hypotheses, this scheme is difficult to implement,

C. Pizzi (✉)
Department of Economics, University Ca' Foscari of Venezia, Sestiere Cannaregio 873, 30121 Venice, Italy
e-mail: pizzic@unive.it

F. Parpinel
Department of Economics, University Ca' Foscari of Venezia, Sestiere Cannaregio 873, 30121 Venice, Italy
e-mail: parpinel@unive.it

M. Corazza, C. Pizzi (eds.), *Mathematical and Statistical Methods for Actuarial Sciences and Finance*, DOI 10.1007/978-3-319-02499-8_24, © Springer International Publishing Switzerland 2014

because the complexity increases remarkably in reason of making the selection of the order to be preceded by the choice between linear or nonlinear models and, in this case, we should also select the kind of nonlinear model.

In the last decades, several nonlinear models have been proposed in the literature, each of which is able to catch a certain kind of nonlinearity. Among these, an interesting class of model is the Threshold AutoRegressive one (Tong e Lim [13]) that performs a local linearization of the DGP by means of a piecewise linear structure. More precisely this class of models is characterized by regimes in which a particular linear model works. So their specification requires at first to estimate the number of regimes and of the thresholds that define the regimes; moreover for each regimes we have to identify the orders and to estimate the parameters of the linear autoregressive models. Let note that the regimes define a partition of the real space $\mathbb{R}$. The whole procedure is complex and different iterative procedures have been proposed, with the main drawback that it is the difficult to identify the number of regimes as they depend on both thresholds and delayed variables. In a recent work Battaglia and Protopapas [1] introduce a new approach in the identification and estimation of a TAR model, using the evolutive paradigm in the simultaneous estimation of all the parameters of the model. More precisely they use the genetic algorithm showing that their procedure enables us to overcome the previously mentioned drawbacks.

In this paper we propose a technique to identify and estimate a threshold model on which the non-overlapping structure of the regimes are relaxed. Thus the resulting DGP is characterized by asymmetric behaviour with respect of the switching rule from one regimes to another one. In other word, if we consider two contiguous regimes the transition from the first regime to the second one depends on a threshold different from that defining the transition from the second regimes to the first one. In order to identify and estimate the asymmetric TAR we propose to use the Particle Swarm Optimization algorithm (PSO), following the idea by Battaglia and Protopapas. The application of PSO algorithms in time series analysis is present in some recent works, for example Wang and Zhao [15], that introduce the use of PSO in ARIMA models, showing its excellent forecast ability. The PSO, born in the eighties, replicates the behaviour of natural flocks and swarm of animals having one specific objective and its main feature is the simplicity of the algorithm, in fact there is no need of computing the gradient of the objective function, starting from random positions of swarm particles and moving the particles in the space of m dimension according to some speed rule. Among all the variants of the procedure proposed in the literature, in particular we use the PSO in the "Inertia weight" variant to improve the velocity rule.

The aim of this work is to evaluate the capability of an evolutive algorithm to improve some classical estimation procedures that we can use even in this context, namely we wish to estimate simultaneously all the parameters of the model. The remainder of this paper is organized as follows. Section 2 presents the asymmetric threshold model and describes the PSO algorithm that we use in our applications. Section 3 presents the methodology and discusses the empirical findings, finally Sect. 4 concludes.

2 The Asymmetric SETAR Model

It is well-known that the different nonlinear behaviour of the DGP may be hardly described by a model featured by unique linear structure. To overcome this limit Tong and Lim [13] proposed, in the eighties, a piecewise linear model that may be viewed as a generalization of the linear model and able to capture nonlinear behaviours such as jump resonance, amplitude–frequency dependency, limit cycle, subharmonics and higher harmonics. This class of model, characterized by the presence of one or more thresholds that define the regimes and that involve the time series itself, is denoted as Self-Exciting Threshold AutoRegressive model (SETAR). The seminal paper by Tong and Lim captured the attention of the scientific community; on one hand many articles tackle the problem to test the threshold hypothesis, as Petruccelli and Davies [8], Tsay [14] and Chang and Tong [2]. On the other hand, several authors face the parameters estimation problem both looking for new procedures and applying the procedures to real data. For example, Battaglia and Protopapas [1], Wu and Chang [16] use genetic algorithms, and Gonzalo and Wolf [4] propose to use some subsampling techniques. The idea of the threshold mechanism has been also extended to the study of financial time series, in particular in modeling volatility. Zakoian [17] introduced the TGARCH model to capture the asymmetry in the conditional standard deviation and variance of a process. In finance for example, the stock market is intrinsically asymmetric. Time series referred to an asset often show asymmetry both on level and variance and this seems due to the behaviour of the market with respect to a threshold that discriminates the bull market from the bear market. In particular the threshold at which the bull market switches to the bear market is different from that needed for the contrary movement, that is from bear to bull market. Li and Lam [6] showed that for *Hong Kong data* the return series could have a conditional mean structure which depends on the rise and fall of the market on a previous day. This asymmetric behaviour of the time series of stock prices during bear and bull markets has been efficiently modelled by a threshold-like model with conditional heteroscedasticity.

More recently Pizzi [9] proposed the Asymmetric–SETAR (AsyTAR) model in order to capture more stringently these asymmetric behaviours. The main difference of the AsyTAR models with respect to the SETAR ones is the partitioning of the thresholds domain, namely the AsyTAR model allowed to regimes to be overlapped. More precisely, whereas in the SETAR model each threshold may be viewed as the boundary point that separates two contiguous regions in each of which a different linear model may locally approximate the unknown DGP, in the AsyTAR model the boundary point between two regions may change in reason of the state of the process. In SETAR model the space $\mathbb{R}$ is partioned in l intervals $(-\infty, r_1), [r_1, r_2), \ldots,$ $[r_{l-1}, \infty)$ by a set of thresholds $\{r_i\}, i = 1, \ldots, l-1$ in such a way that the unknown data generating process is locally, that is in each interval, approximated by a different linear model, from which follows the definition of piecewise linear model. Partitioning space $\mathbb{R}$ implies that the generating process crosses from the first regime to the second one when the delayed value of the time series increases until the threshold is exceeded whereas the passage from the second regime to the first one occurs when the delayed value decreases until it becomes less than the threshold. In some sense, there is a symmetric transition behaviour with respect to the threshold. Conversely we are interested on a different threshold model that allows us to consider an asymmetric transition mechanism. The simplest AsyTAR model has two regimes with AR(1) structure for each regimes.

More generally following Pizzi [9], let $\{y_t\}_{t=1,\dots,T}$, be a times series and let $r_1 > r_2$ be two thresholds such that $(-\infty, r_1), (r_2, \infty)$ are two overlapping intervals. So we can define an Asymmetric Self-Exciting Threshold Autoregressive model, in short AsyTAR(p,d), with two regimes each of order p with delay $d = (d_1, d_2)$, as follows

$$
y_t =
\begin{cases}
\phi_0^{(1)} + \sum\limits_{i=1}^{p} \phi_i^{(1)} y_{t-i} + \varepsilon_t^{(1)} & \text{if } y_{t-d_1} \leq r_1 \text{ and } J_{t-d_1} = 1 \\[2mm]
\phi_0^{(2)} + \sum\limits_{i=1}^{p} \phi_i^{(2)} y_{t-i} + \varepsilon_t^{(2)} & \text{if } y_{t-d_1} > r_1 \text{ and } J_{t-d_1} = 1 \\[2mm]
\phi_0^{(2)} + \sum\limits_{i=1}^{p} \phi_i^{(2)} y_{t-i} + \varepsilon_t^{(2)} & \text{if } y_{t-d_2} > r_2 \text{ and } J_{t-d_2} = 2 \\[2mm]
\phi_0^{(1)} + \sum\limits_{i=1}^{p} \phi_i^{(1)} y_{t-i} + \varepsilon_t^{(1)} & \text{if } y_{t-d_2} \leq r_2 \text{ and } J_{t-d_2} = 2
\end{cases}
\tag{1}
$$

where J_{t-d_i} is an unobservable variable denoting the regime in which the system operates at time $t - d_i$, d_1, d_2 are the delay at which the system reacts and changes regime, $\phi_0^{(1)}, \phi_0^{(2)}$ and $\phi_j^{(1)}, \phi_j^{(2)}$ with $j = 1,\dots,p$ are the model parameters. The exponent between brackets of the parameters ϕ_i denotes the regimes in which the system operates. This means that when the system work in regime *1* the activating threshold is different with respect to the one working in regime *2*. Figure 1 represents the behaviour of a SETAR in a) against an AsyTAR in b).

Another difference in the behaviour of the SETAR and AsyTAR models can be revealed applying, in both cases, the classical estimation procedures for the SETAR.

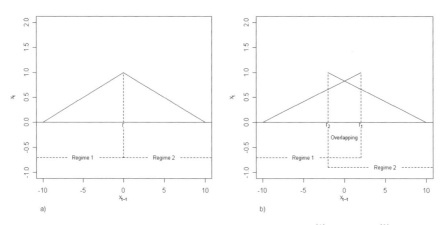

Fig. 1 Theorical behaviour a) SETAR(2;1,1) with $d = 2$, $r = 0$, $\phi_1^{(1)} = -0.8$, $\phi_1^{(2)} = 0.8$, b) AsyTAR(1,2) with $r_1 = 1, r_2 = -1, d = 2$, $\phi_1^{(1)} = -.8$ and $\phi_1^{(2)} = .8$

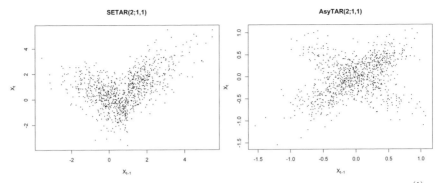

Fig. 2 Simulated data behaviour: on the left side SETAR(2;1,1) with $d = 1$, $r = 1$, $\phi_1^{(1)} = -0.8$, $\phi_1^{(2)} = 0.8$, on the right side AsyTAR(1,2) with $r_1 = 1, r_2 = -1$, $d = 1$, $\phi_1^{(1)} = -0.8$ and $\phi_1^{(2)} = 0.8$

In fact, in order to get the estimates, the classical techniques suggest to use the Standardized Forecast Error (in short SFE) as proposed by Petruccelli and Davies [8], employed also by Tong and Yeung [12] and Hansen [3], and described in Pizzi and Parpinel [10].

2.1 Simultaneous Parameter Estimation

Following some recent works that try to overcome the problem of simultaneous parameter estimation, we suggest to use an evolutive approach, i.e. the Particle Swarm Optimization, born in the eighties and theorized by Kennedy and Eberhart [5] and Shi and Eberhart [11], that replicates the behaviour of natural flocks and swarms of animals having one specific objective.

This general optimization procedure may be applied in very different problems and is based on the creation of a *population* of agents, called *particles*, which are uniformly distributed over some space $\mathscr{X}$. In our context each particle represents a model. More precisely a particle is a vector formed by the values of the parameters involved in the time series model, in such a way that the dimension of space $\mathscr{X}$ is equal to number of unknown DGP parameters. In other words the parameters determine the position of each particle in the space $\mathscr{X}$.

Then it is performed an updating process on the positions of the particles until some stopping criterion is satisfied. Tipically each particle's position is evaluated according to an objective function and it must be updated only if a particle's current position is better, following the rule

$$\mathbf{x}_i^{\tau+1} = \mathbf{x}_i^{\tau} + \mathbf{v}_i^{\tau+1}. \tag{2}$$

The PSO variant that we use is based on the velocity-update rule proposed by Shi and Eberhart [11], who defined the *Inertia weight* variant adding in the velocity rule a parameter **w** called inertia weight as indicated below

$$\mathbf{v}_i^{\tau+1} = \mathbf{w}\mathbf{v}_i^{\tau} + \varphi_1 \mathbf{U}_1^{\tau}(\mathbf{pb}_i^{\tau} - \mathbf{x}_i^{\tau}) + \varphi_2 \mathbf{U}_2^{\tau}(\mathbf{lb}_i^{\tau} - \mathbf{x}_i^{\tau})$$ (3)

where $\mathbf{v}_i^{\tau}$ is the velocity of particle i at iteration τ, **w** is an inertia weight, $\mathbf{U}_j^{\tau}$ are values from uniform random variables, φ_j are weights on the attraction towards the particle's own best known position, $\mathbf{pb}_i^{\tau}$, and the swarm's best known position, $\mathbf{lb}^{\tau}$.

In details, in the procedure applied for this work, we create a *population* of P particles with dimension equal to the number of parameters involved in the proposed model ($\mathbf{x}_i^{\tau}, i = 1, \ldots, P$) and uniformly distributed over the parametric vectorial space and we choose to evaluate each particle's position according to the following objective function

$$MSE_i^{(\tau)} = \frac{1}{T} \sum_{t=1}^{T} (y_t - \hat{y}_t^{(i,\tau)})^2$$ (4)

where $\{y_t\}_{t=1,\ldots,T}$ is the observed time series and $\{\hat{y}_t^{(i,\tau)}\}_{t=1,\ldots,T}$ is the series computed by the model defined by a specific particle. So at each step we have P values of $MSE_i^{(\tau)}$.

If a particle's current position is better than its previous best position ($\mathbf{pb}_i$), it is updated. Then we determine the best particle ($\mathbf{lb}_i$), according to the particle's previous best position and we update particles' velocities according to rule (3). Then the particles are moved to their new positions according to (2).

The stopping rule we adopted ends the algorithm when there are no improvements in fitness function, that is if at least one particle has improved its fitness function or until N iteration are reached (typically we fixed $N = 100$).

This PSO technique has many advantages if compared to other optimization procedures, first for its simpleness, furthermore it does not require the gradient of the objective functions. On the other hand, the performance of the procedure is strongly influenced by the tuning of its behavioural parameters (ω, φ_1, φ_2), and to overcome such a problem Pedersen [7] studies and proposes some optimal combinations for them, depending on the particle dimension, that in our case the number of the estimating parameters, and on the replication of the procedure. In order to arrange an estimate for one series we consider the mean vector of the particles each getting the best fitness.

3 Some Results

In the following we show the results obtained in the estimation procedure, based both on simulated and real data.

3.1 Simulation Results

In our applications we consider different data generating processes, some in the class of the autoregressive linear model, the other ones in a non linear class, considering different AsyTAR models class. To obtain some Monte Carlo comparisons, we simulate $B = 100$ series each of $T = 500$ observations.

The fitness function is computed with respect to different structure, in order to evaluate a sort of bias.

For problems in which the simulated series were autoregressive we consider the dimension of each particle equal to 10, and we see that, applying the PSO procedure, the values of the estimates were not influenced by the tuning of procedure parameters. The number of parameters is set equal to 10 because we suppose that, in the PSO procedure, $\hat{y}_t^{(i\tau)}$ of formula (4) is computed by a model of type (1) with two regimes and autoregressive structure, each of order $p = 3$, there are two thresholds, r_1 and r_2, and the lags of delay, d_1 and d_2.

Table 1 reports the mean values and the standard deviations (in brackets) of the one hundred estimates of each statistical parameter in the case with simple linear DGP, simulated with AR of order 1 and 2. Here we see that the first three parameters, corresponding to the first regime, are typically estimated closed to the second three, that are those of the second regime, and this means that the two regimes are just the same. In reason of the absence of difference between the two regimes the thresholds

Table 1 Linear Models: estimates means over 100 replications (standard deviation)

Mean on 100 simulations	AR(1), $\phi =$			AR(2) $(\phi_1, \phi_2) =$		
	-0.95	-0.05	0.95	$(0.6, 0.3)$	$(1, -0.3)$	$(0.7, -0.3)$
$\hat{\phi}_1^{(1)}$	-0.95	-0.03	0.95	0.61	0.99	0.70
(sd)	(0.08)	(0.15)	(0.09)	(0.10)	(0.13)	(0.18)
$\hat{\phi}_2^{(1)}$	-0.01	0.01	-0.02	0.28	-0.29	-0.31
(sd)	(0.13)	(0.24)	(0.13)	(0.09)	(0.18)	(0.26)
$\hat{\phi}_3^{(1)}$	-0.00	-0.01	0.01	-0.01	-0.02	-0.02
(sd)	(0.09)	(0.26)	(0.11)	(0.11)	(0.13)	(0.28)
$\hat{\phi}_1^{(2)}$	-0.94	-0.07	0.93	0.58	0.99	0.71
(sd)	(0.08)	(0.12)	(0.09)	(0.08)	(0.08)	(0.09)
$\hat{\phi}_2^{(2)}$	0.02	-0.01	0.01	0.30	-0.31	-0.30
(sd)	(0.11)	(0.11)	(0.13)	(0.09)	(0.11)	(0.10)
$\hat{\phi}_3^{(2)}$	0.02	-0.01	-0.01	-0.00	0.01	-0.00
(sd)	(0.08)	(0.20)	(0.09)	(0.09)	(0.09)	(0.10)
$\hat{r}_1$	-0.07	-0.11	-0.04	-0.05	0.02	0.04
(sd)	(0.69)	(0.72)	(0.72)	(0.77)	(0.78)	(0.68)
$\hat{r}_2$	0.24	-0.19	0.06	0.09	-0.23	-0.25
(sd)	(0.86)	(0.82)	(0.77)	(0.83)	(1.06)	(0.97)
$\hat{d}_1$	1.13	1.17	1.22	1.21	1.17	1.19
(sd)	(0.33)	(0.38)	(0.43)	(0.40)	(0.36)	(0.39)
$\hat{d}_2$	1.15	1.13	1.13	1.10	1.20	1.17
(sd)	(0.35)	(0.30)	(0.36)	(0.32)	(0.37)	(0.37)

are negligible although they appeare both approximately equal to zero. In all the case for each estimate of dimension 10, the distributions can be easily depicted by the 100 simulated trajectories and we have seen that the Gaussian hypothesis of estimates are typically accepted according D'Agostino normality test.

So we see that even if we estimate an overparametrized model, the real linear structure is well captured. Some considerations must be done at this regard; in fact, following a sort of parsimony criterion, between two models with the same performance, we prefer the one requiring fewer parameters, so when we find that some model parameters are statistically equal to zero, we can re-identify a more parsimoniously model by dropping the non-significant delayed variables.

The results shown in Tables 2 and 3, for simulations from the AsyTAR models proposed in Sect. 2 respectively with $p = 1$ and $p = 2$, are based on the *Parsimony criterion*. This means that if the estimated model has some parameter statistically non-significant, that is if we use an autoregressive order, say $p + m$, greater than the needed one, p, we may reduce the number of delayed variables taken into account by the model. We remind, in fact, that the estimation may be affected by the presence of unnecessary parameters.

Furthermore, in order to control the attainment of the global minimum of the fitness function by the PSO procedure, we performed an estimate for some trajectories

Table 2 AsyTAR(1,1) above with $\phi^{(1)} = -0.5$, $\phi^{(2)} = 0.8$, $r_1 = 1.5$, $r_2 = 0$, $d_1 = d_2 = 1$, below with $\phi^{(1)} = -0.8$, $\phi^{(2)} = 0.8$, $r_1 = 2$, $r_2 = 0$, $d_1 = d_2 = 1$

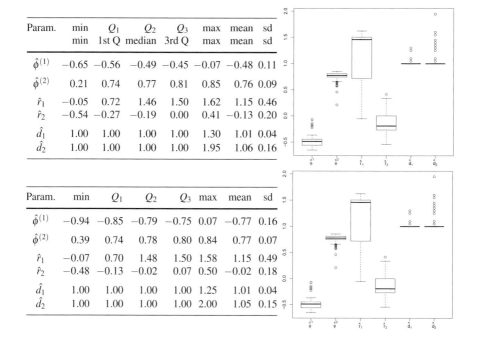

Param.	min	Q_1	Q_2	Q_3	max	mean	sd
	min	1st Q	median	3rd Q	max	mean	sd
$\hat{\phi}^{(1)}$	−0.65	−0.56	−0.49	−0.45	−0.07	−0.48	0.11
$\hat{\phi}^{(2)}$	0.21	0.74	0.77	0.81	0.85	0.76	0.09
$\hat{r}_1$	−0.05	0.72	1.46	1.50	1.62	1.15	0.46
$\hat{r}_2$	−0.54	−0.27	−0.19	0.00	0.41	−0.13	0.20
$\hat{d}_1$	1.00	1.00	1.00	1.00	1.30	1.01	0.04
$\hat{d}_2$	1.00	1.00	1.00	1.00	1.95	1.06	0.16

Param.	min	Q_1	Q_2	Q_3	max	mean	sd
$\hat{\phi}^{(1)}$	−0.94	−0.85	−0.79	−0.75	0.07	−0.77	0.16
$\hat{\phi}^{(2)}$	0.39	0.74	0.78	0.80	0.84	0.77	0.07
$\hat{r}_1$	−0.07	0.70	1.48	1.50	1.58	1.15	0.49
$\hat{r}_2$	−0.48	−0.13	−0.02	0.07	0.50	−0.02	0.18
$\hat{d}_1$	1.00	1.00	1.00	1.00	1.25	1.01	0.04
$\hat{d}_2$	1.00	1.00	1.00	1.00	2.00	1.05	0.15

Table 3 AsyTAR(2,1) above with $\phi^{(1)} = (0.6, 0.3)$, $\phi^{(2)} = (-0.6, -0.3)$, $r_1 = 0$, $r_2 = -1.5$, $d_1 = d_2 = 1$, below with $\phi^{(1)} = (0.3, 0.6)$, $\phi^{(2)} = (-0.3, -0.6)$, $r_1 = 0$, $r_2 = -1.5$, $d_1 = d_2 = 1$

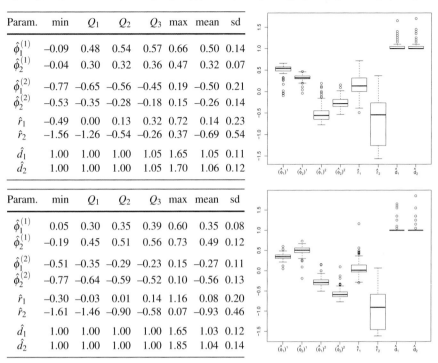

Param.	min	Q_1	Q_2	Q_3	max	mean	sd
$\hat{\phi}_1^{(1)}$	−0.09	0.48	0.54	0.57	0.66	0.50	0.14
$\hat{\phi}_2^{(1)}$	−0.04	0.30	0.32	0.36	0.47	0.32	0.07
$\hat{\phi}_1^{(2)}$	−0.77	−0.65	−0.56	−0.45	0.19	−0.50	0.21
$\hat{\phi}_2^{(2)}$	−0.53	−0.35	−0.28	−0.18	0.15	−0.26	0.14
$\hat{r}_1$	−0.49	0.00	0.13	0.32	0.72	0.14	0.23
$\hat{r}_2$	−1.56	−1.26	−0.54	−0.26	0.37	−0.69	0.54
$\hat{d}_1$	1.00	1.00	1.00	1.05	1.65	1.05	0.11
$\hat{d}_2$	1.00	1.00	1.00	1.05	1.70	1.06	0.12

Param.	min	Q_1	Q_2	Q_3	max	mean	sd
$\hat{\phi}_1^{(1)}$	0.05	0.30	0.35	0.39	0.60	0.35	0.08
$\hat{\phi}_2^{(1)}$	−0.19	0.45	0.51	0.56	0.73	0.49	0.12
$\hat{\phi}_1^{(2)}$	−0.51	−0.35	−0.29	−0.23	0.15	−0.27	0.11
$\hat{\phi}_2^{(2)}$	−0.77	−0.64	−0.59	−0.52	0.10	−0.56	0.13
$\hat{r}_1$	−0.30	−0.03	0.01	0.14	1.16	0.08	0.20
$\hat{r}_2$	−1.61	−1.46	−0.90	−0.58	0.07	−0.93	0.46
$\hat{d}_1$	1.00	1.00	1.00	1.00	1.65	1.03	0.12
$\hat{d}_2$	1.00	1.00	1.00	1.00	1.85	1.04	0.14

Table 4 AsyTAR(1,1) with $\phi^{(1)} = -0.8$, $\phi^{(2)} = 0.8$, $r_1 = 2$, $r_2 = 0$, $d_1 = 1$, $d_2 = 1$, one time series on 5000 PSO-replications

Param.	min	Q_1	Q_2	Q_3	max	mean	sd
$\hat{\phi}^{(1)}$	−1.07	−0.86	−0.83	−0.71	0.33	−0.73	0.23
$\hat{\phi}^{(2)}$	−0.13	0.73	0.79	0.81	0.92	0.73	0.16
$\hat{r}_1$	−0.24	1.75	1.97	1.99	2.13	1.63	0.63
$\hat{r}_2$	−0.95	−0.25	−0.18	−0.00	0.69	−0.14	0.21
$\hat{d}_1$	1.00	1.00	1.00	1.00	1.60	1.04	0.08
$\hat{d}_2$	1.00	1.00	1.00	1.10	2.80	1.07	0.14

simulated by AsyTAR process with $\phi_1^{(1)} = -0.8$, $\phi_1^{(2)} = 0.8$, $r_1 = 2$, $r_2 = 0$, $d_1 = 1$, $d_2 = 1$ using 5000 replications, that means 5000 different starting points (parameters). The results of the estimation procedure are summarized in Table 4. As concern the comparison with the procedure with lesser iterations, we can underline that the differences among the estimates are negligible.

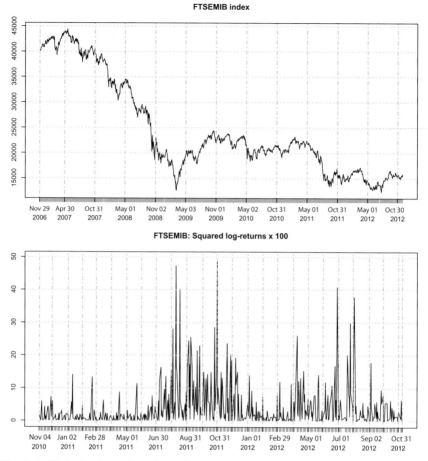

Fig. 3 FTSE MIB: above the price series (from November 2006 to October 2012), in the bottom the squared log-returns series (from November 2010 to October 2012)

3.2 Example in Stock Market Data

In order to apply the proposed procedure to real dataset, we consider the FTSE MIB index (Fig. 3) from 5th November 2010 to 9th November 2012. The analysis of the log-return of the index, as expected, doesn't reveal any particular structure. So we pay attention to the volatility of the time series and considered the squared log-return as its proxy. The AsyTAR(2,1) model is estimated using the PSO algorithm with 60 particles and stopping criterion fixed to 10^{-6} and the procedure parameters are set as following: $\omega = 0.6571$, $\varphi_1 = 1.6319$ and $\varphi_2 = 0.6239$. As our parameters are subject to some constrains in order to obtain a model for positive time series, we consider a version of the PSO procedure in which the velocity is null if the particle goes out of the eligibility region.

Table 5 summarizes the results of the estimation procedure and we can see that the procedure reveals the presence of two overlapped regimes; in fact, the threshold r_1 is greater than r_2. Furthermore both the regimes are estimated with autoregressive model of order two, for the first regime with $\hat{\phi}_1^{(1)} = 0.369$ and $\hat{\phi}_2^{(1)} = 0.324$ and for the second one with $\hat{\phi}_1^{(2)} = 0.175$ and $\hat{\phi}_2^{(2)} = 0.038$, the two thresholds are estimated $\hat{r}_1 = 0.350$ and $\hat{r}_2 = 0.167$. This result highlights the ability of the AsyTAR model to capture the asymmetric structure of the DGP. In the case of the FTSE MIB the behaviour of the two regimes is quite different. This fact reveals that the efficiency of the market sometimes fails showing the presence of a structure in squared logarithmic returns. Moreover in our model the sequences of each regimes are relative short and the two regimes switch at relatively high frequency.

4 Conclusions

The contribution seems of relevant interests as it provides new perspectives along three different patterns. Firstly, the proposed technique seems to obtain interesting results in the parameters estimation. In particular the use of the PSO algorithm makes it possible to identify the best model. Moreover, an appropriate definition of the fitness function enables us to select the most suitable model.

Secondly, the preliminary results show robustness with respect to model misspecification. Thirdly, in the cases of threshold model the technique allows to estimate simultaneously thresholds, delay and autoregressive parameters.

In particular, it seems that an appropriate dimension of the particles let us apply the same procedure to time series generated by different kind of processes, although this requires the definition of a fitness function that should cope with different DGPs. In fact, as the algorithm estimates the redundant parameters equal to zero, we improve the identification of the underlying process.

Eventually, we think that this model might be a useful tool to analyze real time series such as finance data that likely show asymmetric features.

Table 5 Estimation FTSMIB with AsyTAR structure with two regressors and 100 replications

Param.	min	Q_1	Q_2	Q_3	max	mean	sd
$\hat{\phi}_1^{(1)}$	−0.316	0.365	0.396	0.490	0.593	0.396	0.140
$\hat{\phi}_2^{(1)}$	0.111	0.269	0.329	0.374	0.447	0.324	0.070
$\hat{\phi}_1^{(2)}$	−0.396	0.157	0.210	0.261	0.676	0.175	0.175
$\hat{\phi}_2^{(2)}$	−0.781	−0.041	0.123	0.142	0.642	0.038	0.228
$\hat{r}_1$	0.045	0.108	0.238	0.282	1.581	0.350	0.375
$\hat{r}_2$	0.048	0.084	0.110	0.219	0.620	0.167	0.131
$\hat{d}_1$	1.000	1.000	1.000	1.017	1.517	1.025	0.085
$\hat{d}_2$	1.000	1.000	1.000	1.017	2.033	1.073	0.227

References

1. Battaglia, F., Protopapas, M.K.: Multi-regime models for nonlinear nonstationary time series. Computational Statistics **47**, 277–295 (2011)
2. Chan, K.S., Tong, H.: On Likelihood Ratio Tests for Threshold Autoregression. Journal of the Royal Statistical Society **B**(52) (Methodological), 469–476 (1990)
3. Hansen, B.E.: Sample Splitting and Threshold Estimation. Econometrica **68**, 575–603 (2000)
4. Gonzalo, J., Wolf, M.: Subsampling inference in threshold autoregressive models. Journal of Econometrics **127**, 201–224 (2005)
5. Kennedy, J., Eberhart, R.C.: Subsampling inference in threshold autoregressive models. In: In Proceedings of IEEE International Conference on Neural Networks, pp. 1942–1948 (1995)
6. Li, W.K., Lam, K.: Modelling Asymmetry in Stock Returns by a Threshold Autoregressive Conditional Heteroscedastic Model. The Statistician **44**, 333–341 (1995)
7. Pedersen, M.E.H.: Good Parameters for Particle Swarm Optimization. Hvass Laboratories. Technical Report HL1001 (2010)
8. Petruccelli, J.D., Davies, N.: A Portmanteau Test for Self-Exciting Threshold Autoregressive-Type Nonlinearity in Time Series. Biometrika **73**, 687–694 (1986)
9. Pizzi, C.: The asymmetric threshold model AsyTAR(2;1,1). In: Sco2007 (2007)
10. Pizzi, C.; Parpinel, F.: Evolutionary computational approach in TAR model estimation. In: University of Venice "Ca' Foscari" (edc.) Working Papers, vol. 26/2011. Department of Economics, Venezia (2011)
11. Shi, Y., Eberthart, R.: A modified particle swarm optimizer. In: Proceedings of the IEEE International Conference on Evolutionary Computation, pp. 69–73 (1998)
12. Tong, H., Yeung, I.: Threshold Autoregressive Modelling in Continuous Time. Statistica Sinica **1**, 411–430 (1991)
13. Tong, H., Lim, K.S.: Threshold Autoregression, Limit Cycles and Cyclical Data. Journal of the Royal Statistical Society B **42** (Methodological), 245–292 (1980)
14. Tsay, R.S.: Testing and Modeling Multivariate Threshold Models, Journal of the American Statistical Association **93**, 1188–1202 (1998)
15. Wang, H., Zhao, W.: ARIMA Model Estimated by Particle Swarm Optimization Algorithm for Consumer Price Index Forecasting. In: Deng, H., Wang, L., Wang, F.H., Lei, J. (eds.) Artificial Intelligence and Computational Intelligence. Lecture Notes in Computer Science 5855, pp. 48–58. Springer (2009)
16. Wu, B., Chang, C.L.: Using genetic algorithm to parameters (d,r) estimation for threshold autoregressive models. Comput. Stat. Data Anal. **17**, 241–264 (2002)
17. Zakoian, J.M.: Threshold Heteroskedastic Models. Journal of Economic Dynamics and Control **18**, 931–955 (1994)

Valuation of Collateralized Funds of Hedge Fund Obligations: A Basket Option Pricing Approach

Gian Luca Tassinari and Corrado Corradi

Abstract The purpose of the present contribution is to provide an extension to a model developed by Tassinari and Corradi [7] to price equity and debt tranches of collateralized funds of hedge fund obligations (CFOs). The key idea is to price each CFO liability as an option on the underlying basket of hedge funds. The proposed model is able to reproduce the empirical characteristics observed in the distribution of hedge funds' returns: skewness, excess kurtosis and dependence in the tails. Additionally, it can be easily calibrated to the empirical correlation matrix and it requires only historical information to be estimated and implemented. The result is a scheme that can be useful in structuring a CFO. In particular, we believe that the approach described in this work can be helpful to rating agencies and to deal structures to evaluate various capital structures, test levels, liquidity profiles, coupons and equity distribution rules.

1 Introduction

CFOs are structured finance products created by using a standard securitization approach. A special purpose vehicle issues multiple tranches of senior and subordinated notes that pay interest at fixed or floating rates and an equity tranche, and invests the proceeds in a portfolio of hedge funds. CFOs typically have a stated term of three to seven years at the end of which the collateral portfolio is sold and all of

G.L. Tassinari (✉)
Department of Mathematics, *Alma Mater Studiorum*, University of Bologna, Porta San Donato 5, 40126 Bologna, Italy
e-mail: gianluca.tassinari2@unibo.it

C. Corradi
Department of Mathematics, *Alma Mater Studiorum*, University of Bologna, Porta San Donato 5, 40126 Bologna, Italy
e-mail: corrado.corradi@unibo.it

M. Corazza, C. Pizzi (eds.), *Mathematical and Statistical Methods for Actuarial Sciences and Finance*, DOI 10.1007/978-3-319-02499-8_25, © Springer International Publishing Switzerland 2014

the securities must be redeemed. Redemptions before maturity are only possible if some predetermined events happen. The CFO manager periodically checks the net asset value (NAV) of the collateral portfolio. If this value falls below an established threshold, the CFO structure goes into bankruptcy and debts are repaid in sequential order according to their degree of subordination with the liquidation proceeds of the collateral portfolio. Following the approach developed in [7], in this contribution we consider each CFO liability as an option written on the underlying pool of hedge funds and we compute its fair price as its risk neutral expected payoff, discounted at the risk free rate. Since the value of every liability is linked to the dynamics of the collateral portfolio's NAV during the life of the contract, it is necessary to model the joint risk neutral evolution of the underlying hedge funds and at the same time any CFO's structural features like coupon payments, over collateralization test, liquidity profile, equity distribution rules, management fees have to be taken into account. Due to use of derivatives, leverage and short selling, hedge funds' log-returns distribution usually exhibits negative skewness and heavier tails than the normal distribution. Furthermore, hedge funds log-returns presents a higher degree of dependence during high volatility periods and severe market crashes (see [7] and references therein). Since multivariate Brownian model is not able to describe these empirical phenomena, a more general Lévy process is needed. In [7] dependence among hedge fund log-returns is introduced through a gamma stochastic time-change of a multivariate Brownian motion, with uncorrelated components. The idea is that the economy is driven by only one common factor, whose dynamics is described by a gamma subordinator, and there are no other sources of dependence (see [5] and [7]). In this contribution we get a more flexible hedge funds' log-returns model by time-changing a multivariate arithmetic Brownian motion with correlated components. Since we assume dependent Brownian motions, jump sizes are correlated (see [1] and [4]). This model allows to get a richer dependence structure than the one in [7] and it can be calibrated to the empirical correlation matrix. The market model presented in this work is not complete, because the risk due to jumps cannot be hedged. Therefore, the equivalent martingale measure is not unique. Additionally, we cannot change probability measure by using the so called mean correcting martingale technique, because the log-returns process we propose has no diffusion component (see [1]). Among the possible candidates we select the Esscher equivalent martingale measure (EEMM) (see [3]) and we find explicit relations among physical and risk neutral processes and distributions at both marginal and joint levels. These results are very important in the pricing procedure because only historical data are available. Furthermore, this approach can be very useful to price derivatives on equity baskets when no traded options on the underlying assets are available for calibration purpose. The work is organized as follows. In Sect. 2 we present the model applied to describe the physical evolution of hedge funds' log-returns. In Sect. 3 we discuss the change of measure. In Sect. 4 equity and debt tranches of a theoretical CFO are priced. Section 5 concludes.

2 Hedge Funds' Log-Returns P-Dynamics

In this section we model the dynamics of hedge funds' log-returns under the physical probability measure P as a multivariate variance gamma (MVG) process with a linear drift. The pure jump part of the process is got by time changing an arithmetic multivariate Brownian motion with an independent one-dimensional gamma process. We refer to this last process as a subordinator or a stochastic clock. Let $G = \{G_t, \ t \geq 0\}$ be a gamma process, i.e., a process which starts at zero and has stationary and independent increments which follows a gamma distribution. G is a Lévy process in which the defining law of G_1 is a gamma distribution with parameters $\alpha > 0$ and $\beta > 0$. For normalization reasons, we work with a gamma process such that $E(G_t) = t$, which in terms of the parameters implies that $\alpha = \beta = 1/v$. The density function of the random variable G_1 is

$$f(g; 1/v, 1/v) = \frac{v^{-1/v}}{\Gamma(1/v)} g^{1/v-1} \exp(-g/v), \ \ g > 0, \tag{1}$$

where

$$\Gamma(1/v) = \int_0^\infty g^{1/v-1} \exp(-g/v) dg, \tag{2}$$

and its characteristic function is

$$\Psi_{G_1}(\omega) = (1 - i\omega v)^{-1/v}. \tag{3}$$

We also assume that the components of the multivariate Brownian motion are correlated, i.e, $W^j = \{W_t^j, \ t \geq 0\}$ and $W^k = \{W_t^k, \ t \geq 0\}$ are Wiener processes with correlation coefficient ρ_{jk}, for $j = 1, \ldots, n$ and $k = 1, \ldots, n$. Under this assumptions, the log-return of hedge fund j over the interval $[0, t]$ can be written as

$$Y_t^j = \mu_j t + \theta_j G_t + \sigma_j W_{G_t}^j = \mu_j t + X_t^j, \tag{4}$$

where $W_G^j = \{W_{G_t}^j, \ t \geq 0\}$ is the j-th Wiener process subordinated by the common gamma process, $X^j = \{X_t^j, \ t \geq 0\}$ is pure jump process component of $Y^j = \{Y_t^j, \ t \geq 0\}$. μ_j, θ_j, and $\sigma_j > 0$ are constants. Modelling dependence in this way allows to introduce two sources of co-movement among different hedge funds. In particular, the use of a common subordinator generates a new business time in which all the market operates (see [2]). This means all prices jump simultaneously (see [5,7]). Since we assume dependent Brownian motions, jump sizes are correlated (see [1,4]). Furthermore, Brownian motion subordination allows to produce margins able to describe the skewness and excess kurtosis[1] that characterize the empirical distribution of hedge funds' log-returns.

By composition of the Laplace exponent of the gamma subordinator

$$\Phi_{G_1}(\omega) = -\frac{\ln(1 - \omega v)}{v} \tag{5}$$

---

[1] See [1] or [6].

with the characteristic exponent of the multivariate Brownian motion

$$\Lambda(u) = \sum_{j=1}^{n} iu_j\theta_j - \frac{1}{2}\sum_{j=1}^{n}\sum_{k=1}^{n} u_j u_k \sigma_j \sigma_k \rho_{jk}, \quad u \in \mathbf{R}^n, \tag{6}$$

we get the characteristic exponent of the pure jump process (see [1]):

$$\Lambda_{X_1}(u) = -\frac{1}{v}\left[1 - v\left(\sum_{j=1}^{n} iu_j\theta_j - \frac{1}{2}\sum_{j=1}^{n}\sum_{k=1}^{n} u_j u_k \sigma_j \sigma_k \rho_{jk}\right)\right]. \tag{7}$$

Now, it is easy to find the characteristic function of Y_1:

$$\Psi_{Y_1}(u) = \exp\left(i\sum_{j=1}^{n} u_j\mu_j\right)\left[1 - v\left(\sum_{j=1}^{n} iu_j\theta_j - \frac{1}{2}\sum_{j=1}^{n}\sum_{k=1}^{n} u_j u_k \sigma_j \sigma_k \rho_{jk}\right)\right]^{-1/v}. \tag{8}$$

From (8) it is possible to compute marginal and joint moments of hedge funds' log-returns for $t = 1$:

$$E\left(Y_1^j\right) = \mu_j + \theta_j, \tag{9}$$

$$Var\left(Y_1^j\right) = \sigma_j^2 + v\theta_j^2, \tag{10}$$

$$Skew\left(Y_1^j\right) = \theta_j v\left(3\sigma_j^2 + 2v\theta_j^2\right)/\left(\sigma_j^2 + v\theta_j^2\right)^{3/2}, \tag{11}$$

$$Kurt\left(Y_1^j\right) = 3\left(1 + 2v - v\sigma_j^4\left(\sigma_j^2 + v\theta_j^2\right)^{-2}\right), \tag{12}$$

$$Corr\left(Y_1^j; Y_1^k\right) = \frac{\theta_j\theta_k v + \sigma_j\sigma_k\rho_{jk}}{\sqrt{\sigma_j^2 + v\theta_j^2}\sqrt{\sigma_k^2 + v\theta_k^2}}, \tag{13}$$

for $j = 1, \ldots, n$ and $k = 1, \ldots, n$. From (13) and (11) it follows that pairs of hedge funds with skewness of the same sign could be negatively correlated and hedge funds with skewness of opposite sign could be positively correlated. Furthermore, pairs of assets have null correlation if and only if at least one of them has a symmetric distribution and their underlying Brownian motions are uncorrelated. Thus, due to jumps size correlation, this process is more flexible in modelling dependence compared to the one presented in [7]. Finally, note that due to the common stochastic clock null correlation doesn't imply independence.

3 Change of Measure and Hedge Funds' Log-Returns Q_h-Dynamics

Assuming the existence of a bank account which provides a continuously compounded risk free rate r, this market model is arbitrage free, since the price process of every asset has both positive and negative jumps (see [1]). This ensures the existence of an equivalent martingale measure. However, the model is not complete, because the risk due to jumps cannot be hedged. Therefore, the equivalent martingale measure is not unique. Among the possible risk neutral probability measures we choose the EEMM, which we denote by Q_h (see [3]). The Q_h measure associated with the multivariate log-returns process Y is defined by the following Radon-Nikodym derivative:

$$\frac{dQ_h}{dP}|\Im_t = \frac{\exp(\sum_{j=1}^{n} h_j Y_t^j)}{E\left[\exp(\sum_{j=1}^{n} h_j Y_t^j)\right]}, \tag{14}$$

where $\Im_t$ is the filtration originated by the log-returns process. To get the Esscher risk neutral dynamics of Y:

1. find a vector h such that the discounted NAV process of every hedge fund is a martingale under the new probability measure Q_h solving the system

$$E\left[\exp(\sum_{j=1}^{n} h_j Y_t^j + Y_t^1)\right] / E\left[\exp(\sum_{j=1}^{n} h_j Y_t^j)\right] = \exp(rt)$$

$$\vdots$$

$$E\left[\exp(\sum_{j=1}^{n} h_j Y_t^j + Y_t^n)\right] / E\left[\exp(\sum_{j=1}^{n} h_j Y_t^j)\right] = \exp(rt); \tag{15}$$

2. find the characteristic function of the process Y under the measure Q_h as

$$\Psi_{Y_t}^{Q_h}(u) = \frac{E\left[\exp \sum_{j=1}^{n} (h_j + iu_j) Y_t^j\right]}{E\left[\exp \sum_{j=1}^{n} h_j Y_t^j\right]} = \frac{\Psi_{Y_t}(u - ih)}{\Psi_{Y_t}(-ih)}. \tag{16}$$

Taking into account (8), the system (15) may be written as:

$$\frac{1}{v} \ln\left[1 - \frac{v(\theta_1 + 0.5\sigma_1^2 + \sum_{j=1}^{n} h_j \sigma_1 \sigma_j \rho_{1j})}{1 - v\left(\sum_{j=1}^{n} h_j \theta_j + \frac{1}{2}\sum_{j=1}^{n}\sum_{k=1}^{n} h_j h_k \sigma_j \sigma_k \rho_{jk}\right)}\right] = \mu_1 - r$$

$$\vdots$$

$$\frac{1}{v} \ln\left[1 - \frac{v(\theta_n + 0.5\sigma_n^2 + \sum_{j=1}^{n} h_j \sigma_n \sigma_j \rho_{nj})}{1 - v\left(\sum_{j=1}^{n} h_j \theta_j + \frac{1}{2}\sum_{j=1}^{n}\sum_{k=1}^{n} h_j h_k \sigma_j \sigma_k \rho_{jk}\right)}\right] = \mu_n - r \tag{17}$$

with the following constraints

$$\left[1 - v\left(\sum_{j=1}^{n} h_j \theta_j + \frac{1}{2}\sum_{j=1}^{n}\sum_{k=1}^{n} h_j h_k \sigma_j \sigma_k \rho_{jk}\right)\right] > 0 \tag{18}$$

and

$$1 - v(\sum_{j \neq q}^{n} h_j \theta_j + (h_q + 1)\theta_q + \frac{1}{2} \sum_{j \neq q}^{n} \sum_{k \neq q}^{n} h_j h_k \sigma_j \sigma_k \rho_{jk})$$

$$- \frac{1}{2} v(\sum_{j \neq q}^{n} h_j (h_q + 1)\sigma_j \sigma_q \rho_{jq} + (h_q + 1)^2 \sigma_q^2) > 0, \ q = 1, \ldots, n. \quad (19)$$

It can be shown that this system has at most two solutions.[2] Furthermore, in all our experiments we found that system (17) possesses a unique solution and therefore a unique vector h exists satisfying the constraints (18) and (19). This ensures the existence and the uniqueness of the EEMM. From (16) and taking into account (8), we get the risk neutral characteristic function of the process for $t = 1$, i.e.,

$$\Psi_{Y_1}^{Q_h}(u) = \exp\left(\sum_{j=1}^{n} iu_j \mu_j\right) \times$$

$$\left[1 - \frac{v(\sum_{j=1}^{n} u_j(\theta_j + \sum_{k=1}^{n} h_j \sigma_j \sigma_k \rho_{jk}) - \frac{1}{2}\sum_{j=1}^{n}\sum_{k=1}^{n} u_j u_k \sigma_j \sigma_k \rho_{jk})}{1 - v\left(\sum_{j=1}^{n} h_j \theta_j + \frac{1}{2}\sum_{j=1}^{n}\sum_{k=1}^{n} h_j h_k \sigma_j \sigma_k \rho_{jk}\right)}\right]^{-1/v}. \quad (20)$$

Comparing (20) with (8), we note that the Esscher change of probability measure does not modify the nature of the log-returns process. In particular, we find the following relations among risk neutral and statistical parameters:

$$\mu_j^{Q_h} = \mu_j, \quad (21)$$

$$v^{Q_h} = v, \quad (22)$$

$$\theta_j^{Q_h} = \frac{\theta_j + \sum_{k=1}^{n} h_k \sigma_j \sigma_k \rho_{jk}}{1 - v\left(\sum_{j=1}^{n} h_j \theta_j + \frac{1}{2}\sum_{j=1}^{n}\sum_{k=1}^{n} h_j h_k \sigma_j \sigma_k \rho_{jk}\right)}, \quad (23)$$

$$(\sigma_j^{Q_h})^2 = \frac{\sigma_j^2}{1 - v\left(\sum_{j=1}^{n} h_j \theta_j + \frac{1}{2}\sum_{j=1}^{n}\sum_{k=1}^{n} h_j h_k \sigma_j \sigma_k \rho_{jk}\right)}, \quad (24)$$

$$\sigma_{jk}^{Q_h} = \frac{\sigma_{jk}}{1 - v\left(\sum_{j=1}^{n} h_j \theta_j + \frac{1}{2}\sum_{j=1}^{n}\sum_{k=1}^{n} h_j h_k \sigma_j \sigma_k \rho_{jk}\right)}, \quad (25)$$

$$\rho_{jk}^{Q_h} = \rho_{jk}, \quad (26)$$

for $j = 1, \ldots, n$ and $k = 1, \ldots, n$. Under the Q_h measure the log-returns process can be obtained by time-changing a multivariate Brownian motion with correlated components, with an independent Gamma process. We emphasize that the underlying dependence structure is not affected by the change of measure. Precisely, the Brownian motions have the same correlation matrix and the gamma process has the same parameters. However, joint and marginal moments change.

---

[2] To be more precise, this is true under a certain hypothesis. See the Appendix for the proof.

4 Pricing CFOs Tranches

In this section we price debt and equity securities of a theoretical CFO as options written on a portfolio of hedge funds. The fair price of each tranche is computed as its expected discounted payoff under the EEMM. The payoff of every tranche depends on CFO structural features as over-collateralization tests, priority of payment waterfall and liquidity profile, and it is linked to the risk-neutral evolution of the portfolio NAV, which depends on the temporal behaviour of all its underlying hedge funds. This section is organized as follows. In Subsect. 4.1 we describe the data. In Sect. 4.2 we illustrate how to estimate the parameters of the model described in Sects. 2 and 3. In Sect. 4.3 we explain how to simulate a simple path of the collateral portfolio NAV. In Sect. 4.4 we discuss the pricing application and the results.

4.1 Data and Summary Statistics

We get NAV monthly data from the Credit Suisse/Tremont Hedge Index for the following hedge fund indices: convertible arbitrage (CA), dedicated short bias (DSB), emerging markets (EM), equity market neutral (EMN), event driven (ED), distressed (DST), multi-strategy (MS), risk arbitrage (RA). The sample covers the period from January 1994 through May 2008. Table 1 reports some descriptive statistics. A brief examination of the last two columns of this table indicates that hedge fund returns are clearly not Gaussian. Six hedge fund indices over eight exhibit a negative skewness. All indices display excess kurtosis. However, the degree of asymmetry and fat tails is quite different among hedge funds.

Table 1 Summary statistics of monthly log-returns for CS/Tremont indices, period January 1994–May 2008

Index	Mean %	Median %	Max %	Min %	Std.Dev. %	Skew.	Kurt.
CA	0,58	0,86	3,45	−5,80	1,38	−1,64	7,64
DSB	−0,21	−0,36	20,2	−9,36	4,75	0,56	4,11
EM	0,70	1,38	15,3	−26,2	4,50	−1,18	10,4
EMN	0,71	0,67	3,19	−1,27	0,76	0,36	3,90
ED	0,83	1,02	3,84	−12,6	1,61	−3,58	30,1
DST	0,93	1,11	4,08	−13,4	1,78	−3,15	26,1
MS	0,78	0,86	4,29	−12,3	1,74	−2,65	20,9
RA	0,55	0,55	3,58	−6,48	1,16	−1,29	10,4

4.2 *Physical and Risk Neutral Parameters Estimation*

Physical parameters are estimated using a two steps procedure. We estimate marginal parameters imposing the equality among the first three empirical moments of log-returns and their theoretical variance gamma counterparts and requiring a fitted mean kurtosis equal to the sample one. Then, we compute the correlation between j-th and k-th underlying Brownian motions as

$$\rho_{jk} = \frac{\sqrt{\sigma_j^2 + v\theta_j^2}\sqrt{\sigma_k^2 + v\theta_k^2}\rho_{y_j y_k} - \theta_j \theta_k v}{\sigma_j \sigma_k}, \tag{27}$$

where $\rho_{y_j y_k}$ is the sample correlation between hedge fund j and k. Then, we compute the vector h by solving the system (17). Finally, from the relations among risk neutral and statistical parameters of Sect. 3, we get the risk neutral ones.

4.3 *Simulation*

To simulate the paths of n dependent hedge fund NAVs under the EEMM we can proceed as follows.

Let $NAV_{t_0}^j$ the NAV of hedge fund j at time 0 for $j = 1, \ldots, n$.

Divide the time-interval $[0, T]$ into N equally spaced intervals $\Delta t = T/N$ and set $t_k = k\Delta t$, for $k = 0, \ldots, N$.

For every hedge fund repeat the following steps for k from 1 to N:

1. sample a random number g_k out of the Gamma$(\Delta t/v, 1/v)$ distribution;
2. sample n independent standard normal random numbers $w_{t_k}^j$;
3. convert these random numbers $w_{t_k}^j$ into correlated random numbers $v_{t_k}^j$ by using the Cholesky decomposition of the implied correlation matrix of the underlying Brownian motions;
4. compute

$$NAV_{t_k}^j = NAV_{t_{k-1}}^j \exp\left[\mu_j \Delta t + \theta_j^{Q_h} g_k + \sigma_j^{Q_h} \sqrt{g_k} v_{t_k}^j\right]. \tag{28}$$

To simulate a simple trajectory of the collateral portfolio NAV it is sufficient to compute

$$NAV_{t_k} = \sum_{j=1}^n NAV_{t_k}^j, \quad k = 1, \ldots, N. \tag{29}$$

In the next subsection we will also explain how to adapt this procedure to take into account CFO structural features such as coupon payments, equity distribution rules, over-collateralization tests, liquidity profile and management fees.

4.4 Applications and Results

In this section we price debt and equity securities of a theoretical CFO.[3] Consider a CFO structure with a scheduled maturity $T = 5$ years. The collateral portfolio has a current value of 1000 currency units allocated to the funds of hedge funds as follows: CA 175, DSB 50, EM 50, EMN 250, ED 100, DST 50, MS 100 and RA 225. The liabilities are structured in the following way: a debt tranche A with a nominal value of 570 and annual coupon rate of 4%, a debt tranche B with a nominal value of 150 and annual coupon rate of 4.05%, a debt tranche C with a nominal value of 100 and annual coupon rate of 4.5%, and a paying dividend equity tranche with a nominal value of 180. If the NAV of the collateral at the end of a year is greater than 1000 after the payment of coupons to bondholders, the 50% of annual profits is distributed to equityholders. We assume the CFO has enough liquidity to pay coupons and dividends. In particular, a part of each hedge fund, proportional to its NAV, is sold at the payment date. The CFO manager is assumed to make the over collateralization test every three months. In the simulation procedure we consider a barrier equal to 1,05 times the total nominal value of the debt tranches. If the NAV of the fund of hedge funds falls below this level, when its value is checked by the CFO manager, then the collateral portfolio will be sold in order to redeem the rated notes. In the default event, tranche A is redeemed first. In particular, we assume that both capital and current coupon have to be paid. Then, tranche B has to be repaid in the same way and so on. In the event of default, we model the sale of the assets by assuming this simple liquidity profile: 30% after three months, 30% after six months, all the residual collateral portfolio value after nine months. For simplicity, we assume that hedge funds are liquidated proportionally to their NAV. Additionally, we assume the existence of an initial lock out period of two years. This means that redemptions before two years are not admitted. The annual management fee is assumed to be equal to 0.5% of the total nominal amount of CFO tranches. Finally, we assume the existence of a risk free asset with a constant annual log-return $r = 4\%$. Table 2 shows the price of each CFO tranche computed by using three different models to describe the evolution of hedge funds log-returns: multivariate Brownian motion (MBM), multivariate variance gamma process with independent underlying Brownian motions (MVG IND), multivariate variance gamma process with dependent underlying Brownian motions (MVG DEP). This table shows that the choice of the model has a significant impact on the pricing of debt tranches. This effect is greater the smaller is the degree of protection offered by the CFO structure. On the contrary, the price of the equity tranche seems not particularly sensitive with respect to this choice. The prices of all bonds decrease from the MBM to the MVG DEP model. This is reasonable because the normal distribution is symmetric and is not able to capture the kurtosis present in hedge funds' returns. Also, the tail events of a normal distribution are asymptotically independent. Consequently, the Gaussian model underestimates the risk of default, and thus it overestimates prices. The degree of dependence generated by the MVG DEP model is greater because the underlying Brownian motions are almost all positively

---

[3] This CFO has the same structure of the third CFO considered in [7].

Table 2 Asset side 1000: funds of hedge funds. Liability side 1000: one equity security and three coupon bonds (CB). CFO tranche prices with barrier (105%) and management fees $(0,5\%)$

MODEL	EQUITY TRANCHE	CB A TRANCHE	CB B TRANCHE	CB C TRANCHE
MBM				
Prices with fees	154,977	569,912	149,994	101,439
(Prices with no fees)	(177,282)	(569,974)	(150,226)	(102,014)
MVG IND				
Prices with fees	154,894	567,517	146,788	92,873
(Prices with no fees)	(176,443)	(568,073)	(147,725)	(95,291)
MVG DEP				
Prices with fees	154,837	566,788	145,762	90,356
(Prices with no fees)	(176,199)	(567,475)	(146,921)	(93,294)

correlated.[4] A comparison between the prices MVG IND and MVG DEP allows to evaluate the effect on prices due only to the different dependence structures since the physical margins are the same. More intense the link between the margins is, greater the risk of failure is and lower the prices are. From these prices it is possible to infer the importance of a correct modeling of the dependence structure for pricing purpose.

5 Conclusions

In this contribution, we provided an extension to a model developed by Tassinari and Corradi [7] to price equity and debt tranches of a CFO. The model presented here is able to reproduce the empirical characteristics observed in the distribution of hedge funds' returns: skewness, excess kurtosis and dependence in the tails. Additionally, it can be easily calibrated to the empirical correlation matrix and it requires only historical information to be estimated and implemented, thanks to the existence of explicit relations among physical and risk neutral processes and distributions at both marginal and joint levels. The result is a scheme that can be useful in structuring a CFO. In particular, we believe that the approach described can be helpful to rating agencies and to deal structures to evaluate various capital structures, test levels, liquidity profiles, coupons and equity distribution rules.

---

[4] These correlations are not reported in this contribution, but they can be provided on request.

Appendix

In this appendix we discusse the existence and uniqueness of the solution of system (17). From (17) we easily get:

$$\left(a_1 + \sum_{j=1}^{n} h_j b_{1j}\right)/A_1 = 1 - \sum_{j=1}^{n} c_j h_j - \frac{1}{2} \sum_{j=1}^{n} \sum_{k=1}^{n} h_j h_k b_{jk}$$

$$\vdots$$

$$\left(a_n + \sum_{j=1}^{n} h_j b_{nj}\right)/A_n = 1 - \sum_{j=1}^{n} c_j h_j - \frac{1}{2} \sum_{j=1}^{n} \sum_{k=1}^{n} h_j h_k b_{jk} \qquad (30)$$

where $a_j = v(\theta_j + 0.5\sigma_j^2)$, $b_{jk} = v\sigma_{jk}$, $c_j = v\theta_j$, $A_j = 1 - \exp[v(\mu_j - r)]$, $j = 1,\ldots,n$ and $k = 1,\ldots,n$. Bringing the first member of the first equation to the second member in the subsequent equations of system (30), the last $n-1$ equations can be written as follows:

$$\sum_{j=2}^{n} h_j F_{2j} = h_1 D_2 + E_2$$

$$\vdots$$

$$\sum_{j=2}^{n} h_j F_{nj} = h_1 D_n + E_n \qquad (31)$$

where $D_j = A_1 b_{j1} - A_j b_{11}$, $E_j = A_1 a_j - A_j a_1$, and $F_{kj} = A_k b_{1j} - A_1 b_{kj}$ for $j = 2,3,\ldots,n$ and $k = 2,3,\ldots,n$. Under the assumption that the matrix of the coefficients F_{kj} is not singular, we can express the solution of (31) as a linear function of h_1, using Cramer's method:

$$h_2 = \frac{\det F_1(D)}{\det F} h_1 + \frac{\det F_1(E)}{\det F}$$

$$\vdots$$

$$h_n = \frac{\det F_{n-1}(D)}{\det F} h_1 + \frac{\det F_{n-1}(E)}{\det F} \qquad (32)$$

where F is the coefficients matrix F_{kj}, $F_k(D)$ is the matrix obtained substituting its k-th column with vector D, $F_k(E)$ is the matrix obtained substituting its k-th column with vector E. Substituting (32) in the first equation of the system (30), after simple calculations we get a quadratic equation in only one unknown h_1:

$$d h_1^2 + e h_1 + f = 0, \qquad (33)$$

where

$$d = \frac{A_1}{2} \sum_{k=1}^{n} \sum_{j=1}^{n} I_k I_j b_{kj}, \tag{34}$$

$$e = A_1 \left[\sum_{j=1}^{n} I_j c_j + \sum_{j=1}^{n} \sum_{k=2}^{n} I_j L_k b_{jk} \right] + \sum_{j=1}^{n} I_j b_{1j}, \tag{35}$$

$$f = A_1 \left[\sum_{j=2}^{n} L_j c_j + \frac{1}{2} \sum_{j=2}^{n} \sum_{k=2}^{n} L_j L_k b_{jk} - 1 \right] + a_1 + \sum_{j=2}^{n} L_j b_{1j}, \tag{36}$$

and with $I_1 = 1$, $I_k = \frac{\det F_{k-1}(D)}{\det F}$, $L_k = \frac{\det F_{k-1}(E)}{\det F}$, $k = 2, 3, \ldots n$.
The analysis of the existence of solutions of Eq. (33), although simple in principle, is a very hard task in practice. However, in all our experiments we found that Eq. (33) possesses a unique solution and therefore a unique vector

$$h = [h_1; D_2 h_1 + E_2; \ldots; D_n h_1 + E_n] \tag{37}$$

exists satisfying the constraints (18) and (19), where

$$h_1 = \frac{-e - \sqrt{e^2 - 4df}}{2d}. \tag{38}$$

References

1. Cont, R., Tankov, P.: Financial Modelling with Jump Processes. Chapman & Hall/CRC Press, London (2004)
2. Geman, H., Madan, D.D., Yor, M.: Time Changes for Lévy Processes. Mathematical Finance **11**(1), 79–96 (2001)
3. Gerber, H.U., Shiu, E.S.W.: Option Pricing by Esscher Transforms. Transactions of the Society of Actuaries **46**, 99–144 (1994)
4. Leoni, P., Schoutens, W.: Multivariate Smiling.Wilmott Magazine (March 2008)
5. Luciano, E., Shoutens, W.: A Multivariate Jump-Driven Asset Pricing Model. Quantitative Finance **6**(5), 385–402 (2006)
6. Schoutens, W.: Lévy processes in Finance: Pricing Financial derivatives. Wiley, Chichester, New York (2003)
7. Tassinari, G.L., Corradi, C.: Pricing Equity and Debt Tranches of Collateralized Funds of Hedge Fund Obligations: an Approach Based on Stochastic Time Change and Esscher-Transformed Martingale Measure. Quantitative Finance (2013)

Valuation of R&D Investment Opportunities Using the Least-Squares Monte Carlo Method

Giovanni Villani

Abstract In this paper we show the applicability of the Least Squares Monte Carlo (LSM) in valuing R&D investment opportunities. As it is well known, R&D projects are made in a phased manner, with the commencement of subsequent phase being dependent on the successful completion of the preceding phase. This is known as a sequential investment and therefore R&D projects can be considered as compound options. Moreover, R&D investments often involve considerable cost uncertainty so that they can be viewed as an exchange option, i.e. a swap of an uncertain investment cost for an uncertain gross project value. In this context, the LSM method is a powerful and flexible tool for capital budgeting decisions and for valuing R&D investments. In fact, this method provides an efficient technique to value complex real investments involving a set of interacting American-type options.

1 Introduction

Real options analysis has become a well-know R&D project valuation technique that values managerial flexibility to adjust decisions under uncertainty. For instance, a project that is started now may be abandoned or expanded in the future, making the project conditional. As the investment decision is conditional, it can be regarded as an "option" that is acquired by making the prior investment. The fundamental difference between real options and traditional net present value (NPV) is the flexibility to adapt when circumstances change. Whereas NPV assumes that investments are fixed, an option will be exercised if future opportunities are fovourable, otherwise the option will expire without any further cost.

G. Villani (✉)
Department of Economics, University of Foggia, Largo Papa Giovanni Paolo II 1, 71100 Foggia, Italy
e-mail: giovanni.villani@unifg.it

M. Corazza, C. Pizzi (eds.), *Mathematical and Statistical Methods for Actuarial Sciences and Finance*, DOI 10.1007/978-3-319-02499-8_26, © Springer International Publishing Switzerland 2014

Among the option pricing methods, the [4] model is mostly influential. Using financial theory, [12] was the first to described real options as the opportunities to purchase real assets on possibility favorable terms. In the R&D real option literature, [13] shows empiracally that flexibility under uncertainty allows firms to continuously adapt to change and improve products; [8] find that real options theory is used as an auxiliary valuation tool in pharmaceutical investment valuation and so on.

As it is well known, R&D projects are by their nature sequential and the start of a phase depends on the success of the preceding phase. Each phase represents an option on a new phase of process. In this context, [9] view the R&D process and subsequent discoveries as compound exchange options; [2] consider N-phased investment opportunities where the time evolution of project value follows a jump-diffusion process; [6] provide an analytical model for valuing the phased development of a pharmaceutical R&D project; [7] describe a methodology for evaluating R&D investment projects using Monte Carlo method.

However, most real investments opportunities are American-type options, to capture the manager flexibility to realize an investment before the maturity time. So Monte Carlo simulation, as pointed out in [7], is an attractive tool to solve complex real option models. One of the most new approaches in this environment is the Least-Squares Monte Carlo (LSM) proposed by [10], as it is witnessed by [11] and [3].

Aim of this paper is to value an R&D project through a Compound American Exchange option (CAEO). In particular way we assume that, during the commercialization phase, the firm can realize the respective investment cost before the maturity T benefiting of underlying project value. Moreover, R&D investments often involve considerable cost uncertainty so that they can be viewed as an exchange option, i.e. a swap of an uncertain investment cost for an uncertain gross project value.

We value the CAEO applying the LSM method. This approach presents several advantages with respect to the basic Monte Carlo, as analysed in our previous paper (see [7]). In fact, the major drawbacks of simulations are the high computation requirements and also the low speed.

The paper is organized as follows. Section 2 analyses the structure of an R&D investment and the valuation of a CAEO using the LSM method. In Sect. 3 we value four R&D projects using the CAEO and we presents also a sensitivity analysis. Finally, Sect. 4 concludes.

2 The Basic Model

In this model, we assume a two-stage R&D investment with the following structure:

- R is the Research investment spent at initial time $t_0 = 0$;
- IT is the Investment Technology to develop innovation payed at time t_1. We further suppose that $IT = qD$ is a proportion q of asset D, so it follows the same stochastic process of D;

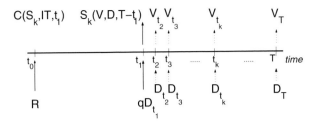

Fig. 1 R&D structure

- D is the production and the commmercialization investment in order to receive the R&D project's value. We assume that D can be realized between t_1 and T;
- V is the R&D project value.

In particular way, investing R at time t_0, the firm obtains a first investment opportunity that can be value as a Compound American Exchange Option (CAEO) denoted by $C(S_k, IT, t_1)$. This option allows to realize the Investment Techology IT at time t_1 and to obtain, as underlying asset, the option to realize the market launch; let denote by $S_k(V, D, T - t_1)$ this option value at time t_1, with maturity date $T - t_1$ and exercisable k times. In detail, during the market launch, the firm has got another investment opportunity to invest D between t_1 and T and to receive the R&D project value V. Specifically, using the LSM model, the firm must decide at any discrete time $\tau_k = t_1 + k\Delta t$, for $k = 1, 2, \cdots h$ with $\Delta t = \frac{T - t_1}{h}$ and h is the number of discretizations, whether to invest D or to wait, and so to delay the decision at next time. In this way we capture the managerial flexibility to invest D before the maturity T and so to realize the R&D cash flows. Figure 1 shows the R&D investment structure.

2.1 Assumptions and General Computations

We assume that V and D follow a geometric Brownian motion:

$$\frac{dV}{V} = (\mu_v - \delta_v)dt + \sigma_v dZ_t^v \tag{1}$$

$$\frac{dD}{D} = (\mu_d - \delta_d)dt + \sigma_d dZ_t^d \tag{2}$$

$$cov\left(\frac{dV}{V}, \frac{dD}{D}\right) = \rho_{vd}\sigma_v\sigma_d \, dt \tag{3}$$

where μ_v and μ_d are the expected rates of return, δ_v and δ_d are the corresponding dividend yields, σ_v^2 and σ_d^2 are the respective variance rates, ρ_{vd} is the correlation between changes in V and D, $(Z_t^v)_{t \in [0,T]}$ and $(Z_t^d)_{t \in [0,T]}$ are two Brownian processes defined on a filtered probability space $(\Omega, \mathcal{A}, \{\mathcal{F}_t\}_{t \geq 0}, \mathbb{P})$, where Ω is the space of all possible outcomes, $\mathcal{A}$ is a sigma-algebra, $\mathbb{P}$ is the probability measure and $\{\mathcal{F}_t\}_{t \geq 0}$ is a filtration with respect to Ω space.

Assuming that the firm keeps a portafolio of activities which allows it to value activities in a risk-neutral way, the dynamics of the assets V and D under the risk-neutral martingale measure $\mathbb{Q}$ are given by:

$$\frac{dV}{V} = (r - \delta_v)dt + \sigma_v dZ_v^* \tag{4}$$

$$\frac{dD}{D} = (r - \delta_d)dt + \sigma_d dZ_d^* \tag{5}$$

$$Cov(dZ_v^*, dZ_d^*) = \rho_{vd}\, dt \tag{6}$$

where r is the risk-free interest rate, Z_v^* and Z_d^* are two Brownian standard motions under the probability $\mathbb{Q}$ with correlation coefficient ρ_{vd}. After some manipulation, we get the equations for the price ratio $P = \frac{V}{D}$ and D_T under the probability $\mathbb{Q}$:

$$\frac{dP}{P} = \left(-\delta + \sigma_d^2 - \sigma_v\sigma_d\rho_{vd}\right)dt + \sigma_v dZ_v^* - \sigma_d dZ_d^* \tag{7}$$

$$D_T = D_0 \exp\left\{(r - \delta_d)T\right\} \cdot \exp\left(-\frac{\sigma_d^2}{2}T + \sigma_d Z_d^*(T)\right) \tag{8}$$

where D_0 is the value of asset D at initial time.

We can observe that $U \equiv \left(-\frac{\sigma_d^2}{2}T + \sigma_d Z_d^*(T)\right) \sim N(-\frac{\sigma_d^2}{2}T, \sigma_d\sqrt{T})$ and therefore $\exp(U)$ is log-normal distributed whose expection value $E_{\mathbb{Q}}[\exp(U)] = 1$. By Girsanov's theorem, we define a new probability measure $\widetilde{\mathbb{Q}}$ equivalent to $\mathbb{Q}$ whose Radon-Nikodym derivative is:

$$\frac{d\widetilde{\mathbb{Q}}}{d\mathbb{Q}} = \exp\left(-\frac{\sigma_d^2}{2}T + \sigma_d Z_d^*(T)\right). \tag{9}$$

Hence, substituing in (8) we can write:

$$D_T = D_0\, e^{(r-\delta_d)T} \cdot \frac{d\widetilde{\mathbb{Q}}}{d\mathbb{Q}}. \tag{10}$$

By the Girsanov's theorem, the processes:

$$d\hat{Z}_d = dZ_d^* - \sigma_d dt \tag{11}$$

$$d\hat{Z}_v = \rho_{vd}d\hat{Z}_d + \sqrt{1 - \rho_{vd}^2}\, dZ' \tag{12}$$

are two Brownian motions under the risk-neutral probability space $(\Omega, \mathscr{A}, \mathscr{F}, \widetilde{\mathbb{Q}})$ and Z' is a Brownian motion under $\widetilde{\mathbb{Q}}$ independent of $\hat{Z}_d$.

By using equations (11) and (12), we can now obtain the risk-neutral price simulation P:

$$P(t) = P_0 \exp\left\{\left(\delta_d - \delta_v - \frac{\sigma^2}{2}\right)t + \sigma Z^P(t)\right\} \tag{13}$$

where $\sigma = \sqrt{\sigma_v^2 + \sigma_d^2 - 2\sigma_v\sigma_d\rho_{vd}}$ and Z^P is a Brownian motion under $\widetilde{\mathbb{Q}}$.

2.2 Valuation of Compound American Exchange Option Using LSM Method

The value of CAEO can be determined as the expectation value of discounted cash-flows under the risk-neutral probability $\mathbb{Q}$:

$$C(S_k, IT, t_1) = e^{-rt_1} E_{\mathbb{Q}}[\max(S_k(V_{t_1}, D_{t_1}, T - t_1) - IT, 0)]. \tag{14}$$

Assuming the asset D as numeraire and using Eq. (10) we obtain:

$$C(S_k, IT, t_1) = D_0 e^{-\delta_d t_1} E_{\widetilde{\mathbb{Q}}}[\max(S_k(P_{t_1}, 1, T - t_1) - q, 0)] \tag{15}$$

where $IT = q D_{t_1}$.

The market launch phase $S_k(P_{t_1}, 1, T - t_1)$ can be analyzed using the LSM method. Like in any American option valuation, the optimal execise decision at any point in time is obtained as the maximun between immediate exercise value and expected continuation value.

The LSM method allows us to estimate the conditional expection function for each exercise date and so to have a complete specification of the optimal exercise strategic along each path. The method starts by simulating n price paths of asset P_{t_1} defined by Eq. (13) using Matlab code:

```
Pt1=P0*exp(norminv(rand,-d*t1-sig^2*t1/2,sig*sqrt(t1)));
```

with $\delta = \delta_v - \delta_d$. Let $\hat{P}^i_{t_1}, i = 1 \cdots n$ the simulated prices. Starting from each ith simu-lated-path, we begin by simulating a discretization of Eq. (13) for $k = 1 \cdots h$:

```
Pt(:,k)=Pt(:,k-1).*exp((-d-0.5*sig^2)*dt+sig*dBt(:,k));
```

where dBt is a random variable with a standard normal distribution. The process is repeated m times over a time horizont T. Starting with the last j^{th} price $\hat{P}^{i,j}_T$, for $j = 1 \cdots m$, the option value in T can be computed as $S_0(\hat{P}^{i,j}_T, 1, 0) = \max(\hat{P}^{i,j}_T - 1, 0)$:

```
S(:,h)=max(PPit(:,h)-1,0);
```

Working backward, at time τ_{h-1}, the process is repeated for each jth path. In this case, the expected continuation value may be computed using the analytic espres-sion for an European option $S_1(\hat{P}^{i,j}_{\tau_{h-1}}, 1, \Delta t)$. Moving backwards, at time τ_{h-1}, the management must decide whether to invest or not. The value of the option is maxi-mized if the immediate exercise exceeds the continuation value, i.e.:

$$\hat{P}^{i,j}_{\tau_{h-1}} - 1 \geq S_1(\hat{P}^{i,j}_{\tau_{h-1}}, 1, \Delta t). \tag{16}$$

We can find the critical ratio $P^*_{\tau_{h-1}}$ that solve the inequality (16):

$$P^*_{\tau_{h-1}} - 1 = S_1(P^*_{\tau_{h-1}}, 1, \Delta t)$$

and so the condition (16) is satisfied if $\hat{P}^{i,j}_{\tau_{h-1}} \geq P^*_{\tau_{h-1}}$. But it is very heavy to compute the expected continuation value for all previous time and so to determine the critical price $P^*_{\tau_k}$, $k = 1 \cdots h - 2$, as it is shown in [5].

The main contribution of the LSM method is to determine the expected continuation values by regressing the subsequent discounted cash flows on a set of basis functions of current state variables. As described in [1], a common choice of basis functions are the weighted Power, Laguerre, Hermite, Legendre, Chebyshev, Gegenbauer and Jabobi polynomials. In our paper we consider as basis function a three weighted Power polynomial. Let be L^w the basis of functional forms of the state varibable $\hat{P}^{i,j}_{\tau_k}$ that we use as regressors. We assume that $w = 1, 2, 3$. At time τ_{h-1}, the least square regression is equivalent to solve the following problem:

$$\min_{\mathbf{a}} \sum_{j=1}^{m} \left[S_0(\hat{P}^{i,j}_T, 1, 0)e^{-r\Delta t} - \sum_{w=1}^{3} a_w L^w(\hat{P}^{i,j}_{\tau_{h-1}}) \right]^2. \tag{17}$$

The optimal $\hat{\mathbf{a}} = (\hat{a}_1, \hat{a}_2, \hat{a}_3)$ is then used to extimate the expected continuation value along each path $\hat{P}^{i,j}_{\tau_{h-1}}$, $j = 1 \cdots m$:

$$\hat{S}^i_1(\hat{P}^{i,j}_{\tau_{h-1}}, 1, \Delta t) = \sum_{w=1}^{3} \hat{a}_w L^w(\hat{P}^{i,j}_{\tau_{h-1}}).$$

After that, the optimal decision for each price path is to choose the maximum between the immediate exercise and the expected continuation value.

Proceding recursively until time t_1, we have a final vector of continuation values for each price-path $\hat{P}^{i,j}_{\tau_k}$ that allows us to build a stopping rule matrix in Matlab that maximises the value of american option:

```
%Find when the option is exercised:
IStop=find(PPit(:,j-1)-1>=max(XX2*BB,0));
%Find when the option is not exercised:
ICon=setdiff([1:m],IStop);
%Replace the payoff function with the value of the option
%(zeros when not exercised and values when exercised):
S(IStop,j-1)=PPit(IStop,j-1)-1;
S(IStop,j:h)=zeros(length(IStop),h-j+1);
S(ICon,j-1)=zeros(length(ICon),1);
```

As consequence, the ith option value approssimation $\hat{S}^i_k(\hat{P}^i_{t_1}, 1, T - t_1)$ can be determined by averaging all discounted cash flows generated by option at each date over all paths $j = 1 \cdots m$.

Finally, it is possible to implement Monte Carlo simulation to approssimate the CAEO:

$$C(S_k, IT, t_1) \approx D_0 e^{-\delta_d t_1} \left(\sum_{i=1}^{n} \frac{\max(\hat{S}^i_k(\hat{P}^i_{t_1}, 1, T - t_1) - q, 0)}{n} \right). \tag{18}$$

The Appendix illustrates the complete Matlab algorithm to value CAEO. We conclude that, applying real option methodology, the R&D project will be realized at time t_0 if $C(S_k, IT, t_1) - R$ is positive, otherwise the investment will be rejected.

3 Numerical R&D Applications

Table 1 summarizes the input parameters about four ipotetical R&D investments. The R&D project value V_0 is the current value of the underlying project cash flows appropriately discounted. We assume that V_0 ranges from 210 000 to 750 000.

For simplicity, we consider a two-staged R&D investment. The projects start with the research phase that is expected to end at time t_1 with the discovery of a new good. We consider that $t_1 = 1$ for projects I and II, $t_1 = 2$ for III and $t_1 = 3$ for IV. On evarage, the research phase is about one year for software-technological investments, two years for motor-telecommunication industries and three years for pharmaceutical one.

At time t_1, the firm realizes a second investment in technologies to develop innovation. Its current value is IT_0 and we assume that IT is a proportion q of asset D. the indentical stochastic process of D, except that it occurs at time $t_1 = 1$.

After that, we have the production and the commercialization phase in which the new product is ready for the market launch. We assume that this phase starts in t_1 and ends at time T. After time T each business opportunity disappears. During the commercialization phase, the firm realizes the investment cost D and receives the project value V. The investment D can be realize at any time between t_1 and T and its current value is D_0. In this way we value the decision flexibilities to capture the R&D cash flows before the maturity T. The length of commercialization phase depends on typology of R&D investment: this is shorter for software-technological

Table 1 Input values for R&D valuation

Project		I (Software)	II (Tech.)	III (Motor)	IV (Pharma.)
R&D Project Value	V_0	250 000	210 000	750 000	410 000
Development Cost	D_0	140 000	200 000	950 000	310 000
Investment Technology	IT_0	70 000	120 000	171 000	46 500
Research Investment	R	50 000	40 000	35 000	100 000
Exchange Comp. ratio	q	0.50	0.60	0.18	0.15
Dividend-Yield of V	δ_v	0.20	0.15	0.15	0.15
Dividend-Yield of D	δ_d	0.05	0.05	0	0
Time to Maturity	t_1	1 year	1 year	2 year	3 year
Time to Maturity	T	2 year	3 year	5 year	7 year
Correlation	ρ_{vd}	0.38	0.26	0.08	0.12
Volatility of V	σ_v	0.83	0.64	0.54	0.88
Volatility of D	σ_d	0.32	0.41	0.15	0.31

R&D than motor-pharmaceutical one. So we assume that $T = 2, 3, 5, 7$ for projects I,II,III and IV, respectively.

Appropriately, in order to value the volatility of asset V and D, we take into account the quoted shares and traded options of similar companies. Moreover, as an R&D investment presents a high uncertainty about its result, we assume that σ_v ranges from 0.54 to 0.88 and σ_d from 0.15 to 0.41.

According to financial options, δ denotes the dividends paid on the stock that are foregone by option holder. In real option theory, δ_v is the opportunity cost of deferring the project and δ_d is the "dividend yield" on asset D.

To compute the value of CAEO we assume that $m = 20\,000$, $n = 10\,000$ with $x = 20$ steps for year. Moreover, the Standard Error $\varepsilon_n = \frac{\hat{\sigma}}{\sqrt{n}}$ is a measure of simulation accurancy and it is estimated as the realised standard deviation of simulations divided by the square root of simulations.

Table 2 contains the Monte Carlo numerical results. In particular way, we have computed four simulated values for each R&D project and then we have considered the average among them to determine the CAEO. For each simulated value we have also computed the standard error (SE). We can observe that this value increases when the variances σ_v and σ_d go up. The last two culumns of Table 2 show the comparison between the NPV and the Real option methodology. The NPV is given by the difference between the receipts and expenses in t_0, namely NPV$= V_0 - (D_0 + IT_0 + R)$ while the real option value (RO) is the CAEO minus the investment R. As we can observe, the NPV of each project is always negative and so, according to the NPV, the firm should reject all projects. On the other hand, considering the real option approach, the investment opportunities I, III, and IV are remuneratives since we take into account both the sequential frame of an *R&D*, i.e. the possibility that the project may be abandoned in the future, and the managerial flexibility to realize the investment D before the maturity T and so to benefit of R&D cash flows.

In real options valuation, many times the binomial method can be unusable owing for instance to the dimensionality of problem, or when we take into account discrete dividends or with the valuation of compound options. In this context, Monte Carlo

Table 2 Simulated Values of CAEO

	1st Sim	2nd Sim	3rd Sim	4th Sim	CAEO	NPV	RO
Project I (Soft.)	64 444	64 516	64 803	64 592	64 589	−10 000	14 589
SE I	0.0111	0.0111	0.0114	0.0113			
Project II (Tech.)	19 775	19 923	20 020	20 261	19 995	−150 000	−20 005
SE II	0.0038	0.0038	0.0039	0.0040			
Project III (Motor)	66 908	65 326	67 973	66 207	66 603	−406 000	31 603
SE III	0.0030	0.0030	0.0030	0.0032			
Project IV (Pharm.)	150 910	152 120	147 410	148 353	149 698	−46 500	49 698
SE IV	0.0329	0.0230	0.0169	0.0240			

Table 3 Comparison between basic MC and LSM approach

	1st Sim	2nd Sim	3rd Sim	4th Sim
Least-Square Monte Carlo	64 444	64 516	64 803	64 592
Standard Error LSM	0.0111	0.0111	0.0114	0.0113
Basic Monte Carlo	65 821	63 380	64 215	65 152
Standar Error MC	0.0423	0.0420	0.0422	0.0423

simulation provides an easy way of valuing options but its disadvantage is that it is computationally intensive and inefficient. To relieve this problem, it is possible to reduce the standard error and to improve the accuracy of simulation estimates by increasing the number of simulated paths or by using variance reduction techniques. About the simple and compound exchange options, this analysis has been illustrated in [14]. Comparing the basic Monte Carlo (MC) approach used in our previous paper (see [7]) and in particular way the Second Matlab Algorithm for the Pseudo Compound American Exchange option with the LSM methodology, we improve both the computation time and the accurancy of simulations. Specifically, using a 2 GHz Intel Core 2 Due processor and assuming that $m = 20\,000, n = 10\,000$ and $x = 20$, we obtain for the project I the results listed in Table 3 .

To give an idea of CPU computation time, it takes about two hours for LSM and three hours for basic MC. So the LSM is a little faster than basic MC. Moreover, the standard errors for simulated values are lower for LSM, ranging between 0.0111 and 0.0114, than basic MC, ranging between 0.0420 and 0.0423. The variance of four simulations is 18 040 for LSM and 861 754 for basic MC and so we can state that LSM methodology improves the simulation accuracy.

Finally we examine the sensitivity of our results with respect to the parameters V, σ_v and the maturity time $\tau = T - t_1$.

As it is shown in Fig. 2a, it is obvious that the R&D real option values increase when the asset V goes up. In particular way, project I can be placed first and it is remunerative when $V > 240\,000$, after that we have project II that it is positive starting from $V > 270\,000$. Projects IV and III are remuneratives from $V > 340\,000$ and $V > 560\,000$, respectively.

Figure 2b displays the effects of volatility σ_v on the real option selection. We can observe that the R&D real option values increase when the volatility σ_v grows. In this case, under the same volatility value, project III is the best. After that the investors will pick out in order of remuneration projects IV, I and II, respectively.

Finally, Table 4 summarises the sensibility of R&D real option values when the maturity time τ changes. In this case we can observe that project IV is the top considering the same maturity time. Successively we have projects III, I and II, respectively.

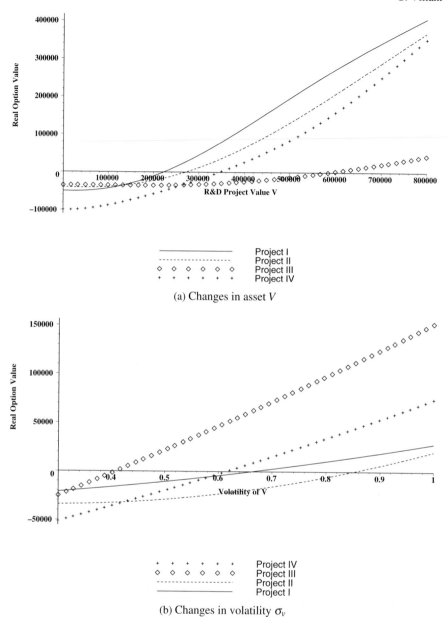

(a) Changes in asset V

(b) Changes in volatility σ_v

Fig. 2 Comparison among projects

Table 4 Comparison among projects when the maturity τ changes

τ	1	2	3	4
Project I (Soft.)	14 589	18 901	21 831	12 944
Project II (Tech.)	−20 541	−20 005	−18 169	−22 038
Project III (Motor)	32 309	36 282	31 603	18 572
Project IV (Pharm.)	83 270	61 160	54 840	49 698

4 Conclusions

In this paper we have shown how the Least Squares Monte Carlo can be used to evaluate R&D projects. In particular way, an R&D opportunity is a sequential investment and therefore can be considered as a compound option. We have assumed the managerial flexibility to realize the investment D before the maturity T in order to benefit of R&D cash flows. So an R&D project can be view as a Compound American Exchange option, that allows us to couple both the sequential frame and the managerial flexibility of an R&D investment.

Moreover, using the asset D as numeraire, we have reduced the bidimensionality problem of valuing the CAEO to one variable P. As we have analyzed, the main contribution of the LSM method is to determine the expected continuation value by regressing the discounted cash flows on the simple powers of variable P, and so to overcome the effort to compute the critical prices $P^*_{\tau_k}, k = 1 \cdots h - 2$. In this way we have improved both the computation time and the accurancy of simulations with respect to our previous paper (see [7].) Nevertheless, the accurancy of LSM method can be improved testing and comparing different basis functions on real option values. Finally, we have studied four R&D projects. We have observed that the NPV of each project is always negative while, according to real option approach, the investment opportunities I,III and IV are remuneratives. Moreover, we have examined how the R&D real option value changes according to parameters V, σ_v and τ of the model.

Appendix

In this appendix we present the Matlab algorithm of LSM method.

```
function LMSCAEO= LMSCAEO(V,D,q,sigv,sigd,rho,...
dV,dD,t1,T,x,m,n);
sig=sqrt(sigv^2+sigd^2-2*sigv*sigd*rho);
P0=V/D;
d=dV-dD;
dt =1/x; %Lenght of the interval of time
t=T-t1;
h=(t/dt); %Number of periods to simulate the price
```

```
for i=1:n
Pt1=P0*exp(norminv(rand,-d*t1-sig^2*t1/2,sig*sqrt(t1)));
dBt=sqrt(dt)*randn(m,h); %Brownian motion
Pt=zeros(m,h);  %Initialize matrix
Pt(:,1)=Pt1*ones(m,1); %Vector of initial stock price
for k=2:h;
Pt(:,k)=Pt(:,k-1).*exp((-d-0.5*sig^2)*dt+sig*dBt(:,k));
end
PPit=Pt; %Change the name
%Work Backwards; Initialize CashFlow Matrix
S=NaN*ones(m,h);
S(:,h)=max(PPit(:,h)-1,0);
for j=h:-1:3;
% Step 1: Select the path in the money at time j-1
I=find(PPit(:,j-1)-1>0);
ISize=length(I);
% Step 2: Project CashFlow at time j onto basis...
% function at time j-1
if j==h;
YY=(ones(ISize,1)*exp(-dD*[1:h-j+1]*dt)).*S(I,j:h);
else
YY=sum(((ones(ISize,1)*exp(-dD*[1:h-j+1]*dt)).*S(I,j:h))')';
end
PPb=PPit(I,j-1);
XX=[ones(ISize,1),PPb,PPb.^2,PPb.^3];
BB=pinv(XX'*XX)*XX'*YY;
PPb2=PPit(:,j-1);
XX2=[ones(m,1),PPb2,PPb2.^2,PPb2.^3];
%Find when the option is exercised:
IStop=find(PPit(:,j-1)-1>=max(XX2*BB,0));
%Find when the option is not exercised:
ICon=setdiff([1:m],IStop);
%Replace the payoff function with the option value:
S(IStop,j-1)=PPit(IStop,j-1)-1;
S(IStop,j:h)=zeros(length(IStop),h-j+1);
S(ICon,j-1)=zeros(length(ICon),1);
end
YY=sum(((ones(m,1)*exp(-dD*[1:h-1]*dt)).*S(:,2:h))')';
AEOSim(i)=mean(YY);
PAYOFF(i)=max(AEOSim(i)-q,0);
end
CAEO=D*exp(-dD*t1)*mean(PAYOFF)
```

References

1. Abramowitz, M., Stegun, I.A.: Handbook of Mathematical Functions. Dover Pubblications, New York (1970)
2. Andergassen, R., Sereno, L.: Valuation of N-stage Investments Under Jump-Diffusion Processes. Computational Economics **39**(3), 289–313 (2012)
3. Areal, N., Rodrigues, A., Armada, M.J.R.: Improvements to the Least Squares Monte Carlo Option Valuation Method. Review of Derivative Research **11**(1–2), 119–151 (2008)
4. Black, F., Scholes, M.: The Pricing of Options and Corporate Liabilities. Journal of Political Economy **81**, 637–659 (1973)
5. Carr, P.: The Valuation of American Exchange Options with Application to Real Options. In: Trigeorgis, L. (eds) Real Options in Capital Investment: Models, Stratigies and Applications. Praeger, Westport Connecticut, London (1995)
6. Cassimon, D., Engelen, P.J., Thomassen, L., Van Wouwe, M.: The valuation of a NDA using a 6-fold compound option. Research Policy **33**(1), 41–51 (2004)
7. Cortelezzi, F., Villani, G.: Valuation of R&D sequential exchange options using Monte Carlo approach. Computational Economics **33**(3), 209–236 (2009)
8. Hartmann, M., Hassan, A.: Application of real options analysis for pharmaceutical R&D project valuation-Empirical results from a survey. Research Policy **35**(3), 343–354 (2006)
9. Lee, J., Paxson, D.A.: Valuation of R&D real American sequential exchange option. R&D Management **31**(2), 191–201 (2001)
10. Longstaff, F.A., Schwartz, E.S.: Valuing American options by simulation: a simple least-squares approach. The Review of Financial Studies **14**(1), 113–147 (2001)
11. Moreno, M., Navas, J.F.: On the Robustness of Least-Squares Monte Carlo (LSM) for Pricing American Derivatives. Review of Derivative Research **6**(2), 107–128 (2003)
12. Myers, S.C.: Determinants of corporate borrowing. Journal of Financial Economics **5**(2), 147–175 (1977)
13. Thomke, S.H.: The role of flexibility in the development of a new products: An empirical study. Research Policy **26**(1), 105–119 (1997)
14. Villani, G.: Generalization of stratified variance reduction methods for monte carlo exchange options pricing. In: Perna Cira e Marilena Sibillo (eds.) Mathematical and Statistical Methods for Actuarial Sciences and Finance, pp. 379–387. Springer (2012)

The Determinants of Interbank Contagion: Do Patterns Matter?

Stefano Zedda, Giuseppina Cannas and Clara Galliani[*]

Abstract The recent financial crisis highlighted that interconnectedness between banks has a crucial role, and can push the effects of bank defaults to extreme levels. Interconnectedness in banking systems can be modelled trough the interbank market structure. As only data on interbank credits and debts aggregated at bank level are publicly available, one common hypothesis is to assume that banks maximise the dispersion of their interbank credits and debts, so that the interbank matrix is approximated by its maximum entropy realisation.

The aim of this paper is to test the influence of this approximation on simulations, and verifying if variations in the structure of the interbank matrix systematically change the magnitude of contagion.

Numerical experiments on samples of banks from four European countries, showed that different interbank matrices produce small changes in the point estimation. Nevertheless, they significantly affect variability and confidence interval for the estimates, in particular in banking systems when contagion effects are more intense.

S. Zedda (✉)
Joint Research Centre, European Commission, Ispra (VA), Italy, and Department of Business and Economics, University of Cagliari, Via S. Ignazio 74, 09123 Cagliari, Italy
e-mail: szedda@unica.it

G. Cannas
Joint Research Centre, European Commission, Ispra (VA), Italy
e-mail: giuseppina.cannas@jrc.ec.europa.eu

C. Galliani
Joint Research Centre, European Commission, Ispra (VA), Italy
e-mail: clara.galliani@ec.europa.eu

[*] The opinions expressed are those of the authors and do not imply any reference to those of the European Commission.

M. Corazza, C. Pizzi (eds.), *Mathematical and Statistical Methods for Actuarial Sciences and Finance*, DOI 10.1007/978-3-319-02499-8_27, © Springer International Publishing Switzerland 2014

1 Introduction

Interbank markets are important for the proper functioning of modern financial systems. They therefore need to be considered in any banking model aiming at estimating the probability of a systemic banking crisis. One of the effects of interbank connections is that one initial bank failure could have domino effects on the whole system: interbank markets can be a major carrier of contagion among banks, as problems affecting one bank may spread to others.

Contagion results from two risks: first, the risk that at least one component of the system could default (probability of a bank defaulting) and, second, the risk that this shock could propagate through the system (potential impact of the default). As the former can stem from a variety of unexpected situations, and is driven mainly by assets' riskiness and solvency, this research focuses on the latter. In particular, the goal of this paper is to assess how a hypothesis on the structure of the interbank market (i.e. the matrix of credit and debts among banks) affects the magnitude of a systemic banking crisis.

One common problem in dealing with interbank market structures is that only partial data are available, as balance sheets report only aggregated interbank assets and liabilities. Maximum entropy approximation offers a way to proxy interbank bilateral exposures, assuming that banks maximise the dispersion of their interbank credits and debts. But what is the cost of such an approximation?

This paper assesses the influence of the maximum entropy hypothesis by verifying if variations in the matrix structure lead to significantly different results in systemic excess losses, i.e. losses that exceed capital requirements. The model generates losses in the banking systems of four countries (Belgium, Ireland, Italy and Portugal) via Monte Carlo simulations, as performed in the model recently developed by De Lisa et al. [4].

Interbank exposures are initially modelled using a matrix that maximises the dispersion of banks' bilateral exposures. Contagion results obtained from this scenario are then compared with those achieved with a more concentrated interbank matrix, in order to evaluate if contagion is influenced by hypotheses on interbank exposures.

This paper is structured as follows. Section 2 gives an overview of the literature on interbank market contagion. Section 3 explains the maximum entropy matrix approximation, the algorithm to adjust the interbank exposures matrix and the scenario generation procedure. Section 4 presents data used to perform the numerical analysis. Section 5 shows results. Conclusions are drawn in Sect. 6.

2 Literature Review

It is well-known that if a failing bank does not repay its obligations in the interbank market, this could compromise the solvency of its creditor banks and lead to a domino effect in the banking system. Hence, contagion occurs when the financial distress of a single bank affects one bank's ability to pay debts to other financial institutions. Therefore, interlinkages between banks could eventually have an impact on the whole financial system and, beyond that, on the state of the entire economy.

Moreover, the pattern of the interbank linkages could affect the way a crisis propagates through the system.

Theoretical studies like Allen and Gale [1] and Freixas et al. [7] often apply network theory to the banking system and, in particular, focus on the completeness and connectedness of the interbank matrix. Looking at empirical approaches, Upper [10] provides a summary of the numerous papers that have focused on the role played by the interbank market in spreading financial contagion. Among them, only few contributions have detailed information about the interbank matrix (see for example Upper and Worms [11], van Lelyveld and Liedorp [12] and Mistrulli [9]). When data about single exposures are not available, it's necessary to make some assumptions about the structure of the matrix: the most common one is maximum entropy, used for example in Wells [13], van Lelyveld and Liedorp [9, 12].

It's also possible to have only partial data referred, in the majority of the cases, to large interbank exposures (see Degryse and Nguyen [3]). In these cases the analysis uses a mixed approach that refers both to real data and to simplifying assumption when necessary.

About the scenarios generating method, there are two methodologies currently used to model defaults: the fictitious default algorithm provided by Eisenberg and Noe [5] and the sequential default algorithm provided by Furfine [8]. Both methods start from the artificial failure of a bank and then count losses of sequential failures. The main difference among them is that the first method takes into account the simultaneity problem (defaults occurring after the trigger may increase losses at the banks that have failed previously), whereas the second one does not.

3 Methodology

3.1 Interbank Matrix Structure

Available data at single bank level only cover total credits and debts to other banks, and information on bilateral exposures between banks is not publicly available. For this reason, the interbank matrix must be inferred by making assumptions on how interbank debts and credits are spread over the system. This analysis aims to assess the uncertainty of simulated bank losses due to the approximation of the interbank matrix.

Following Upper and Worms [11], the first step is to approximate the interbank matrix with the maximum entropy one, i.e. to assume that banks maximise the dispersion of their interbank credits and debts. Individual interbank exposures in the sample are assumed to display maximum dispersion, so that each bank lends to each of the others in proportion to its share of the total interbank credit. In this way the largest lender will be the largest creditor for all other banks, and banks with no debts will result in a column of zeros. The maximum entropy matrix is obtained numerically via the ENTROP algorithm (see Blien and Graef [2]).

In order to test the robustness of the maximum entropy assumption, variations were introduced in the interbank matrix to evaluate if these changes induce a significant variation in results. At each step, the process introduces one zero more in a random cell, but preserving the totals per rows and per columns (for a complete description of the methodology see Zedda et al. [14]). Starting from the maximum entropy matrix, we produce more concentrated interbank matrices with 20%, 35%, 50%, 65% and 80% more elements set to zero other than the diagonal elements or elements already set at zero. We implement this mechanism for banking systems in four different countries (Belgium, Ireland, Italy and Portugal).

3.2 Generating Scenarios

To verify the effectiveness of contagion, we have considered as fundamental to generate scenarios as close as possible to the real banking system situation. To do this, a Monte Carlo simulation coherent with a Basel II framework and based on balance-sheet data was performed, with banks' correlated assets. The correlation between banks' assets is a key point: in this way we do not have only cases with just one primary default, but also cases with few or more contemporaneous primary defaults, typically rounded by some other cases of near-to-default banks that are more likely to start financial contagion. These values are recorded for reference before applying the contagion mechanism, as NO CONTAGION scenario.

Contagion is then looped up until the cycle where no more banks default, and net losses are recorded when at least one bank defaults. Simulations were performed in order to have 10,000 cases with at least one default for each considered country and for each differently concentrated interbank matrix. To reach this goal it was necessary to run a total number of simulations ranging from 220,501 (for Italy) to 6,273,040 (for Belgium). This is far larger from the number of simulated scenarios performed in the recent literature employing Monte Carlo methods (see for example Elsinger et al. [6]).

To perform a *ceteris paribus* analysis, in each simulation the variation in the interbank matrix is set randomly, whereas the internal losses suffered by each bank are always the same. In this way different results related to the same country can only be due to variations in the interbank matrix.

4 Data

Our analysis has been conducted on four banking systems showing different features: Belgium (BE), Italy (IT), Ireland (IE) and Portugal (PT). This makes it possible to evaluate if changes on interbank matrix have an impact on simulations depending on the countrys specific characteristics.

Table 1 Key features of the sample used for simulations

	BE	IE	IT	PT
Number of banks	23	24	473	14
Sample (% population)	82.26%	101.91% *	81.81%	66.49%
Capital (m euro)	48,401	65,392	270,876	26,341
Total assets (m euro)	878,336	1,221,181	2,827,051	323,762
Interbank debts (m euro)	97,493	276,738	188,375	43,561
Interbank credits (m euro)	84,727	148,729	195,958	34,504
Capitalization	5.50%	5.40%	9.60%	8.10%
Interbank debts/total assets	11.10%	22.70%	6.70%	13.50%
Interbank credits/total assets	9.6%	12.2%	6.9%	10.7%
Herfindhal index (on total assets)	29.3%	15.4%	5.4%	25.9%
Herfindhal index (on IB debts)	30.0%	17.7%	9.2%	22.8%
Herfindhal index (on IB credits)	25.6%	21.4%	11.7%	34.5%

* The imperfect coherence between the two sources used to construct the sample for Ireland (ECB and Irish Central Bank) generates a percentage of sample rescaled to population that is above 100%.

Data are based on the Bankscope dataset, as of December 2009, integrated with European Central Bank (ECB) and Central Banks' single countries data.

Table 1 shows the data used in our analysis aggregated at country level.

The sample to population percentage is the ratio of total assets for all banks in the sample to the total in each country reported by ECB. Capitalisation levels, measured by the capital to total assets ratio, roughly approximate how much banks are resilient to defaults of their own assets. This capacity also depends on the riskiness of the assets, which is taken into account in the scenario-generating process anyway. We report rows containing interbank credits and debts (in Euro and as percentage of total assets). In addition, the Herfindal index on total assets and interbank volumes has been calculated to give an idea of the concentration in each considered banking system.

Data in Table 1 lead to some considerations on the key features of the considered sample. Belgium has a small number of banks and, according to its Herfindhal index, a highly concentrated banking system in terms of total assets and interbank exposures. The Irish banking system is not so highly concentrated but is made up of a small number of banks highly exposed in the interbank market. Italy has the largest number of banks, high capitalisation, low interbank exposures and low Herfindhal indices. Portugal has the smallest number of banks, a high capitalisation level and, as for Belgium, the highest level of concentration in terms of both total assets and interbank exposures.

5 Results

The interbank matrix evidently plays a central role in default contagion risks. What is not evident a priori is how and how much results differ changing the matrix structure. On one hand, the maximum entropy assumption could lead to underestimation of contagion risk, as the consequences of a default are actually spread across all the other banks, limiting the effects on each single entity. On the other hand, this assumption reflects the connectedness between all banks, even where no real interbank links exist, thus possibly creating fictitious ways of propagating contagion. For this reason, the influence of variations in the interbank matrix is verified for the whole probability distribution of estimated losses.

It must be reminded that simulations are run in order to have 10,000 scenarios with at least one default in each country. For each scenario 20 different interbank matrices were considered for each concentration level, so that contagion in each country can be monitored by five matrices (one for each concentration level) with dimensions 10,000 × 20, for both losses and defaults.

Table of results report estimates both in case of NO CONTAGION and CONTAGION. Maximum entropy assumption (our base scenario) and modified interbank matrices are proposed within the CONTAGION framework. NO CONTAGION refers to a case where no interbank linkages are considered, meaning that only primary defaults are taken into account.

The average results (point estimation of the mean excess loss in each country) clearly indicate that higher concentration in the interbank matrix do not significantly influence the expected value of excess losses with respect to what results in the maximum entropy case. In fact, the estimates tends to slightly increase the average values with the concentration level and contagion dimension, the only relevant difference being for IRELAND +80%, where the amount of losses is significantly higher when the extreme concentration level is reached.

To comment on these results, a reference to network analysis can be relevant, assessing that the completeness of the matrix does not affect point estimates. On the contrary, important effects are evident on the variability of estimates.

Considering variability in each banking system, quantified by the interquartile range and 10% to 90% range calculated for each row of the five matrices, interesting results are found.

As expected, concentration in the interbank matrix does affect variability and confidence intervals. In particular, the higher the concentration in interbank connections (number of zeros in the interbank matrix), the higher is the variability in results.

This is due to the fact that the maximum entropy hypothesis leads to only one possible realisation of the estimated interbank matrix, while more concentrated interbank matrices offer multiple possible solutions, and the higher is the number of zeros introduced in the matrices, the higher can be the gap between two different matrices with the same concentration (and completeness).

The general trend is an increase in variability as the simulation moves up from a situation with 20% of zeros added in the interbank matrices to 80% more. This trend is confirmed in all four considered countries, but different countries result in deeply different results in the magnitude of variability (see, for example, the comparison between Ireland and Italy).

On one side, Italy has almost no variability in results (even considering the most concentrated interbank matrices), while Ireland's estimates are rather unstable: interquartile ranges reveal high variability even with low levels of concentration (+20% of zeros). It can also be noticed that higher interbank values (Ireland) result in higher variability, while a higher number of banks (Italy) possibly induces more stability.

These differences are evidently related with contagion dimension. In fact, comparing NO CONTAGION estimates so the values obtained before applying the contagion mechanism, with BASE SCENARIO, obtained applying contagion on the base of the maximum entropy hypothesis, we can have a proxy of the contagion dimension. Ireland is more affected by contagion (in terms of average losses) than the other countries.

When considering contagion we have a mean value of losses ten times higher than the same value obtained without contagion (16,998,231 vs. 1,707,946). Italy is in the opposite situation, results are almost not affected by contagion (167,925 average losses without contagion vs. 171,048 in the base scenario).

Table 2 reports the number of primary defaults before contagion in each of the considered country.

Table 2 Number of primary defaults (before contagion)

	BE	IE	IT	PT
1	8,663	8,931	6,696	8,855
2	959	806	1,493	840
3	252	183	696	197
4	73	51	330	69
5	29	16	185	21
>5	24	13	600	18
Total	10,000	10,000	10,000	10,000

Table 3 Average losses distribution by scenario – Belgium (th euro)

BE	Mean	1st–3rd Quartile range	10th–90th Quantile range
No contagion	1,536,509		
Base	2,696,176		
+20% zeros	2,693,177	0.1%	1.2%
+35% zeros	2,693,778	0.6%	1.8%
+50% zeros	2,699,382	1.5%	5.8%
+65% zeros	2,710,214	3.5%	11.2%
+80% zeros	2,761,571	12.1%	23.4%

Table 4 Average losses distribution by scenario – Ireland (th euro)

IE	Mean	1st–3rd Quartile range	10th–90th Quantile range
No contagion	1,707,946		
Base	16,998,231		
+20% zeros	17,049,103	2.6%	11.7%
+35% zeros	16,968,441	10.1%	32.9%
+50% zeros	17,206,620	26.2%	53.1%
+65% zeros	17,321,954	41.6%	78.3%
+80% zeros	19,941,129	71.2%	116.5%

Table 5 Average losses distribution by scenario – Italy (th euro)

IT	Mean	1st–3rd Quartile range	10th–90th Quantile range
No contagion	167,925		
Base	171,048		
+20% zeros	171,046	0.0%	0.0%
+35% zeros	171,045	0.0%	0.1%
+50% zeros	171,052	0.0%	0.1%
+65% zeros	171,047	0.1%	0.2%
+80% zeros	171,042	0.2%	0.5%

Table 6 Average losses distribution by scenario Ñ Portugal (th euro)

PT	Mean	1st–3rd Quartile range	10th–90th Quantile range
No contagion	549,885		
Base	881,506		
+20% zeros	881,520	1.9%	6.2%
+35% zeros	878,946	4.0%	12.9%
+50% zeros	884,939	6.2%	13.0%
+65% zeros	887,535	11.1%	20.9%
+80% zeros	898,572	17.2%	29.1%

Column 2 in Tables 3–6 show the average magnitude of systemic excess losses, whereas columns 3 and 4 contain the reference ranges of variability in results for the matrices with 20%, 35%, 50%, 65% or 80% more interbank elements set to zero. Figures 1–4 show, for each country, the distribution of losses in the 10,000 simulated scenarios.

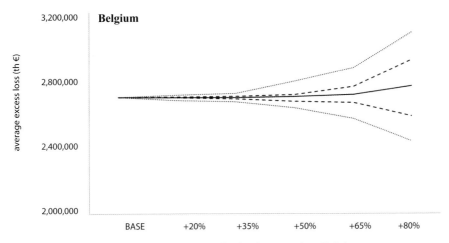

Fig. 1 Average losses distribution by scenario – Belgium

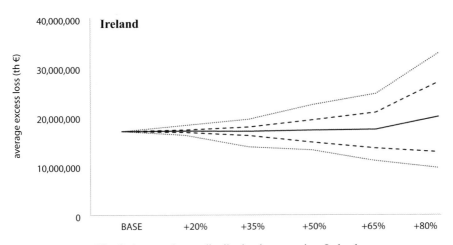

Fig. 2 Average losses distribution by scenario – Ireland

6 Conclusions

This paper compares the effects on contagion obtained assuming that interbank market patterns are maximally diffused with those coming up from more concentrated interbank matrices. To realise this, an algorithm was developed that allows obtaining interbank matrices with higher degrees of concentration in bilateral exposures, but respecting totals constraints on rows and columns. A Monte Carlo method was then applied to generate banking crises scenarios that were used to test contagion effects. We implemented this mechanism for banking systems in Belgium, Ireland, Italy and Portugal.

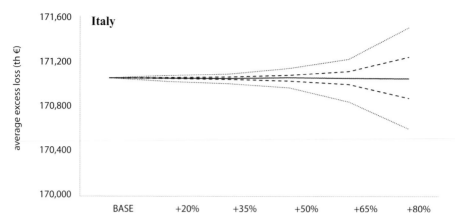

Fig. 3 Average losses distribution by scenario – Italy

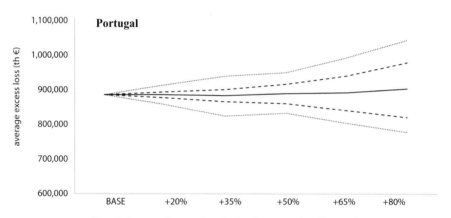

Fig. 4 Average losses distribution by scenario – Portugal

Results show that the expected value of losses is rather stable even in case of maximum concentration (with 80% more zeros in the matrix with respect to the maximum entropy one). Conversely, when considering the variability in estimates, the higher the concentration the higher the variability in results and in confidence intervals.

More specifically, variability seems to be deeply affected by the specific features of the banking system: high levels of capitalisation, low interbank exposure and large samples seem to produce more stable results, whereas low capitalisation, high interbank exposure and a small number of banks in the system seems to lead to higher variability in results.

Summing up, results suggest that structural values (capitalisation, interbank volumes, concentration and dimension) are possibly more important in determining the point estimation of contagion risk than variations in the interbank matrix concentration.

The results presented in this paper represents a preliminary part of a sensitivity analysis aiming to assess the role on contagion of some important variables, such as interbank exposures levels, capitalisation levels, correlation among banks results, concentration, granularity, etc.

Different methods for estimating the interbank matrix will be developed, also considering cross borders relations, central bank role and banking groups' structure.

References

1. Allen, F., Gale, D.: Financial Contagion. Journal of Political Economy **108**, 1–33 (2000)
2. Blien, U., Graef, F.: Entropy Optimisation Methods for the Estimation of Tables. In Balderjahn, I., Mathar, R., Schader, M. (eds.) Classification, Data Analysis and Data Highways, pp. 3–15. Springer Verlag, Berlin (1997)
3. Degryse, H., Nguyen, G.: Interbank Exposures: An Empirical Examination of Contagion Risk in the Belgian Banking System. International Journal of Central Banking **3**, 123–171 (2007)
4. De Lisa, R., Zedda, S., Vallascas, F. , Campolongo, F., Marchesi, M.: Modelling Deposit Insurance Scheme Losses in a Basel 2 Framework. Journal of Financial Services Research (2011)
5. Eisenberg, L., Noe, T.H.: Systemic Risk in Financial Systems. Management Science **47**(2), 236–249 (2001)
6. Elsinger, H., Lehar, A., Summer, M. : Using market information for banking system risk assessment. International Journal of Central Banking **2**, 137–165 (2006)
7. Freixas, X., Parigi, B., Rochet, J.C.: Systemic Risk. Interbank Relations and Liquidity Provision by the Central Bank. Journal of Money. Credit and Banking **32**, 611–638 (2000)
8. Furfine, C.H.: Interbank Exposures: Quantifying the Risk of Contagion. Journal of Money. Credit and Banking **35**, 111–128 (2003)
9. Mistrulli, P.E.: Assessing Financial Contagion in the Interbank Market: Maximum Entropy versus Observed Interbank Lending Patterns. Journal of Banking & Finance **25**, 1114–1127 (2010)
10. Upper, C.: Simulation methods to assess the danger of contagion in interbank markets. Journal of Financial Stability **7**, 111–125 (2011)
11. Upper, C., Worms, A.: Estimating Bilateral Exposures in the German Interbank Market: Is There a Danger of Contagion? European Economic Review **48**, 827–849 (2004)
12. Van Lelyveld, I., Liedorp, F.: Interbank Contagion in the Dutch Banking Sector: A Sensitivity Analysis. International Journal of Central Banking **2**, 99–133 (2006)
13. Wells, S.: Financial Interlinkages in the United Kingdom's Interbank Market and the Risk of Contagion. Working Paper 230. Bank of England (2004)
14. Zedda, S., Cannas, G. , Galliani, C., De Lisa, R.: The role of contagion in financial crises: an uncertainty test on interbank patterns. EUR report 25287. Luxembourg (2012)